R. VOIGT

Birkhäuser Advanced Texts
Basler Lehrbücher

Edited by
Herbert Amann, Zürich
Hanspeter Kraft, Basel

Helmut Hofer
Eduard Zehnder
Symplectic Invariants and Hamiltonian Dynamics

Birkhäuser Verlag
Basel · Boston · Berlin

Authors:
Helmut Hofer and
Eduard Zehnder
Mathematik
ETH Zentrum
8092 Zürich
Switzerland

Deutsche Bibliothek Cataloging-in-Publication Data

Hofer, Helmut:
Symplectic invariants and Hamiltonian dynamics / Helmut Hofer ; Eduard Zehnder. – Basel ; Boston ; Berlin : Birkhäuser, 1994
(Birkhäuser advanced texts)
ISBN 3-7643-5066-0 (Basel...)
ISBN 0-8176-5066-0 (Boston)
NE: Zehnder, Eduard:

This work is subject to copyright. All rights are reserved, whether the whole or part of the material is concerned, specifically the rights of translation, reprinting, re-use of illustrations, broadcasting, reproduction on microfilms or in other ways, and storage in data banks. For any kind of use permission of the copyright owner must be obtained.

© 1994 Birkhäuser Verlag, P.O. Box 133, CH-4010 Basel, Switzerland
Printed on acid-free paper produced of chlorine-free pulp
Printed in Germany
ISBN 3-7643-5066-0
ISBN 0-8176-5066-0

9 8 7 6 5 4 3 2

Contents

1 Introduction
1.1 Symplectic vector spaces 1
1.2 Symplectic diffeomorphisms and Hamiltonian vector fields 6
1.3 Hamiltonian vector fields and symplectic manifolds 9
1.4 Periodic orbits on energy surfaces 18
1.5 Existence of a periodic orbit on a convex energy surface 23
1.6 The problem of symplectic embeddings 31
1.7 Symplectic classification of positive definite quadratic forms . . . 35
1.8 The orbit structure near an equilibrium, Birkhoff normal form . . . 42

2 Symplectic capacities
2.1 Definition and application to embeddings 51
2.2 Rigidity of symplectic diffeomorphisms 58

3 Existence of a capacity
3.1 Definition of the capacity c_0 69
3.2 The minimax idea 77
3.3 The analytical setting 82
3.4 The existence of a critical point 91
3.5 Examples and illustrations 98

4 Existence of closed characteristics
4.1 Periodic solutions on energy surfaces 105
4.2 The characteristic line bundle of a hypersurface 113
4.3 Hypersurfaces of contact type, the Weinstein conjecture 119
4.4 "Classical" Hamiltonian systems 127
4.5 The torus and Herman's Non-Closing Lemma 137

5 Compactly supported symplectic mappings in $\mathbb{R}^{2n}$
5.1 A special metric d for a group $\mathcal{D}$ of
    Hamiltonian diffeomorphisms 143
5.2 The action spectrum of a Hamiltonian map 151
5.3 A "universal" variational principle 154
5.4 A continuous section of the action spectrum bundle 161
5.5 An inequality between the displacement energy
    and the capacity 165
5.6 Comparison of the metric d on $\mathcal{D}$ with the C^0-metric 173
5.7 Fixed points and geodesics on $\mathcal{D}$ 182

6 The Arnold conjecture, Floer homology and symplectic homology
- 6.1 The Arnold conjecture on symplectic fixed points 194
- 6.2 The model case of the torus 202
- 6.3 Gradient-like flows on compact spaces 217
- 6.4 Elliptic methods and symplectic fixed points 222
- 6.5 Floer's appraoch to Morse theory for the action functional 250
- 6.6 Symplectic homology 265

Appendix
- A.1 Generating functions of symplectic mappings in $\mathbb{R}^{2n}$ 273
- A.2 Action-angle coordinates, the Theorem of Arnold and Jost 278
- A.3 Embeddings of $H^{1/2}(S^1)$ and smoothness of the action 286
- A.4 The Cauchy-Riemann operator on the sphere 291
- A.5 Elliptic estimates near the boundary and an application 298
- A.6 The generalized similarity principle 302
- A.7 The Brouwer degree 305
- A.8 Continuity property of the Alexander-Spanier cohomology 314

Index 321

Bibliography 327

We dedicate this book to our friend

Andreas Floer

1956–1991

Preface

The discoveries of the past decade have opened new perspectives for the old field of Hamiltonian systems and led to the creation of a new field: symplectic topology. Surprising rigidity phenomena demonstrate that the nature of symplectic mappings is very different from that of volume preserving mappings which raised new questions, many of them still unanswered. On the other hand, due to the analysis of an old variational principle in classical mechanics, global periodic phenomena in Hamiltonian systems have been established. As it turns out, these seemingly different phenomena are mysteriously related. One of the links is a class of symplectic invariants, called symplectic capacities. These invariants are the main theme of this book which grew out of lectures given by the authors at Rutgers University, the RUB Bochum and at the ETH Zürich (1991) and also at the Borel Seminar in Bern 1992. Since the lectures did not require any previous knowledge, only a few and rather elementary topics were selected and proved in detail. Moreover, our selection has been prompted by a single principle: the action principle of mechanics. The action functional for loops in the phase space, given by

$$F(\gamma) = \int_\gamma pdq - \int_0^1 H\big(t, \gamma(t)\big) dt \, ,$$

differs from the old Hamiltonian principle in the configuration space defined by a Lagrangian. The critical points of F are those loops γ which solve the Hamiltonian equations associated with the Hamiltonian H and hence are the periodic orbits. This variational principle is sometimes called the least action principle. However, there is no minimum for F. Indeed, the action principle is very degenerate. All its critical points are saddle points of infinite Morse index, and at first sight, the principle appears quite useless for existence proofs. But surprisingly it is very effective. This will be demonstrated using several variational techniques starting from minimax arguments due to P. Rabinowitz and ending with A. Floer's homology. The book includes the following subjects:

The introductory chapter presents in a rather unsystematic way some background material. We give the definitions of symplectic manifolds and symplectic mappings and briefly recall the Hamiltonian formalism. For convenience, Cartan's calculus is used. The classification of 2-dimensional symplectic manifolds by the Euler-characteristic and the total volume is proved. Some questions dealt with later on in detail are raised and discussed in special examples. We illustrate the so-called direct method of the calculus of variations in order to establish a periodic orbit on a convex energy surface of a Hamiltonian system in $\mathbb{R}^{2n}$. The Birkhoff invariants are introduced in order to describe without proofs the intricate orbit structure of a Hamiltonian system near an equilibrium point or near a periodic solution. These local results are quite in contrast to the global questions dealt with in the following chapters.

In a systematic way the symplectic invariants, called symplectic capacities, are introduced axiomatically in Chapter 2. Considering the family of all symplectic manifolds of fixed dimension $2n$, a capacity c is a map associating with every symplectic manifold (M, ω) a positive number $c(M, \omega)$ or ∞ satisfying these axioms: a monotonicity axiom for symplectic embeddings, a conformality axiom for the symplectic structure, and a normalization axiom which rules out the volume in higher dimensions. For subsets of $\mathbb{R}^{2n}$, the capacity extends a familiar linear symplectic invariant for positive quadratic forms to nonlinear symplectic mappings. If M and N are symplectically diffeomorphic then $c(M, \omega) = c(N, \tau)$. In view of its monotonicity property a capacity represents, in particular, an obstruction to certain symplectic embeddings and it will be used in order to explain rigidity phenomena for symplectic embeddings, discovered by Ya. Eliashberg and M. Gromov. In particular, Gromov's squeezing theorem is deduced using capacities as well as Eliashberg's C^0-stability of symplectic diffeomorphisms. We introduce a notion of a symplectic homeomorphism, a concept which raises many questions. There are many different capacity functions. For example, the size of the largest ball in $\mathbb{R}^{2n}$ which can be symplectically embedded into a symplectic manifold (M, ω) leads to a special capacity called the Gromov width. It is the smallest capacity function originally introduced by M. Gromov. There are many other "embedding" capacities.

Chapter 3 is devoted entirely to a very detailed construction of a distinguished symplectic capacity c_0. It is dynamically defined by means of Hamiltonian systems. It measures the minimal C^0-oscillation of a Hamiltonian function $H : M \to \mathbb{R}$ which allows to conclude the existence of a fast periodic solution of the corresponding Hamiltonian vector field X_H on M. In the special case of a connected 2-dimensional symplectic manifold, the capacity c_0 agrees with the total area. The existence proof is based on the above action principle which is introduced from scratch in its proper functional analytic framework. The interesting aspect of this principle is that it is bounded neither from below nor from above so that standard variational techniques do not apply directly. Techniques going back to P. Rabinowitz permit us to establish effectively distinguished saddle points of the functional representing special periodic solutions of the system. In the special case of a convex, bounded and smooth domain $U \subset \mathbb{R}^{2n}$, the capacity is represented by a distinguished closed characteristic of its boundary ∂U: it has minimal reduced action equal to $c_0(U)$. But, in general, it is rather difficult to compute the invariant c_0. Some of the recent computations based on more advanced techniques of first order elliptic systems and Fredholm theory are presented without proofs. With the construction of the capacity c_0, the proofs of the rigidity phenomena described in Chapter 2 are complete. Due to its special properties this invariant turns out to be useful also for the dynamics of Hamiltonian systems.

PREFACE

In Chapter 4 the dynamical capacity c_0 is applied to an old question of the qualitative theory of Hamiltonian systems originating in celestial mechanics: does a compact energy surface carry a periodic orbit? We shall demonstrate that many well-known global existence results previously obtained by technically intricate proofs emerge immediately from this invariant. The phenomenon is simply this: if a compact hypersurface in a symplectic manifold possesses a neighborhood of finite capacity c_0, then there are always uncountably many closed characteristics nearby. If one poses, in addition, symplectically invariant restrictions, such as of "contact type", then the hypersurface itself carries a closed characteristic. We shall prove, in particular, the seminal solution of the Weinstein conjecture in $\mathbb{R}^{2n}$ due to C. Viterbo. A nonstandard symplectic torus shows that, in contrast to the Gromov width mentioned above, not every compact symplectic manifold is of finite capacity c_0. Our special example is related to M. Herman's celebrated counterexample to the closing lemma which answers a longstanding open question in dynamical systems. M. Herman's "non-closing-lemma" is proved at the end of the chapter.

In Chapter 5 we study the subgroup $\mathcal{D}$ of symplectic diffeomorphisms of $\mathbb{R}^{2n}$ which are generated by time dependent Hamiltonian vector fields of compact support. The distance from the identity map or the energy $E(\varphi)$ of such a symplectic diffeomorphism φ will be measured by means of the oscillation of its generating Hamiltonian function. This will lead to a surprising bi-invariant metric on $\mathcal{D}$ called the Hofer metric and defined by $d(\varphi, \psi) = E(\varphi^{-1} \circ \psi)$. The definition does not involve derivatives of the Hamiltonian and is of C^0-nature. The verification of the metric property requiring that $d(\varphi, \psi) = 0$ if and only if $\varphi = \psi$ is the difficult aspect. It is based on more refined minimax arguments for the action functional valid simultaneously for a large class of Hamiltonians. We shall investigate the relations of this distinguished metric to the dynamical symplectic invariant c_0 introduced in Chapter 3 and also to another symplectic invariant which is defined for subsets of $\mathbb{R}^{2n}$ and called the displacement energy. The displacement energy of a subset U measures the minimal energy $E(\varphi)$ needed in order to dislocate a given set U from itself in the sense that $U \cap \varphi(U) = \emptyset$. The bi-invariant metric will also be compared with the standard sup-metric. Geodesic arcs associated with the metric will be defined and described in detail. A special example of a geodesic arc is the flow generated by an autonomous Hamiltonian. An important role in our approach is played by the action spectrum of a Hamiltonian mapping $\varphi \in \mathcal{D}$, which turns out to be a nowhere dense subset of the real numbers. Our minimax principle singles out a nontrivial continuous section of the action spectrum bundle over $\mathcal{D}$ called the γ-invariant. This invariant is the main technical tool in this chapter. It allows the characterization of the geodesics and is used also in the existence proof of infinitely many nontrivial periodic points for compactly supported Hamiltonian mappings.

The subject of Chapter 6 is the fixed point theory for Hamiltonian mappings on compact symplectic manifolds (M, ω). It differs from topological fixed point theories. A Hamiltonian map is a special symplectic map: it is homotopic to the identity and the homotopy is generated by the flow of a time dependent Hamiltonian vector field. Prompted by H. Poincaré's last geometric theorem, V.I. Arnold conjectured in the sixties that such a Hamiltonian map possesses at least as many fixed points as a real-valued function on M possesses critical points. Reformulated in terms of dynamical systems, the conjecture asks for a Ljusternik-Schnirelman theory respectively for a Morse theory of forced oscillations solving a time periodic Hamiltonian system on M. We shall first prove the conjecture for the special case of the standard torus T^{2n}. The proof is again based on the action principle. But this time the aim is to find all its critical points. Our strategy is inspired by C. Conley's topological approach to dynamical systems: we shall study the topology of the set of all bounded solutions of the regularized gradient equation belonging to the action functional defined on the set of contractible loops on the manifold M. This way the study of the gradient flow in the infinite dimensional loop space is reduced to the study of a gradient like continuous flow of a compact metric space, whose rest points are the desired critical points. Their number is then estimated by Ljusternik-Schnirelman theory presented in 6.3. A reinterpretation will then lead us to the proof of the Arnold conjecture for the larger class of symplectic manifolds satisfying $[\omega]|\pi_2(M) = 0$. In this general case there is no natural regularization and we are forced to investigate in 6.4 the set of bounded solutions of the non regularized gradient system which now are smooth solutions of a special system of first order elliptic partial differential equations of Cauchy Riemann type. These solutions are related to M. Gromov's pseudoholomorphic curves in M. The compactness of the solution set will be based on an analytical technique which is sometimes called bubbling off analysis. Following this procedure, we shall arrive at the high point of these developments: A. Floer's new approach to Morse theory and Floer homology. We shall merely outline Floer's beautiful ideas in 6.5. A combination of Floer's approach with the construction of the dynamical capacity c_0 results in a symplectic homology theory which is not yet in its final form and which will be sketched without proofs in the last section. The technical requirements of these theories are quite advanced and beyond the scope of this book. Floer's ideas and further related developments will be presented in detail in a sequel. Chapter 6 illustrates, in particular, that old problems emerging from celestial mechanics still lead to powerful new techniques useful also in other branches of mathematics. We should point out that the Arnold conjecture for a general symplectic manifold is still open in the dimensions ≥ 8.

The Appendix contains some technical topics presented for the convenience of the reader. In A.1 we show that a symplectic diffeomorphism can be locally represented in terms of a single function, the so-called generating function. This classical fact is used in Chapter 5. Appendix A.2 illustrates the generating functions in the construction of action-angle coordinates for integrable systems occurring in Chapter 4. A special Sobolev embedding theorem required in the analysis of the action functional (Chapter 5) is proved in A.3. We derive some basic estimates

PREFACE   xiii

for the Cauchy-Riemann operator on the sphere (A.4), elliptic estimates near the boundary (A.5) and prove the generalized Carleman similarity principle (A.6); all these results for special partial differential equations are important in Chapter 6. While the analytical tools required in the first five chapters are introduced in detail, we make use of topological tools without explanations: we use the Brouwer mapping degree (Chapter 2), the Leray-Schauder degree (Chapter 3), the Smale degree (mod2) and (co-) homology theories (Chapter 6). References concerning these topological topics are given in A.7 and A.8 where we explain the Brouwer degree and the continuity property of the Alexander-Spanier cohomology. This continuity property is important to us for the proof of the Arnold conjecture in the general case.

Acknowledgements

We are very grateful for many helpful comments, ideas and suggestions, for encouragement to write and to finish this book. In particular, we would like to thank M. Chaperon, Y. Eliashberg, S. Kuksin, J. Moser, D. Salamon, J.C. Sikorav, C. Viterbo and K. Wysocki. We are indebted to M. Bialy and L. Polterovich for giving us their results on geodesics prior to publication. For carefully reading the manuscript, discovering numerous mistakes, correcting them and improving the presentation, we are very grateful to C. Abbas, K. Cieliebak, M. Flucher, A. Göing, B. Gratza, L. Kaas, M. Kriener, M. Schwarz and K.F. Siburg. We enjoyed the interest of the participants of the Borel seminar in Bern and appreciated their questions. We thank R. McLachlan for his valuable help in improving our English, and Claudia Flepp for her efficiency and patience typing and retyping the manuscript.

E.T.H. Zürich, March 28, 1994.

Chapter 1
Introduction

We shall introduce the concepts of symplectic manifolds, symplectic mappings and Hamiltonian vector fields. It is not the intention to give a systematic treatment of the Hamiltonian formalism, because it is already presented in many books. Rather we shall ask some questions related to these concepts which recently lead to new phenomena and interesting open problems. The question: "What can be done with a symplectic mapping?" leads, for example, to new symplectic invariants different from the volume and discussed in detail in subsequent chapters. We shall illustrate that a seemingly very different and old problem originating in celestial mechanics is related to these invariants. Namely, prompted by the Poincaré recurrence theorem, we ask whether a compact energy surface of a Hamiltonian vector field possesses a periodic orbit. For the very special case of a convex hypersurface in $\mathbb{R}^{2n}$, historically one of the landmarks in this qualitative problem of Hamiltonian systems, we shall give an existence proof in order to illustrate the so-called direct method of the calculus of variation. This classical method is in contrast to the more recent methods introduced in the following chapters in order to establish global periodic solutions. At the end of the introduction we shall illustrate without proofs the rich and intricate orbit structure to be expected near a given periodic orbit. The considerations are based on the local, nonlinear Birkhoff-invariants presented in detail.

1.1 Symplectic vector spaces

Definition. A symplectic vector space (V, ω) is a finite dimensional real vector space V equipped with a distinguished bilinear form ω which is antisymmetric and nondegenerate, i.e.,

$$\omega(u,v) = -\omega(v,u), \quad u, v \in V \tag{1.1}$$

and, for every $u \neq 0 \in V$, there is a $v \in V$ satisfying $\omega(u,v) \neq 0$. This nondegeneracy is equivalent to the requirement that the map

$$V \to V^*, \quad v \mapsto \omega(v, \cdot) \tag{1.2}$$

is a linear isomorphism of V onto its dual vector space V^*. An example is the so-called standard symplectic vector space $(\mathbb{R}^{2n}, \omega_0)$ with

$$\omega_0(u,v) = \langle Ju, v \rangle \quad \text{for all} \quad u, v \in \mathbb{R}^{2n}, \tag{1.3}$$

where the bracket denotes the Euclidean inner product in $\mathbb{R}^{2n}$, and where the $2n \times 2n$ matrix J is defined by

$$(1.4) \qquad J = \begin{pmatrix} 0 & 1 \\ -1 & 0 \end{pmatrix}$$

with respect to the splitting $\mathbb{R}^{2n} = \mathbb{R}^n \times \mathbb{R}^n$. Clearly $\det J \neq 0$ and since $J^T = -J$ the form ω_0 is nondegenerate and antisymmetric. We note that

$$(1.5) \qquad J^T = J^{-1} = -J.$$

In particular $J^2 = -1$ and

$$\omega_0(u, Jv) = \langle u, v \rangle.$$

Therefore, J is a complex structure on $\mathbb{R}^{2n}$ compatible with the Euclidean inner product. Recall that a complex structure on a real vector space V is a linear transformation $J: V \to V$ satisfying $J^2 = -1$. It makes V into an n-dimensional complex vector space by defining

$$(\alpha + i\beta)v = \alpha v + \beta J v$$

for $\alpha, \beta \in \mathbb{R}$ and $v \in V$. In the example $(\mathbb{R}^{2n}, \omega_0)$ we may identify $\mathbb{R}^{2n}$ with $\mathbb{C}^n$ in the usual way by mapping $z = (x, y) \in \mathbb{R}^n \times \mathbb{R}^n$ onto $x + iy \in \mathbb{C}^n$. The linear map J corresponds to the multiplication by $-i$ in $\mathbb{C}^n$.

In the following we shall call v orthogonal to u and write $v \perp u$ if $\omega(v, u) = 0$. If E is a linear subspace of V, we define its orthogonal complement by

$$(1.6) \qquad E^\perp = \bigl\{ u \in V \bigm| \omega(v, u) = 0 \quad \text{for all} \quad v \in E \bigr\}.$$

$E^\perp$ is a linear subspace and in view of the nondegeneracy of the bilinear form ω, we have

$$(1.7) \qquad \dim E + \dim E^\perp = \dim V.$$

Indeed, choosing a basis $e_1, \ldots, e_d$ in E, the subspace $E^\perp$ is the kernel of the linearly independent functionals $\omega(e_j, \cdot)$ on V such that $\dim E^\perp = \dim V - \dim E$ as claimed. Since $u \perp v$ is equivalent to $v \perp u$ we see that

$$(1.8) \qquad (E^\perp)^\perp = E.$$

The concept of orthogonality in symplectic geometry differs sharply from that in Euclidean geometry: E and $E^\perp$ need not be complementary subspaces. For example, every vector $v \in V$ is orthogonal to itself since $\omega(v, v) = -\omega(v, v)$. Hence if $\dim E = 1$ we have $E \subset E^\perp$.

1.1 Symplectic vector spaces

We can, of course, restrict the bilinear form ω to a linear subspace $E \subset V$. This restricted form will obviously be antisymmetric but, in general, fails to be nondegenerate. It is nondegenerate on E if and only if

$$E \cap E^\perp = \{0\}, \tag{1.9}$$

which follows immediately from the definitions. In view of (1.7) the statement (1.9) holds precisely if E and $E^\perp$ are complementary, i.e.,

$$E \oplus E^\perp = V.$$

We see that (E, ω) is a symplectic vector space if (1.9) is satisfied and we call E a symplectic subspace. Because of the symmetry of (1.9) in E and $E^\perp$, we conclude that E is symplectic if and only if $E^\perp$ is symplectic.

The following proposition shows that every symplectic space looks like the standard space $(\mathbb{R}^{2n}, \omega_0)$.

Proposition 1. The dimension of a symplectic vector space (V, ω) is even. If $\dim V = 2n$ there exists a basis $e_1, \ldots, e_n, f_1, \ldots, f_n$ of V satisfying, for $i, j = 1, 2, \ldots n$,

$$\omega(e_i, e_j) = 0$$
$$\omega(f_i, f_j) = 0$$
$$\omega(f_i, e_j) = \begin{cases} 1 & \text{if } i = j \\ 0 & \text{if } i \neq j \end{cases}.$$

Such a basis is called a symplectic (or canonical) basis of V. Representing $u, v \in V$ in this basis by

$$u = \sum_{j=1}^n \left(x_j\, e_j + x_{n+j}\, f_j \right)$$
$$v = \sum_{j=1}^n \left(y_j\, e_j + y_{n+j}\, f_j \right)$$

one computes readily that

$$\omega(u, v) = \langle Jx, y \rangle, \quad x, y \in \mathbb{R}^{2n},$$

where the matrix J is defined by (1.4). The subspaces $V_j = \text{span}\,\{e_j, f_j\}$ are symplectic and orthogonal to each other if $i \neq j$, so that the vector space V is the orthogonal sum

$$V = V_1 \oplus V_2 \oplus \cdots \oplus V_n \tag{1.10}$$

of 2-dimensional symplectic subspaces. With respect to this splitting the bilinear form ω is, in symplectic coordinates, represented by the matrix

$$\begin{pmatrix} \begin{pmatrix} 0 & 1 \\ -1 & 0 \end{pmatrix} & & & \\ & \begin{pmatrix} 0 & 1 \\ -1 & 0 \end{pmatrix} & & \\ & & \ddots & \\ & & & \begin{pmatrix} 0 & 1 \\ -1 & 0 \end{pmatrix} \end{pmatrix}.$$

Proof of Proposition 1. Choose any vector $e_1 \neq 0$ in V. Since ω is nondegenerate we find $u \in V$ satisfying $\omega(u, e_1) \neq 0$, and we can normalize $f_1 = \alpha u$ such that

$$\omega(f_1, e_1) = 1.$$

Consequently, f_1 and e_1 are linearly independent since ω is antisymmetric so that $E = \text{span}\{e_1, f_1\}$ is a 2-dimensional symplectic subspace of V. If $\dim V = 2$ the proof is finished. If $\dim V > 2$ we apply the same argument to the complementary symplectic subspace $E^\perp$ of V and thus find the desired basis in finitely many steps. ∎

We see that for fixed dimension every symplectic vector space (V, ω) can be put into the same normal form, quite in contrast to the situation of nondegenerate symmetric bilinear forms. The symplectic form ω singles out those linear maps of v which leave the form invariant.

Definition. A linear map $A : V \to V$ of a symplectic vector space (V, ω) is called symplectic (or canonical) if

$$A^*\omega = \omega.$$

By definition, $A^*\omega$ is the so-called pullback 2-form given by $A^*\omega(u,v)=\omega(Au, Av)$. In the standard space $(\mathbb{R}^{2n}, \omega_0)$ a matrix A is, therefore, symplectic if and only if $\langle JAu, Av \rangle = \langle Ju, v \rangle$ for all $u, v \in \mathbb{R}^{2n}$, or equivalently,

(1.11) $$A^T J A = J.$$

In the special case $\mathbb{R}^2$ of two dimensions this is equivalent to the condition that $\det A = 1$. In general we conclude from (1.11) immediately that $(\det A)^2 = 1$. It turns out that

(1.12) $$\det A = 1,$$

so that symplectic matrices in $\mathbb{R}^{2n}$ are volume-preserving. This requires a proof and it is convenient to use the language of differential forms. Recall that, with the

1.1 Symplectic vector spaces

coordinates $z = (z_1, \ldots, z_{2n}) \in \mathbb{R}^{2n}$, the bilinear form $dz_i \wedge dz_j$ on $\mathbb{R}^{2n}$ is defined by
$$(dz_i \wedge dz_j)(u,v) = u_i v_j - u_j v_i,$$
for $u, v \in \mathbb{R}^{2n}$. Introducing the notation $z = (x, y) \in \mathbb{R}^{2n}$, we can, therefore, represent ω_0 in the form
$$\omega_0 = \sum_{j=1}^{n} dy_j \wedge dx_j.$$

Then the $2n$-form
$$\Omega = \omega_0 \wedge \omega_0 \wedge \ldots \wedge \omega_0 \quad (n \text{ times })$$
on $\mathbb{R}^{2n}$ is the volume form
$$\Omega = c\, dx_1 \wedge \ldots \wedge dx_n \wedge dy_1 \wedge \ldots \wedge dy_n$$
with a constant $c \neq 0$. If A is a matrix in $\mathbb{R}^{2n}$ then $A^*\Omega = (\det A)\Omega$ by the definition of a determinant. Assuming that $A^*\omega_0 = \omega_0$ we conclude $A^*\Omega = \Omega$ and, hence, $\det A = 1$ as claimed.

The set of symplectic matrices in $\mathbb{R}^{2n}$, which meet the conditions (1.11), is a group under matrix multiplication. It is one of the classical Lie groups and is denoted by $Sp(n)$.

Proposition 2. *If A and $B \in Sp(n)$ then $A^{-1}, AB \in Sp(n)$. Moreover, $A^T \in Sp(n)$ and $J \in Sp(n)$.*

Proof. By multiplying $A^T J A = J$ with A^{-1} from the right and with $(A^T)^{-1}$ from the left, we find $J = (A^T)^{-1} J A^{-1}$ so that $A^{-1} \in Sp(n)$. Taking now the inverse on both sides of the latter identity we find $J^{-1} = A J^{-1} A^T$, and since $J^{-1} = -J$ we find $(A^T)^T J A^T = J$ and $A^T \in Sp(n)$. ∎

Note that if a $2n$ by $2n$ matrix is written in block form

(1.13)
$$U = \begin{pmatrix} A & B \\ C & D \end{pmatrix}$$

with respect to the splitting $\mathbb{R}^{2n} = \mathbb{R}^n \times \mathbb{R}^n$, it is symplectic if and only if
$$A^T C,\ B^T D \quad \text{are symmetric and} \quad A^T D - C^T B = 1,$$
as is readily verified. For example, a matrix U having $B = 0$ is symplectic if and only if A is nonsingular and U can be written as
$$U = \begin{pmatrix} A & 0 \\ 0 & (A^T)^{-1} \end{pmatrix} \begin{pmatrix} 1 & 0 \\ S & 1 \end{pmatrix},$$
with some symmetric matrix S.

Definition. If (V_1, ω_1) and (V_2, ω_2) are two symplectic vector spaces we call a linear map $A : V_1 \to V_2$ symplectic if
$$A^* \omega_2 = \omega_1,$$
where, by definition, $(A^* \omega_2)(u, v) = \omega_2(Au, Av)$ for all $u, v \in V_1$. Clearly A is injective such that $\dim V_1 \leq \dim V_2$.

Proposition 3. If (V_1, ω_1) and (V_2, ω_2) are two symplectic spaces of the same dimension, then there exists a linear isomorphism $A : V_1 \to V_2$ satisfying $A^* \omega_2 = \omega_1$.

This means that all symplectic vector spaces of the same dimension are, in this sense, equivalent; they are symplectically indistinguishable.

Proof. The proof follows immediately from the normal form in Proposition 1. Choosing symplectic bases (e_j, f_j) in (V_1, ω_1) and $(\hat{e}_j, \hat{f}_j)$ in (V_2, ω_2) we define the linear map $A : V_1 \to V_2$ by
$$A e_j = \hat{e}_j \quad \text{and} \quad A f_j = \hat{f}_j$$
for $1 \leq j \leq n$. Then clearly $A^* \omega_2 = \omega_1$ by definition of a symplectic basis. ∎

Since the choice of e_1 and $\hat{e}_1$ in the construction of the symplectic bases is at our disposal we conclude from the above proof that the group $Sp(n)$ acts transitively in $\mathbb{R}^{2n}$. Moreover, it also acts transitively on the set of symplectic subspaces of $\mathbb{R}^{2n}$ having the same dimension. This follows because a symplectic basis in a subspace E can always be completed to a basis of V by adding a symplectic basis of its complement $E^\perp$, as we did in the proof of Proposition 1.

1.2 Symplectic diffeomorphisms and Hamiltonian vector fields in $(\mathbb{R}^{2n}, \omega_0)$

We now turn to nonlinear maps in the symplectic space $(\mathbb{R}^{2n}, \omega_0)$. A diffeomorphism φ in $\mathbb{R}^{2n}$ is called symplectic if

(1.14) $$\varphi^* \omega_0 = \omega_0,$$

where, by definition, the pullback of any 2-form ω is given by
$$(\varphi^* \omega)_x(a, b) = \omega_{\varphi(x)}(\varphi'(x) a, \varphi'(x) b)$$
for $x \in \mathbb{R}^{2n}$ and for all $a, b \in T_x \mathbb{R}^{2n} = \mathbb{R}^{2n}$. Here $\varphi'(x)$ denotes the derivative of φ at the point x represented by the Jacobian matrix. In view of the definition of ω_0, a symplectic diffeomorphism in $(\mathbb{R}^{2n}, \omega_0)$ is, therefore, characterized by the identity

(1.15) $$\varphi'(x)^T J \varphi'(x) = J, \quad x \in \mathbb{R}^{2n}$$

1.2 Symplectic diffeomorphisms and Hamiltonian vector fields

for the first derivatives of φ. Hence $\varphi'(x)$ is a symplectic matrix and, in particular,

(1.16) $$\det \varphi'(x) = 1,$$

so that symplectic diffeomorphisms are volume-preserving. However, if $n > 1$ the class of symplectic diffeomorphisms is much more restricted than that of volume-preserving diffeomorphisms. This will become clear below when, taking our lead from Gromov, we look at the question: what can be done with symplectic diffeomorphisms?

A symplectic diffeomorphism φ in $(\mathbb{R}^{2n}, \omega_0)$ not only preserves ω_0 and the associated volume form Ω but also the action of closed curves, as we shall see next. The form ω_0 is an exact form, since

(1.17) $$\omega_0 = \sum_{j=1}^{n} dy_j \wedge dx_j = d\lambda,$$

with the 1-form λ defined by

$$\lambda = \sum_{j=1}^{n} y_j \, dx_j.$$

Therefore, $\lambda - \varphi^*\lambda$ is a closed form provided φ is symplectic. Indeed, $d(\lambda - \varphi^*\lambda) = d\lambda - d(\varphi^*\lambda) = d\lambda - \varphi^* d\lambda = \omega_0 - \varphi^*\omega_0 = 0$. Using the Poincaré lemma one finds a function $F : \mathbb{R}^{2n} \to \mathbb{R}$ satisfying

(1.18) $$\lambda - \varphi^*\lambda = dF.$$

If γ is an oriented simply closed curve we can integrate and find in view of (1.18)

$$\int_\gamma \lambda = \int_\gamma \varphi^*\lambda = \int_{\varphi(\gamma)} \lambda$$

since the integral of an exact form over a closed curve vanishes. Defining the action $A(\gamma)$ of a closed curve γ by

(1.19) $$A(\gamma) = \int_\gamma \lambda \in \mathbb{R},$$

we see that

(1.20) $$A\big(\varphi(\gamma)\big) = A(\gamma)$$

provided φ is symplectic; hence φ leaves the action invariant as claimed. Conversely, of course, if a diffeomorphism φ in $\mathbb{R}^{2n}$ satisfies (1.20) for all closed curves in $\mathbb{R}^{2n}$ we conclude that φ is symplectic. Parameterizing γ by $x(t)$, with $0 \le t \le 1$ and $x(0) = x(1)$, the action becomes

$$(1.21) \qquad A(\gamma) = \frac{1}{2} \int_0^1 \langle -J\dot{x}, x \rangle \, dt.$$

Examples of symplectic diffeomorphisms are generated by the so-called Hamiltonian vector fields which we now recall. To the symplectic form ω_0 and to a smooth function

$$H : \mathbb{R}^{2n} \to \mathbb{R},$$

we can associate a vector field X_H on $\mathbb{R}^{2n}$ by requiring

$$(1.22) \qquad \omega_0 \Big(X_H(x), a \Big) = -dH(x)a$$

for all $a \in \mathbb{R}^{2n}$ and $x \in \mathbb{R}^{2n}$. Since ω_0 is nondegenerate, the vector $X_H(x)$ is determined uniquely. The condition (1.22) is equivalent to $\langle JX_H(x), a \rangle = -\langle \nabla H(x), a \rangle$ where the gradient of H is, as usual, defined with respect to the Euclidean inner product. Therefore, $JX_H(x) = -\nabla H(x)$ and in view of $J^2 = -1$ we find the representation

$$(1.23) \qquad X_H(x) = J\nabla H(x), \quad x \in \mathbb{R}^{2n}.$$

Clearly the Hamiltonian vector fields are very special. They differ in particular sharply from vector fields $X = \nabla H(x)$ of gradient type, since J is antisymmetric.

In the following we denote by φ^t the flow of a vector field X. It is defined by

$$\frac{d}{dt}\varphi^t(x) = X\Big(\varphi^t(x)\Big)$$

$$\varphi^0(x) = x, \quad x \in \mathbb{R}^{2n}.$$

The curve $x(t) = \varphi^t(x)$ solves the Cauchy initial value problem for the initial condition $x \in \mathbb{R}^{2n}$. Assume now that $X = X_H$ is the Hamiltonian vector field determined by ω_0 and H. Then every flow map φ^t preserves the form ω_0:

$$(1.24) \qquad (\varphi^t)^* \omega_0 = \omega_0,$$

and is, therefore, a symplectic map. This is easily verified and will be proved in the next section in a more abstract setting.

It is useful for the following to recall the transformation formula for vector fields X on $\mathbb{R}^m$. Assume $x(t)$ is a solution of the differential equation

$$\dot{x} = X(x), \quad x \in \mathbb{R}^m.$$

If $u : \mathbb{R}^m \to \mathbb{R}^m$ is a diffeomorphism we can define the curve $y(t)$ by
$$x(t) = u\big(y(t)\big).$$
Differentiating in t we conclude that $y(t)$ solves the equation
$$\dot y = Y(y), \quad y \in \mathbb{R}^m$$
for the transformed vector field Y defined by
$$Y(y) = du(y)^{-1} \cdot X \circ u(y).$$
In the following, we shall use the notation
$$(1.25) \qquad u^* X := (du)^{-1} \cdot X \circ u.$$
We have demonstrated that the two flows φ^t of X and ψ^t of $u^* X$ are conjugated by the diffeomorphism u, i.e.,
$$\varphi^t \circ u = u \circ \psi^t.$$
If we subject a Hamiltonian vector field X_H in $\mathbb{R}^{2n}$ to an arbitrary transformation u its special form will be destroyed. However, a symplectic transformation preserves the class of Hamiltonian vector fields. Indeed, if $u^* \omega_0 = \omega_0$ then
$$(1.26) \qquad u^* X_H = X_K \quad \text{and} \quad K = H \circ u.$$
This is easily verified: defining the function K as the composition $K = H \circ u$, then by the chain rule, $dK = dH \circ u \cdot du$, and the gradient with respect to the Euclidean scalar product becomes $\nabla K = (du)^T \nabla H \circ u$. By assumption, du is, at every point, a symplectic map and, therefore, also $(du)^T$ so that $du \cdot J \cdot (du)^T = J$. Consequently, in view of the definition (1.23) of a Hamiltonian vector field
$$\begin{aligned} X_K &= J \nabla K = J (du)^T \nabla H \circ u \\ &= (du)^{-1}(J \nabla H) \circ u \\ &= u^* X_H, \end{aligned}$$
as we set out to prove.

1.3 Hamiltonian vector fields and symplectic manifolds

In order to introduce Hamiltonian vector fields on a manifold, we first have to extend the symplectic structure
$$\omega_0 = \sum_{j=1}^{n} dy_j \wedge dx_j \quad \text{on} \quad \mathbb{R}^{2n}$$
to even dimensional manifolds.

Definition. A symplectic structure on an even dimensional manifold M is a 2-form ω on M satisfying

(i) $\quad d\omega = 0$, i.e., ω is a closed form.

(ii) $\quad \omega$ is nondegenerate.

The second condition requires that for every tangent space $T_x M$: if $\omega_x(u, v) = 0$ for all $v \in T_x M$ then $u = 0$. The pair (M, ω) is then called a symplectic manifold. Thus every tangent space $T_x M$ of a symplectic manifold becomes a symplectic vector space with respect to the distinguished antisymmetric and nondegenerate bilinear form ω_x at x. Therefore, M has even dimension.

An example is the symplectic manifold $(\mathbb{R}^{2n}, \omega_0)$; indeed, since ω_0 is a constant form we have $d\omega_0 = 0$. Since the symplectic form ω is assumed to be closed, every symplectic manifold looks, locally, like $(\mathbb{R}^{2n}, \omega_0)$; we shall now prove that there are always local coordinates in which the symplectic form is represented by the constant form ω_0.

Theorem 1. (Darboux) Suppose ω is a nondegenerate 2-form on a manifold of $\dim M = 2n$. Then $d\omega = 0$ if and only if at each point $p \in M$ there are coordinates (U, φ) where $\varphi : (x_1, \ldots, x_n, y_1, \ldots, y_n) \to q \in U \subset M$ satisfies $\varphi(0) = p$ and

$$\varphi^* \omega = \omega_0 = \sum_{j=1}^{n} dy_j \wedge dx_j.$$

Such coordinates are sometimes called symplectic coordinates. They are clearly not determined uniquely; the most general coordinates of this sort are related to (x, y) by symplectic transformation $u^* \omega_0 = \omega_0$ in $\mathbb{R}^{2n}$, as previously introduced. We see that we can define a symplectic manifold alternatively as follows: it is a manifold of $\dim M = 2n$ for which there are local coordinates φ_j mapping open sets $U_j \subset M$ onto open sets of the fixed symplectic space $(\mathbb{R}^{2n}, \omega_0)$ such that the coordinates changes $\varphi_i \circ \varphi_j^{-1}$ defined on $\varphi_j(U_i \cap U_j)$ are symplectic local diffeomorphisms in $(\mathbb{R}^{2n}, \omega_0)$.

Proof. Choosing any local coordinates, we may assume that ω is a 2-form on $\mathbb{R}^{2n}$ depending on $z \in \mathbb{R}^{2n}$ and that p corresponds to $z = 0$. By a linear change of coordinates we can achieve that the form be in normal form at the origin, i.e.,

$$\omega(0) = \sum_{j=1}^{n} dy_j \wedge dx_j \quad \text{at} \quad z = 0.$$

This is precisely the same as the statement that any nondegenerate antisymmetric bilinear form can be brought into normal form (Proposition 1). With ω_0 we shall denote the constant form $\Sigma dy_j \wedge dx_j$ on $\mathbb{R}^{2n}$. The aim is to find a local diffeomorphism φ in a neighborhood of 0 leaving the origin fixed and solving

$$\varphi^* \omega = \omega_0.$$

1.3 Hamiltonian vector fields and symplectic manifolds

We shall solve this equation by a deformation argument. We interpolate ω and ω_0 by a family ω_t of forms defined by

$$\omega_t = \omega_0 + t(\omega - \omega_0), \quad 0 \leq t \leq 1,$$

such that $\omega_t = \omega_0$ for $t = 0$ and $\omega_1 = \omega$, and look for a whole family φ^t of diffeomorphisms satisfying $\varphi^0 = id$ and

(1.27) $$(\varphi^t)^* \omega_t = \omega_0, \quad 0 \leq t \leq 1.$$

The diffeomorphism φ^t for $t = 1$ will then be the solution to our problem. In order to find φ^t we shall construct a t-dependent vector field X_t generating φ^t as its flow. Differentiating (1.27), such a vector field X_t has to satisfy the identity

(1.28) $$0 = \frac{d}{dt}(\varphi^t)^* \omega_t = (\varphi^t)^* \left\{ L_{X_t} \omega_t + \frac{d}{dt} \omega_t \right\}.$$

Here L_Y denotes the Lie derivative of the vector field Y. Now we use Cartan's identity

(1.29) $$L_X = i_X \circ d + d \circ i_X$$

and the assumption that $d\omega_t = 0$ and find

$$0 = (\varphi^t)^* \left\{ d(i_{X_t} \omega_t) + \omega - \omega_0 \right\}.$$

Hence, X_t has to satisfy the linear equation

(1.30) $$d(i_{X_t} \omega_t) + \omega - \omega_0 = 0.$$

In order to solve this equation we observe that $\omega - \omega_0$ is closed, hence, locally exact by the Poincaré lemma and there is a 1-form λ satisfying

$$\omega - \omega_0 = d\lambda \quad \text{and} \quad \lambda(0) = 0.$$

Since $\omega_t(0) = \omega_0$ the 2-forms ω_t are nondegenerate for $0 \leq t \leq 1$ in an open neighborhood of the origin and hence there is a unique vector field X_t determined by

$$i_{X_t} \omega_t = \omega_t(X_t, \cdot) = -\lambda$$

for $0 \leq t \leq 1$ which then solves the equation (1.30). Since we normalized $\lambda(0) = 0$ we have $X_t(0) = 0$ and there is an open neighborhood of the origin on which the flow φ^t of X_t exists for all $0 \leq t \leq 1$. It satisfies $\varphi^0 = id$ and $\varphi^t(0) = 0$. We can follow our arguments backwards: by construction this family φ^t of diffeomorphisms satisfies

$$\frac{d}{dt}(\varphi^t)^* \omega_t = 0, \quad 0 \leq t \leq 1,$$

hence $(\varphi^t)^* \omega_t = (\varphi^0)^* \omega_0 = \omega_0$ for all $0 \leq t \leq 1$, as we wanted to prove. ∎

The method employed in the proof above is the so-called deformation method of J. Moser. Its unusual aspect is that one searches for a differential equation to be solved. J. Moser introduced the method in [160] in order to prove, in particular, that two symplectic structures ω_0 and ω_1 on a compact manifold M are equivalent, in the sense that
$$\varphi^* \omega_1 = \omega_0$$
for a diffeomorphism φ of M, provided the forms can be deformed into each other within the class of symplectic forms having their periods fixed. Incidentally, the classification of symplectic forms up to equivalence is still open.

From Darboux's normal form we conclude that any two symplectic manifolds having the same dimension are locally indistinguishable: symplectic manifolds do not possess any local symplectic invariants other than the dimension. This is in sharp contrast to Riemannian manifolds: two different metrics generally are not locally isometric, e.g., the Gaussian curvature is an invariant. It is our aim later on to construct global symplectic invariants.

Every manifold M carries a Riemannian structure. In contrast, not every even dimensional manifold admits a symplectic structure. For example, spheres S^{2n} do not admit a symplectic structure if $n \geq 2$. Indeed, arguing by contradiction we assume ω is a symplectic structure. Then $\Omega = \omega \wedge \omega \wedge \ldots \wedge \omega$ (n times) is a volume form, since ω is nondegenerate. But $\omega = d\alpha$ for a 1-form α on S^{2n} since the second de Rham cohomology group vanishes: $H^2(S^{2n}) = 0$. Therefore, $\Omega = d\beta$ with $\beta = \omega \wedge \omega \wedge \ldots \wedge \omega \wedge \alpha$ and by Stokes' theorem
$$\int_{S^{2n}} \Omega = \int_{\partial S^{2n}} \beta = 0$$
which is, of course, not possible for a volume form. This argument evidently applies to any compact manifold M without boundary having $H^{2j}(M) = 0$ for some $1 \leq j \leq n-1$.

Next we introduce the analogue of symplectic maps in $(\mathbb{R}^{2n}, \omega_0)$. A differentiable map $f : M_1 \to M_2$ between two symplectic manifolds (M_1, ω_1) and (M_2, ω_2) is called symplectic if
$$f^* \omega_2 = \omega_1,$$
where, by definition of the pullback of a 2-form ω
$$(f^*\omega)_x(u, v) = \omega_{f(x)}(df(x)u, df(x)v) \quad \text{for all } u, v \in T_x M.$$
Since ω_1 is nondegenerate the tangent map $df(x)$ must be injective at every point and hence $\dim M_1 \leq \dim M_2$. If $\dim M_1 = \dim M_2$ then f is a local diffeomorphism. In the case that f maps a symplectic manifold (M, ω) into itself the condition for f to be symplectic becomes
$$f^*\omega = \omega,$$

1.3 Hamiltonian vector fields and symplectic manifolds

i.e., f preserves the symplectic structure. Expressed in the distinguished local symplectic coordinates defined by Darboux's theorem, this condition for f agrees with our previous condition for a map to be symplectic in $(\mathbb{R}^{2n}, w_0)$. It is useful to point out that locally such a symplectic map can be presented in terms of a single function on $\mathbb{R}^{2n}$, a so-called generating function and we refer to the Appendix for details.

The symplectic structure, being nondegenerate, defines an isomorphism between vector fields X and 1-forms on M given by $X \mapsto w(X, \cdot)$. In particular, if
$$H : M \to \mathbb{R}$$
is a smooth function on M, then dH is a 1-form on M and hence together with w determines the vector field X_H by

$$(1.31) \qquad \left(i_{X_H} w\right)(x) = w\left(X_H(x), \cdot\right) = -dH(x),$$

$x \in M$. This distinguished vector field X_H is called the Hamiltonian vector field belonging to the function H. Since $dw = 0$ we deduce from (1.31) using Cartan's formula $L_X = d i_X + i_X d$ and $ddH = 0$ that

$$(1.32) \qquad L_{X_H} w = 0.$$

We conclude that the maps φ^t belonging to the flow of a Hamiltonian vector field X_H leave the symplectic form invariant,

$$(1.33) \qquad (\varphi^t)^* w = w,$$

hence are symplectic. Indeed, the derivative $\frac{d}{dt}(\varphi^t)^* w = (\varphi^t)^* L_X w = 0$ vanishes in view of (1.32) and since $(\varphi^0)^* w = w$ the claim follows. The set of Hamiltonian vector fields is invariant under symplectic transformations as we shall verify next. Recall that $u^* X = (du)^{-1} X \circ u$ for a vector field X and a diffeomorphism u, and, equivalently, $\varphi^t \circ u = u \circ \psi^t$ for the associated flows φ^t of X and ψ^t of $u^* X$.

Proposition 4. If $u : M \to M$ satisfies $u^* w = w$ then for every function $H : M \to \mathbb{R}$
$$u^* X_H = X_K \quad \text{and} \quad K = H \circ u.$$

Proof. In view of the definition of a Hamiltonian vector field
$$\begin{aligned} i_{X_{H \circ u}} w &= -d(H \circ u) = -u^*(dH) \\ &= u^*(i_{X_H} w) = i_{u^* X_H}(u^* w) \\ &= i_{u^* X_H} w \end{aligned}$$
and since w is nondegenerate the vector fields $X_{H \circ u}$ and $u^* X_H$ must be equal. ∎

From the symplectic structure we shall deduce an auxiliary structure which will be convenient later on. Recall that an almost complex structure on a manifold M associates smoothly with every $x \in M$ a linear map $J = J_x : T_x M \to T_x M$ satisfying $J^2 = -1$.

Proposition 5. If (M, ω) is a symplectic manifold there exists an almost complex structure J on M and a Riemannian metric $\langle \cdot, \cdot \rangle$ on M satisfying

$$\omega_x(v, Ju) = \langle v, u \rangle_x \tag{1.34}$$

for $v, u \in T_x M$. From the symmetry of the bilinear form $\langle \cdot, \cdot \rangle$ it follows that

$$\omega_x(Jv, Ju) = \omega_x(v, u), \tag{1.35}$$

i.e., J is a symplectic map of the symplectic vector space $(T_x M, \omega_x)$. Moreover,

$$J^* = J^{-1} = -J, \tag{1.36}$$

where J^* is the adjoint of J in the inner product space $(T_x M, \langle \cdot, \cdot \rangle_x)$.

Proof. We choose any Riemannian metric g on M. Fixing a point $x \in M$ we shall construct $J = J_x$ in $T_x M$. All the constructions will depend smoothly on x and, for notational convenience, the dependence on x will not be explicitly mentioned. Since ω is nondegenerate there exists a unique linear isomorphism $A : T_x M \to T_x M$ satisfying

$$\omega(u, v) = g(Au, v), \quad u, v \in T_x M.$$

Since ω is antisymmetric we infer $g(Au, v) = \omega(u, v) = -\omega(v, u) = -g(Av, u) = -g(v, A^*u) = g(-A^*u, v)$, where A^* is the g-adjoint map of A. Hence

$$A^* = -A.$$

Consequently, $A^*A = AA^* = -A^2$ is a positive definite g-self-adjoint map and we denote by $Q = \sqrt{-A^2}$ the positive square root of $-A^2$. Set

$$J = AQ^{-1}.$$

Since A and A^* do commute, A is a normal operator and consequently A and Q^{-1} commute and we compute

$$J^2 = AQ^{-1}AQ^{-1} = A^2(-A^2)^{-1} = -id.$$

Finally,

$$\begin{aligned} \omega(u, Jv) &= g(Au, Jv) = g(Au, AQ^{-1}v) \\ &= g(A^*Au, Q^{-1}v) = g(Q^2 u, Q^{-1}v) \\ &= g(Qu, v). \end{aligned}$$

Since Q is symmetric and positive definite we conclude that

$$\langle u, v \rangle := g(Qu, v)$$

defines a Riemannian metric on M which, in general, is different from g. It is the desired metric. The remainder of the statement is now readily verified making use

1.3 HAMILTONIAN VECTOR FIELDS AND SYMPLECTIC MANIFOLDS

of the fact that the metric $\langle u,v\rangle = \langle v,u\rangle$ is symmetric. Since the construction depends smoothly on x the proof is completed. ∎

This almost complex structure compatible with ω extends the complex structure in $(\mathbb{R}^{2n},\omega_0)$ considered above. Moreover, if ∇H denotes the gradient of a function H with respect to the Riemannian metric $\langle \cdot,\cdot\rangle$ of Proposition 5, i.e., $\langle \nabla H(x),v\rangle = dH(x)v$ for all $v \in T_x M$, we find for the Hamiltonian vector field X_H the representation

$$(1.37) \qquad X_H(x) = J\nabla H(x) \in T_x M$$

using that $J^2 = -1$. This agrees with the representation of X_H in $(\mathbb{R}^{2n},\omega_0)$.

We should point out that the almost complex structure is not unique. If we denote by $\mathcal{J}_\omega$ the set of almost complex structures compatible with ω in the sense of (1.34), it can easily be shown that this set is contractible. Indeed, for every $J \in \mathcal{J}_\omega$ there exists, by definition, a unique Riemannian metric g_J satisfying $\omega(u,Jv) = g_J(u,v)$. Starting from any Riemannian metric g, we constructed in the proof of the proposition an almost complex structure $J = J_g$ and a metric g_J such that $J_{g_J} = J$. Hence, fixing any metric g^* on M, we can define the contraction in $\mathcal{J}_\omega$ by

$$(t,J) \mapsto J_{(1-t)g_J + tg^*}$$

for $0 \leq t \leq 1$ and $J \in \mathcal{J}_\omega$.

In view of Darboux's theorem there are locally no symplectic invariants other than the dimension. On the other hand, the total volume is a trivial example of a global symplectic invariant. Indeed, if $u: (M_1,\omega_1) \to (M_2,\omega_2)$ is a symplectic diffeomorphism of M_1 onto M_2 then it follows from $u^*\omega_2 = \omega_1$ that the associated volume forms $\Omega_1 = \omega_1 \wedge \ldots \wedge \omega_1$ (n times) on M_1 and similarly Ω_2 on M_2 are related by

$$(1.38) \qquad u^*\Omega_2 = \Omega_1.$$

Since the diffeomorphism $u: M_1 \to M_2$ preserves the orientation we have

$$\int_{M_1} u^*\Omega_2 = \int_{M_2} \Omega_2$$

and in view of (1.38),

$$\int_{M_1} \Omega_1 = \int_{M_2} \Omega_2,$$

so that the total volumes of Ω_1 and Ω_2 have to agree. Consider now the special case of compact, connected and oriented manifolds of dimension 2, i.e., surfaces. The orientation will be given by a volume form denoted by ω. It evidently is a closed form, since every 3-form on a surface vanishes. Therefore, (M,ω) is a symplectic manifold with the volume form as the symplectic structure.

Assume now that (M_1, ω_1) and (M_2, ω_2) are two such surfaces and let $u : M_1 \to M_2$ be a symplectic diffeomorphism $u^*\omega_2 = \omega_1$, which in dimension 2 is the same as a volume-preserving diffeomorphism, then

$$(1.39) \qquad \int_{M_1} \omega_1 = \int_{M_2} \omega_2,$$

so that the volumes agree. Our aim is to prove the converse. We shall prove that if two compact, connected and oriented surfaces (M_1, ω_1) and (M_2, ω_2) are diffeomorphic (which is the case if their Euler characteristics are equal) and if, in addition, their total volumes agree, then there exists a diffeomorphism $u : M_1 \to M_2$, which satisfies $u^*\omega_2 = \omega_1$. This gives a classification of compact connected 2-dimensional symplectic manifolds according to the Euler characteristic and the total volume. More generally we shall prove the following statement for volume preserving diffeomorphisms.

Theorem 2. (Moser) Assume M is a compact, connected and oriented manifold of dimension m without boundary. If α and β are two volume forms such that their total volumes agree, i.e.,

$$(1.40) \qquad \int_M \alpha = \int_M \beta,$$

then there is a diffeomorphism u of M satisfying $u^*\beta = \alpha$.

Consequently the total volume is the only invariant of volume-preserving diffeomorphisms.

Proof. We proceed as in the proof of Darboux's theorem and deform the volume form α into β defining

$$\alpha_t = (1-t)\alpha + t\beta, \quad 0 \le t \le 1.$$

These forms α_t are volume forms since locally α and β are represented by $\alpha = a(x)dx_1 \wedge \ldots \wedge dx_m$ and $\beta = b(x)dx_1 \wedge \ldots \wedge dx_m$ with nonvanishing smooth functions a and b, which, by assumption (1.40), must have the same sign. We shall construct a family φ^t of diffeomorphisms satisfying

$$(1.41) \qquad \left(\varphi^t\right)^* \alpha_t = \alpha, \quad \varphi^0 = id$$

for $0 \le t \le 1$, so that the diffeomorphism $u = \varphi^1$ will solve our problem. Since M is compact, connected and oriented we conclude from $\int_M (\beta - \alpha) = 0$ that

$$(1.42) \qquad \beta - \alpha = d\gamma$$

1.3 HAMILTONIAN VECTOR FIELDS AND SYMPLECTIC MANIFOLDS

for some $(m-1)$-form γ on M. This is a special case of the de Rham theorem. Since α_t is a volume form we find a unique time-dependent vector field X_t on M solving the linear equation
$$i_{X_t}\alpha_t = -\gamma$$
for $0 \leq t \leq 1$. Denote by φ^t the flow of this vector field X_t satisfying $\varphi^0 = id$. Since M is compact it exists for all t. Since $d\alpha_t = 0$ for volume forms we find, again using Cartan's formula,
$$\begin{aligned}\frac{d}{dt}\left(\varphi^t\right)^*\alpha_t &= \left(\varphi^t\right)^*\left(L_{X_t}\alpha_t + \frac{d}{dt}\alpha_t\right)\\ &= \left(\varphi^t\right)^*\left(d(i_{X_t}\alpha_t) + \beta - \alpha\right),\end{aligned}$$
which vanishes since $d(i_{X_t}\alpha_t) + \beta - \alpha = d(i_{X_t}\alpha_t + \gamma) = 0$ by our choice of the vector field X_t. Therefore (1.41) holds and the proof is finished. ∎

We have seen that symplectic diffeomorphisms are volume-preserving diffeomorphisms so that the total volume is a (trivial) global symplectic invariant. Moreover, for volume-preserving diffeomorphisms, this is the only invariant. One of our aims later on is to establish symplectic invariants other than the volume which will prove that symplectic diffeomorphisms are of a different nature than volume-preserving diffeomorphisms. An example of such a global symplectic invariant is prompted by Darboux's theorem. Recall that this theorem states that locally near every point on a symplectic manifold (M,ω) there is a local diffeomorphism φ from a small ball of $\mathbb{R}^{2n}$ into M satisfying $\varphi^*\omega = \omega_0$. This means in particular that there is always a symplectic embedding of a small open ball $B(r)$ into M,
$$\varphi : \left(B(r), \omega_0\right) \to (M, \omega),$$
where $B(r) = \{x \in \mathbb{R}^{2n} \,|\, |x|^2 < r^2\}$.

We should mention that this is, of course, only a local result; even on $\mathbb{R}^{2n}$ with $n \geq 2$ there exist symplectic structures ω of infinite volume for which there is no diffeomorphism φ of $\mathbb{R}^{2n}$ solving $\varphi^*\omega = \omega_0$. The existence of such an exotic symplectic structure is a deep result due to M. Gromov [107].

We now look for the largest ball $B(r) \subset \mathbb{R}^{2n}$ which can be symplectically embedded into a given symplectic manifold (M,ω) of dimension $\dim M = 2n$, and define
$$D(M,\omega) = \sup\left\{\pi r^2 \,\middle|\, \begin{array}{l}\text{there exists}\\ \text{a symplectic embedding}\end{array} \varphi : \left(B(r),\omega_0\right) \to (M,\omega)\right\}.$$

This is a positive number or ∞ which we shall call the Gromov width of (M,ω). At this point we cannot explain why we choose πr^2 in the definition and not, for example, the volume of $B(r)$. We mention, however, that πr^2 agrees with the action $|A(\gamma)|$ for every closed characteristic γ on the boundary $\partial B(r)$ of the standard sphere of radius r in $\mathbb{R}^{2n}$. This will be explained later on.

It is easy to see that $D(M,\omega)$ is a symplectic invariant. We first show that $D(M,\omega)$ has a monotonicity property in the sense that

$$(1.43) \qquad D(M,\omega) \leq D(N,\tau),$$

if there exists a symplectic embedding $\psi : M \to N$. To every symplectic embedding $\varphi : B(r) \to M$ there is a symplectic embedding $B(r) \to N$ defined by $\psi \circ \varphi$. Therefore, the supremum in the definition of $D(N,\tau)$ is taken over a possibly larger set so that indeed $D(N,\tau) \geq D(M,\omega)$ as claimed. If $f : M \to N$ is a symplectic diffeomorphism of M onto N we can apply the monotonicity property to f and also to f^{-1} and conclude:

Proposition 6. If (M,ω) and (N,τ) are symplectically isomorphic, then $D(M,\omega) = D(N,\tau)$.

We see that the Gromov width is a symplectic invariant. It will turn out, and this is not obvious, that this invariant is quite different form the total volume if $n \geq 2$. The Gromov width is merely one example in the class of symplectic invariants called symplectic capacities which will be introduced in Chapter 2.

1.4 Periodic orbits on energy surfaces

The existence problem we shall briefly describe originates in the search for periodic solutions in celestial mechanics and is, as we shall see in Chapter 4, related to special symplectic invariants. For simplicity we consider the standard symplectic manifold $(\mathbb{R}^{2n}, \omega_0)$. We have seen that a function $H : \mathbb{R}^{2n} \to \mathbb{R}$ together with ω_0 determines the Hamiltonian vector field X_H on $\mathbb{R}^{2n}$ by

$$(1.44) \qquad i_{X_H}\omega_0 = -dH.$$

The flow φ^t of the vector field X_H leaves the function H invariant, i.e.,

$$(1.45) \qquad H\bigl(\varphi^t(x)\bigr) = H(x), \quad x \in \mathbb{R}^{2n}$$

for all t for which the flow is defined, hence H is an integral of X_H. Indeed differentiating (1.45) we find, using the definition of the flow

$$\frac{d}{dt}H(\varphi^t) = dH(\varphi^t) \cdot \frac{d}{dt}\varphi^t = dH(\varphi^t) \cdot X_H(\varphi^t)$$
$$= -\omega_0\bigl(X_H, X_H\bigr) \circ (\varphi^t),$$

which vanishes since ω_0 is antisymmetric. Geometrically, the level sets

$$S = \bigl\{x \in \mathbb{R}^{2n}\,\big|\,H(x) = \text{const}\bigr\} \subset \mathbb{R}^{2n}$$

1.4 Periodic orbits on energy surfaces

are invariant under the flow of X_H. We shall assume now that S is a compact regular energy surface, i.e.,

(1.46) $$dH(x) \neq 0 \quad \text{for } x \in S.$$

Then S is a smooth hypersurface, i.e., a submanifold of $\mathbb{R}^{2n}$ of codimension 1, whose tangent space at $x \in S$ is given by

$$T_x S = \left\{ v \in T_x \mathbb{R}^{2n} \mid dH(x)v = 0 \right\}.$$

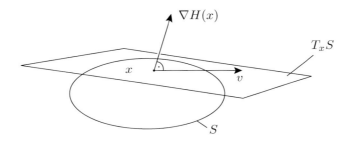

Fig. 1.1

Evidently $X_H(x) \in T_x S$ for $x \in S$ since $dH(x) \cdot X_H(x) = -\omega_0(X_H(x), X_H(x)) = 0$, so that X_H defines a vector field on S which does not vanish. Since S is compact the flow φ^t on S exists for all times. It should be emphasized that this flow is, in general, extremely complicated, since Hamiltonian vector fields describe systems without friction in which oscillations never decrease. However, the flow on S has strong recurrence properties due to the volume-preserving character of Hamiltonian flows. We shall briefly recall this phenomenon known to H. Poincaré.

Let $\Omega = \omega_0 \wedge \ldots \wedge \omega_0$ be the volume form on $\mathbb{R}^{2n}$. Since $dH(x) \neq 0$ in a neighborhood U of S we find a $(2n-1)$-form α on U satisfying

(1.47) $$\Omega = dH \wedge \alpha \quad \text{on } U.$$

Denoting by $j : S \to \mathbb{R}^{2n}$ the inclusion mapping, the $(2n-1)$-form

(1.48) $$\mu = j^*\alpha \quad \text{on } S$$

is clearly a volume form on S. It is uniquely determined by (1.47). Indeed, if $\Omega = dH \wedge \alpha = dH \wedge \beta$ then $dH \wedge (\alpha - \beta) = 0$ so that $\alpha - \beta = dH \wedge \gamma$ for a $(2n-2)$-form γ on U. In view of $j^*dH = 0$ we find $j^*\alpha = j^*\beta + j^*(dH \wedge \gamma) = j^*\beta$ as claimed. The volume form μ is invariant under the flow φ^t of X_H:

(1.49) $$(\varphi^t)^*\mu = \mu.$$

This follows from $(\varphi^t)^*\Omega = \Omega$ and $(\varphi^t)^*dH = dH$. Indeed, in view of (1.47) we first find $\Omega = dH \wedge (\varphi^t)^*\alpha$, and using the uniqueness we conclude $j^*(\varphi^t)^*\alpha = j^*\alpha$. But $\varphi^t \circ j = j \circ \varphi^t$ and hence $(\varphi^t)^*(j^*\alpha) = (j^*\alpha)$, proving the claim (1.49). If we denote the regular measure on S associated to the form μ by m we see that

$$(1.50) \qquad m\big(\varphi^t(A)\big) = m(A), \quad A \subset S.$$

Moreover, $m(S) < \infty$ since S is compact.

Theorem 3. (Poincaré's recurrence theorem) Let S be a compact and regular energy surface of the Hamiltonian vector field X_H with flow φ^t. Then almost every point (with respect to m) on S is a recurrent point, i.e., for almost every $x \in S$ there is a sequence $t_j \uparrow \infty$ satisfying

$$\lim_{j \to \infty} \varphi^{t_j}(x) = x.$$

Proof. The proof is surprisingly easy. By φ we denote the time-1 map of the flow φ^t. We first use the invariance and finiteness of the measure to show that for every $A \subset S$

$$(1.51) \qquad m\Big(A \cap \Big[\bigcap_{k \geq 0} \bigcup_{j \geq k} \varphi^{-j}(A)\Big]\Big) = m(A).$$

Observe that the points x in the above set $[\,\cdot\,]$ are those points $x \in S$ which have the property that for every integer k there is an integer $j \geq k$ such that $x \in \varphi^{-j}(A)$ i.e., $\varphi^j(x) \in A$. The intersection with A consists of the points x in A which return infinitely often to the set A. In order to prove (1.51) we abbreviate

$$A_k := \bigcup_{j \geq k} \varphi^{-j}(A), \quad k = 0, 1, 2, \ldots$$

and have the decreasing sequence of subsets $A_0 \supset A_1 \supset A_2 \supset \ldots$. Clearly, $\varphi^k(A_k) = A_0$ and consequently $m(A_k) = m(A_0)$ since φ preserves the measure. Since $m(A_0) < \infty$ we conclude that $A_0 = A_k$ almost everywhere for $k = 0, 1, 2, \ldots$ and consequently, $\bigcap_{k \geq 0} A_k = A_0$ almost everywhere. Hence, in view of $A \subset A_0$ we find

$$A \cap \bigcap_{k \geq 0} A_k = A \cap A_0 = A \text{ almost everywhere,}$$

which is the desired equation (1.51). Now we use the topological fact that there is a countable basis in S. For every n there are countably many open balls $B_j(\frac{1}{n})$ of radius $\frac{1}{n}$ covering S. Applying the first step to every ball, we find a null set $N \subset S$ having the property that every $x \in S \setminus N$ returns infinitely often to every ball to which it belongs. Since for every n there exists a $j = j(x)$ such that $x \in B_j(\frac{1}{n})$, the theorem is proved. ∎

1.4 Periodic orbits on energy surfaces

We note that the compactness is not relevant: the statement follows if the invariant measure on S is finite and the topology of S has a countable basis. This observation allowed Poincaré to apply his theorem to the restricted 3-body problem.

In view of the recurrence theorem it seems quite natural to search for periodic phenomena on S. One could hope that by perturbing the Hamiltonian system slightly some of the recurrent points do not only return infinitely often but close up in finite time, thus giving rise to a periodic orbit. This is the so-called Closing Problem. In their celebrated Closing-Lemma, C.C. Pugh and C. Robinson ([179] 1983) proved that generically (in the C^2-category of the Hamiltonian functions) the periodic orbits are dense on a compact and regular energy surface. In sharp contrast to this generic phenomenon we are interested in the following global existence question:

Question. Does a compact and regular energy surface S in $(\mathbb{R}^{2n}, \omega_0)$ possess a periodic solution of the Hamiltonian vector field X_H?

The question is still open. It should be emphasized that we are looking for periodic solutions of a very restricted class of vector fields on S and recall that H. Seifert ([193], 1950) raised the question, whether every nonvanishing vector field on the three sphere S^3 has a closed orbit. The problem remained open for many years. In 1974, P.A. Schweitzer [192] constructed a surprising vector field in the class C^1 on S^3 which has no periodic solutions, and only very recently in 1993 K. Kuperberg [132] gave an example of a C^∞ vector field on S^3 with no periodic solutions. For manifolds with dimension higher than three the question had been answered similarly in 1966 by F.W. Wilson [228]. One could ask whether there are vector fields in the more restricted class of measure-preserving smooth vector fields not admitting any periodic solution. For the special class of Reeb vector fields, however, the existence of closed orbits has been established quite recently by H. Hofer [118]: every smooth Reeb vector field on S^3 possesses a periodic orbit. A vector field X on S^3 is called a Reeb vector field, if there exists a 1-form λ on S^3 with $\lambda \wedge d\lambda$ defining a volume form and satisfying

$$i_X d\lambda = 0 \text{ and } i_X \lambda = 1.$$

We shall come back to Reeb vector fields in Chapter 4 in connection with the Weinstein conjecture on contact manifolds.

It will be crucial later on that the above question is independent of the particular choice of the Hamiltonian function representing S. It only depends on the hypersurface and the symplectic structure. This fact is well-known from the regularization of collision singularities in the celestial 2-body problem. Assume that F is a second function defining S, such that

(1.52) $$S = \left\{ x \mid H(x) = c \right\} = \left\{ x \mid F(x) = c' \right\}$$

with $dH \neq 0$ and $dF \neq 0$ on S. Then $\ker dF(x) = \ker dH(x)$ and, therefore, $dF(x) = \rho(x)dH(x)$ for every $x \in S$, with a nowhere-vanishing smooth function ρ. Consequently, the Hamiltonian vector fields are parallel:

(1.53) $$X_F = \rho X_H \quad \text{on} \quad S,$$

with $\rho(x) \neq 0$. It follows that X_H and X_F have the same orbits on S although their parametrization will be different, in general. To be more precise, if φ^t is the flow of X_H on S, then the flow ψ^s of X_F on S is given by

(1.54) $$\psi^s(x) = \varphi^t(x), \quad t = t(s,x), \quad x \in S,$$

where the function t is defined by the differential equation

$$\frac{dt}{ds} = \rho\bigl(\varphi^t(x)\bigr), \quad t(0,x) = 0,$$

depending on the parameter $x \in S$. We see that, in particular, X_H has the same periodic orbits as X_F.

For example, if X_H has the energy surface $S = \{x \in \mathbb{R}^{2n} \mid \frac{1}{2}|x|^2 = R > 0\}$, then all the solutions of X_H on S are periodic. Indeed, also the Hamiltonian F defined by $F(x) = \frac{1}{2}|x|^2$ represents $S = \{x \mid F(x) = R\}$ as a regular energy surface. The vector field X_F is linear and has the flow

$$\psi^s(x) = e^{sJ}x = (\cos s)x + (\sin s)Jx,$$

which evidently is periodic.

We can reformulate the problem more abstractly in terms of a distinguished line bundle. Assume S is a hypersurface in $\mathbb{R}^{2n}$. Then T_xS has dimension $2n-1$ so that the restriction of ω_0 onto T_xS must be degenerate and of rank $(2n-2)$. Its kernel is, therefore, 1-dimensional and we see that ω_0 and S determine the line bundle

(1.55) $$\mathcal{L}_S = \Bigl\{(x,\xi) \in TS \,\Big|\, \omega_0(\xi,v) = 0 \text{ for all } v \in T_xS\Bigr\}.$$

If S is a regular energy surface for X_H, then $\omega_0(X_H(x), v) = -dH(x)v = 0$ for all $v \in T_xS$. Therefore,

$$X_H(x) \in \mathcal{L}_S(x), \quad x \in S$$

for every Hamiltonian vector field having S as a regular energy surface. Consequently, the periodic solutions for X_H on S correspond to the closed characteristics of the line bundle $\mathcal{L}_S$. These are defined as the 1-dimensional submanifold $P \subset S$ diffeomorphic to circles for which the tangent spaces belong to $\mathcal{L}_S$, i.e., $TP = \mathcal{L}_S|P$. Therefore, the existence question of periodic solutions of Hamiltonian equations on regular energy surfaces can be reformulated geometrically as follows:

Question. Does a compact smooth hypersurface $S \subset (\mathbb{R}^{2n}, \omega_0)$ admit a closed characteristic of the distinguished line bundle $\mathcal{L}_S$?

Note that $\mathcal{L}_S$ is defined by the hypersurface and ω_0. It will be demonstrated in Chapter 4 that there are hypersurfaces $S \subset \mathbb{R}^{2n}$ and symplectic structures ω near S and different from ω_0, such that the line bundle of S with respect to ω does not admit closed characteristics.

The breakthrough in this global existence question is due to A. Weinstein [225] and P. Rabinowitz [225] in 1978 who established the existence of a closed characteristic on a convex respectively star-like hypersurface in $\mathbb{R}^{2n}$. The proofs are based on variational principles. In particular P. Rabinowitz demonstrated that the highly degenerate action principle previously used to derive formal transformation properties for Hamiltonian vector fields can be used very effectively for existence proofs. This crucial idea turned out to be decisive in the further development in which the next landmark was C. Viterbo's proof of the A. Weinstein conjecture in $\mathbb{R}^{2n}$ in 1987. The conjecture states that every hypersurface of contact type carries a closed characteristic. This type of symplectically restricted hypersurfaces will be described in Chapter 4, which is devoted to the existence of closed characteristics on hypersurfaces in symplectic manifolds. We should mention that every compact hypersurface S gives rise to an abundance of periodic orbits which, however, are not necessarily on the given surface but on surfaces nearby. We illustrate this with a result which also will be proved in Chapter 4. Assume the hypersurface S belongs to a family defined by

$$S_\varepsilon = \left\{ x \,\middle|\, H(x) = 1 + \varepsilon \right\} \subset \mathbb{R}^{2n}$$

for ε in an interval I around 0, where S corresponds to S_0.

Theorem 4. For almost every $\varepsilon \in I$, the hypersurface S_ε in $\mathbb{R}^{2n}$ possesses a periodic solution for X_H.

In Chapter 4 we shall easily deduce this phenomenon from a distinguished symplectic invariant.

1.5 Existence of a periodic orbit on a convex energy surface

In this section we shall prove a very special qualitative existence result, which historically marked the beginning of a rapid development in global questions of symplectic geometry and Hamiltonian mechanics. Our purpose is to illustrate the classical technique of direct methods of the calculus of variations. This technique is in sharp contrast to the more recent technique introduced in Chapter 3. In Chapter 4 the result itself will be an immediate consequence of the existence of a special symplectic invariant.

Theorem 5. Assume the hypersurface $S \subset (\mathbb{R}^{2n}, \omega_0)$ is the smooth (in the class C^2) boundary of a compact and strictly convex region in $\mathbb{R}^{2n}$. Then S carries a closed orbit.

We conclude, in particular, that every Hamiltonian vector field X_H having S as a regular energy surface possesses a periodic solution on S. The result is due to P. Rabinowitz [180] and A. Weinstein [225], 1978. The proof we shall give below is based on an idea due to F. Clarke [53].

In order to prove the theorem we make use of the freedom to choose a convenient Hamiltonian function and introduce a particular Hamiltonian H which is positively homogeneous of degree 2 and which represents the hypersurface as $S = \{x \in \mathbb{R}^{2n} \mid H(x) = 1\}$. S is the boundary of a compact and strictly convex region C in $\mathbb{R}^{2n}$, and we may assume that C contains the origin in its interior. Then each ray issuing from the origin meets S in exactly one point nontangentially. Thus if $x \neq 0$ is given, there is a unique point $\xi = \lambda^{-1} x$ on S, where $\lambda > 0$.

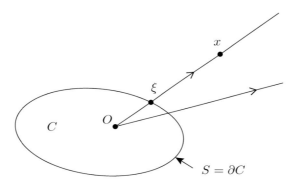

Fig. 1.2

If we define the function F on $\mathbb{R}^{2n}$ by $F(x) = \lambda$ if $\xi = \lambda^{-1} x \in S$ for $x \neq 0$, and $F(0) = 0$, then the manifold S is represented by $S = \{x \mid F(x) = 1\}$. We would like to describe S in terms of a strictly convex function H as the level set $\{x \mid H(x) = 1\}$, i.e., a function whose Hessian $H_{xx}(x)$ is positive definite at every point $x \neq 0$. Since F is homogeneous of degree 1, it is obviously never strictly convex. Indeed, differentiating Euler's formula $\langle F_x, x \rangle = F$ gives $F_{xx} x = 0$. By the convexity assumption $F_{xx} \mid TS > 0$. Therefore, we define the function

$$H(x) = \Big(F(x)\Big)^2$$

which is strictly convex precisely if S is strictly convex as one proves without difficulty. We thus have shown that every strictly convex energy surface S containing

1.5 Existence of a periodic orbit on a convex energy surface

the origin in its interior can be represented by

(1.56) $$S = \{x \mid H(x) = 1\}$$

with a function $H : \mathbb{R}^{2n} \longrightarrow \mathbb{R}$ satisfying

(1.57)
    (i) $H \in C^2(\mathbb{R}^{2n}\setminus\{0\})$, $H(0) = 0$
    (ii) $H_{xx}(x) > 0$ if $x \neq 0$
    (iii) $H(\rho x) = \rho^2 H(x)$ if $\rho \geq 0$.

In view of our discussion in the previous section, it suffices to show that this special Hamiltonian vector field X_H possesses a periodic solution on this energy surface S, i.e., we have to show that there is a periodic solution of

$$\dot{x} = J\nabla H(x) \quad \text{on} \quad S.$$

Normalizing the period we can ask for a periodic solution of period 2π on S for the equation

(1.58) $$\dot{x} = \lambda \, J\nabla H(x) \quad \text{for some} \quad \lambda \neq 0.$$

For example, the variational principle

$$\min \int_0^{2\pi} H\big(x(t)\big)\, dt \quad \text{under} \quad \frac{1}{2}\int_0^{2\pi} \langle Jx, \dot{x}\rangle = 1 \, ,$$

where the functions $x(t)$ are assumed to be 2π-periodic, has the above equations as Euler equations. But one easily shows that neither the infimum nor the supremum is taken on, even if $H(x) = \frac{1}{2}|x|^2$. The trick now is to use an alternative variational principle having the same differential equations for which, however, the infimum is taken on. We shall form the Legendre transformation of H, but in contrast to the usual Legendre transformation of physics, all variables will be transformed. The function $G(y)$ related to $H(x)$ by a Legendre transformation can be defined by

$$G(y) = \max_{\xi \in \mathbb{R}^{2n}} \Big(\langle \xi, y\rangle - H(\xi)\Big) = \langle x, y\rangle - H(x).$$

There is indeed a unique maximum $x \in \mathbb{R}^{2n}$ given by $y = \nabla H(x)$, since $H_{xx}(y)$ is positive definite for $y \neq 0$ and since H satisfies

(1.59) $$\frac{1}{c}|y|^2 \leq H(y) \leq c\,|y|^2, \; y \in \mathbb{R}^{2n}$$

for some constant $c > 1$. The estimate is an immediate consequence of the homogeneity of H. Summarizing, we have

(1.60) $$G(y) + H(x) = \langle x, y\rangle$$

for
(1.61) $$y = \nabla H(x) \quad \text{and} \quad x = \nabla G(y).$$

Clearly $G(0) = 0$, $G \in C^2(\mathbb{R}^{2n}\setminus\{0\})$ and G is positively homogeneous of degree 2. Since $H_{xx}(x) \cdot G_{yy}(y) = \mathbb{1}$ for $x \neq 0$, G is also strictly convex if $y \neq 0$, so that G enjoys the same properties as H:

(i) $\quad G(0) = 0, \; G \in C^2(\mathbb{R}^{2n}\setminus\{0\})$

(ii) $\quad G_{yy}(y) > 0 \quad \text{if} \quad y \neq 0$

(iii) $\quad G(\rho y) = \rho^2 G(y) \quad \text{for} \quad \rho \geq 0.$

We now look at the new, alternate, variational principle for 2π-periodic functions z:

(1.62) $$\min \int_0^{2\pi} G(\dot z)dt \quad \text{under} \quad \frac{1}{2}\int_0^{2\pi} \langle Jz, \dot z\rangle = 1,$$

which we shall solve by the standard direct variational methods. To be more precise, we define the following function space $\mathcal{F}$ of periodic functions $z(t) = z(t+2\pi)$ having mean value 0:

$$\mathcal{F} = \{z \in H_1(S^1) \mid \frac{1}{2\pi}\int_0^{2\pi} z(t)dt = 0\},$$

where $H_1(S^1)$ is the Hilbert space of absolutely continuous 2π-periodic functions whose derivatives are square integrable, i.e., belong to $L_2(S^1)$. Note that with z also $z+$ const. is a solution of (1.62), hence the condition on the mean value fixes the constant. By $\mathcal{A} \subset \mathcal{F}$ we shall denote the subset

$$\mathcal{A} = \{z \in \mathcal{F} \mid \frac{1}{2}\int_0^{2\pi} \langle Jz, \dot z\rangle = 1\}.$$

Following the standard procedure we shall verify that the functional

$$I(z) = \int_0^{2\pi} G\bigl(\dot z(t)\bigr)dt \quad \text{on} \quad \mathcal{A}$$

meets the following properties:

(i) It is bounded from below on $\mathcal{A}$.

1.5 EXISTENCE OF A PERIODIC ORBIT ON A CONVEX ENERGY SURFACE

(ii) It takes on its minimum on $\mathcal{A}$, i.e., there exists $z_* \in \mathcal{A}$ satisfying
$$\int_0^{2\pi} G(\dot{z}_*)dt = \inf_{z \in \mathcal{A}} \int_0^{2\pi} G(\dot{z})dt = \mu > 0.$$

(iii) z_* solves the Euler equations
$$\nabla G(\dot{z}_*) = \alpha J z_* + \beta$$
in $L^2(S^1)$ with constants $\alpha \neq 0$ and β.

(iv) z_* belongs to C^2 and satisfies
$$\dot{z}_* = \nabla H(\alpha J z_* + \beta)$$
pointwise, and $x(t) = c\,(\alpha J z_*(t) + \beta)$ is the desired solution of $\dot{x} = J \nabla H(x)$ on S for an appropriate constant c.

Ad (i): We need some estimates. Since every z belonging to $\mathcal{F}$ has mean value zero, the Poincaré inequality
$$\|z\| \leq \|\dot{z}\|, \quad z \in \mathcal{F}$$
holds, where $\|\cdot\|$ denotes the L_2 norm. This follows simply from the Fourier series. For $z \in \mathcal{A}$ we conclude

(1.63) $$2 = \int_0^{2\pi} \langle Jz, \dot{z}\rangle \leq \|z\|\,\|\dot{z}\| \leq \|\dot{z}\|^2, \quad z \in \mathcal{A}.$$

The function G being strictly convex and positively homogeneous of degree 2 satisfies the estimate $\frac{1}{K}|y|^2 \leq G(y) \leq K|y|^2$ for $y \in \mathbb{R}^{2n}$ with some constant $K \geq 1$. Therefore, by means of (1.63), we find for $z \in \mathcal{A}$

(1.64) $$\int_0^{2\pi} G(\dot{z})dt \geq \frac{1}{K}\|\dot{z}\|^2 \geq \frac{2}{K} > 0,$$

i.e., the functional is bounded from below by a positive constant, in particular, $\mu > 0$.

Ad (ii): We choose a minimizing sequence $z_j \in \mathcal{A}$ such that

(1.65) $$\lim_{j\to\infty} \int_0^{2\pi} G(\dot{z}_j)dt = \mu.$$

In view of (1.63) and (1.64) there is a constant $M > 0$ with $2 \leq \|\dot{z}_j\|^2 \leq M$ and we obtain $\|z_j\| \geq 2\|\dot{z}_j\|^{-1} \geq \frac{2}{\sqrt{M}}$ and hence, by Poincaré's inequality $\frac{2}{\sqrt{M}} \leq$

$\|z_j\| \leq \sqrt{M}$. In particular z_j is a bounded sequence in the Hilbert space $H_1(S^1)$ and, therefore, a subsequence, also denoted by z_j converges weakly in $H_1(S^1)$ to an element $z_* \in H_1(S^1)$:

$$\tag{1.66} z_j \to z_* \quad \text{weakly in } H_1(S^1).$$

We have used the well-known fact that the closed unit ball of a Hilbert space is weakly sequentially compact. In addition, for a function $\hat{z}_* \in C(S^1)$,

$$\tag{1.67} \sup_t |z_j(t) - \hat{z}_*(t)| \to 0,$$

i.e., z_j converges uniformly to $\hat{z}_*$. Indeed the z_j are equicontinuous:

$$\begin{aligned} |z_j(t) - z_j(\tau)| &\leq |\int_\tau^t \dot{z}_j(s) ds| \\ &\leq |t-\tau|^{1/2} \|\dot{z}_j\| \\ &\leq |t-\tau|^{1/2} \sqrt{M}, \end{aligned}$$

and the claim follows from the Arzelà-Ascoli theorem. Clearly $z_*(t) = \hat{z}_*(t)$ almost everywhere. Indeed, with the L^2 inner product we can write $(z_* - \hat{z}_*, z_* - \hat{z}_*) = (z_j - \hat{z}_*, z_* - \hat{z}_*) + (z_* - z_j, z_* - \hat{z}_*)$. The first term on the right hand side converges to 0 in view of the uniform convergence and the second one in view of the weak convergence.

We shall show that $z_* \in \mathcal{A}$. By (1.67) we conclude that the mean value of z_* vanishes. Also,

$$2 = \int_0^{2\pi} \langle J z_j, \dot{z}_j \rangle = \int_0^{2\pi} \langle J(z_j - z_*), \dot{z}_j \rangle + \int_0^{2\pi} \langle J z_*, \dot{z}_j \rangle.$$

The first term on the right hand side tends to 0 by (1.67) since $\|\dot{z}_j\| \leq \sqrt{M}$ is bounded, and the second term converges by (1.66) so that

$$\int_0^{2\pi} \langle J z_*, \dot{z}_* \rangle = 2$$

and therefore $z_* \in \mathcal{A}$. We next show that this z_* is a minimum. From the convexity of G we deduce the pointwise estimate $\langle \nabla G(y_1), y_2 - y_1 \rangle \leq G(y_2) - G(y_1) \leq \langle \nabla G(y_2), y_2 - y_1 \rangle$, which applied to $\dot{z}_*(t) = y_2$ and $y_1 = \dot{z}_j(t)$ gives

$$\int_0^{2\pi} G(\dot{z}_*) dt - \int_0^{2\pi} G(\dot{z}_j) dt \leq \int_0^{2\pi} \langle \nabla G(\dot{z}_*), \dot{z}_* - \dot{z}_j \rangle dt.$$

1.5 EXISTENCE OF A PERIODIC ORBIT ON A CONVEX ENERGY SURFACE

Observing that ∇G, being positively homogeneous of degree 1, satisfies an estimate $|\nabla G(y)| \leq C|y|$ for $y \in \mathbb{R}^{2n}$ and for some constant $C > 0$, we conclude that $\nabla G(\dot{z}_*)$ belongs to L_2 so that the right hand side tends to zero, since $(\dot{z}_* - \dot{z}_j) \to 0$ weakly in L_2. Thus

$$\mu \leq \int_0^{2\pi} G(\dot{z}_*) \leq \liminf_{j \to \infty} \int_0^{2\pi} G(\dot{z}_j) dt = \mu$$

by (1.65) and we have proved that z_* is a minimum of $I(z)$, for $z \in \mathcal{A}$.

Ad (iii): Since z_* is a minimum, we have

(1.68) $$\int_0^{2\pi} \langle \nabla G(\dot{z}_*), \dot{\zeta} \rangle = 0$$

for every test function $\zeta \in \mathcal{F}$ satisfying

(1.69) $$\int_0^{2\pi} \langle J z_*, \dot{\zeta} \rangle = 0.$$

We now choose ζ of the form

$$\dot{\zeta} = \nabla G(\dot{z}_*) - \alpha J z_* - \beta$$

with two constants α und β. In order that ζ be periodic we pick β so that the mean value of $\dot{\zeta}$ is zero:

$$\int_0^{2\pi} \dot{\zeta} dt = \int_0^{2\pi} \nabla G(\dot{z}_*) dt - 2\pi \beta = 0 ,$$

and next determine the constant α so that (1.69) holds true. Using again that z_* has mean value zero we find that the equation

(1.70) $$0 = \int_0^{2\pi} \langle Jz_*, \dot{\zeta} \rangle = \int_0^{2\pi} \langle Jz_*, \nabla G(\dot{z}_*) \rangle - \alpha \int_0^{2\pi} \langle Jz_*, Jz_* \rangle$$

has a unique solution α, since its coefficient does not vanish. With this test function ζ we compute

$$\int_0^{2\pi} |\dot{\zeta}|^2 = \int_0^{2\pi} \langle \nabla G(\dot{z}_*), \dot{\zeta} \rangle - \alpha \int_0^{2\pi} \langle Jz_*, \dot{\zeta} \rangle - \langle \beta, \int_0^{2\pi} \dot{\zeta} \rangle$$

which, in view of the condition (1.68) vanishes. We conclude that z_* satisfies the Euler equations
$$\nabla G(\dot{z}_*) = \alpha J z_* + \beta \tag{1.71}$$
for constants α and β, as was, of course, to be expected. For the constant α we find $\alpha = \mu$, so that $\alpha > 0$. Indeed, using the Euler equations and the homogeneity of G we compute
$$2\mu = 2\int_0^{2\pi} G(\dot{z}_*) = \int_0^{2\pi} \langle \nabla G(\dot{z}_*), \dot{z}_* \rangle = \alpha \int_0^{2\pi} \langle J z_*, \dot{z}_* \rangle = 2\alpha \ .$$

Ad (iv): It is easy to see that z_* belongs to C^2. Recalling that $x = \nabla G(y)$ is inverted by $y = \nabla H(x)$ pointwise we find from the Euler equations that
$$\dot{z}_*(t) = \nabla H(\alpha J z_*(t) + \beta) \tag{1.72}$$
for almost every t. The right hand side is continuous and we conclude that $z_* \in C^1$, and, inserting it again into the right hand side, we find $z_* \in C^2$. Finally, setting
$$x(t) = c\left(\alpha J z_*(t) + \beta\right),$$
with $c > 0$, we obtain from (1.72) using the homogeneity of ∇H
$$\dot{x} = c\,\alpha J\ \nabla H(\alpha J z_* + \beta) = \alpha J\ \nabla H(x)$$
and, since H is of degree 2,
$$\int_0^{2\pi} H(x) dt = \frac{1}{2}\int_0^{2\pi} \langle \nabla H(x), x\rangle = \frac{1}{2}c^2\alpha \int_0^{2\pi} \langle \dot{z}_*, J z_*\rangle = c^2\alpha$$
which is equal to 2π if we choose $c = \sqrt{\frac{2\pi}{\alpha}}$. Therefore, $x(t)$ is a 2π-periodic solution of the equation $\dot{x} = \alpha J\ \nabla H(x)$ and lies on the energy surface $H = 1$. Consequently $y(t) = x(\alpha^{-1}t)$ is the desired periodic solution of $\dot{x} = J\nabla H(x)$ on $H = 1$ which has the period $T = 2\pi\alpha = 2\pi\mu$. But this is unimportant since the period of the periodic solution on S depends on our choice of the Hamiltonian function. This finishes the proof of the theorem. ∎

Recalling the transformation properties of Hamiltonian vector fields, one concludes from the above theorem that every hypersurface in $\mathbb{R}^{2n}$ which is symplectically diffeomorphic to a strictly convex one, always possesses a closed characteristic. Such a situation is, however, hard to recognize and we shall see below that there are hypersurfaces which are not symplectically diffeomorphic to convex ones. It will be our aim in Chapter 4 to describe symplectically invariant properties of a hypersurface which guarantee the existence of a closed characteristic.

1.6 The problem of symplectic embeddings

The existence will, however, be established by means of a very different variational principle.

The idea of the above proof, sometimes called the dual action principle, prompted many existence results for convex Hamiltonians. For example, I. Ekeland [65] 1984 established a Morse theory for periodic solutions on convex hypersurfaces which is analogous to the one for closed geodesics on compact Riemannian manifolds. Using Ekeland's Morse theory he showed that generically (in the C^∞ category) every compact and convex hypersurface $S \subset \mathbb{R}^{2n}$ carries infinitely many periodic orbits [65, 66, 67]. In the special case that the convex surface $S \subset \mathbb{R}^{2n}$ is close to a round sphere in the sense that

$$B(r) \subset S \subset B(R) \quad \text{and} \quad R \leq \sqrt{2}r$$

for two open balls of radius r and R (where the first inclusion means contained in the inside of S), one can prove that S carries at least n closed characteristics [70]. Here the convexity requirement can be relaxed somewhat and we refer to [21] for the precise statement.

For an account of this circle of questions concerning convex Hamiltonians, we refer to the recent book by I. Ekeland: "Convexity Methods in Hamiltonian Mechanics" (1990), [66].

1.6 The problem of symplectic embeddings

We consider two compact connected domains D_1 and D_2 in $\mathbb{R}^m$ having smooth boundaries. If the domains are diffeomorphic by an orientation preserving diffeomorphism, one can ask the question: under what additional conditions on D_1 and D_2 does there exist a volume-preserving diffeomorphism, i.e., a diffeomorphism $\varphi : D_1 \to D_2$ satisfying $\det \varphi'(x) = 1$ for every x in the interior of D_1? Clearly,

$$\text{vol}\,(D_1) = \text{vol}\,(D_2)$$

is a necessary condition. It turns out that this is already the only condition which is also sufficient for the existence of a volume-preserving diffeomorphism. This follows from the following extension of Theorem 2 in Section 3 to the case of manifolds with boundaries.

Theorem 6. (Dacorogna-Moser) Let D_1 and D_2 be two compact connected domains with smooth boundaries in $\mathbb{R}^m$. Assume $\alpha(x) = a(x)\,dx_1 \wedge \ldots \wedge dx_m$ is a smooth volume form on D_1 and $\beta(x) = b(x)dx_1 \wedge \ldots \wedge dx_m$ is a smooth volume form on D_2 with $a > 0$ and $b > 0$. If $\varphi : D_1 \to D_2$ is an orientation preserving diffeomorphism, then there exists a diffeomorphism $\psi : D_1 \to D_2$ satisfying, in addition, $\psi = \varphi$ on ∂D_1 and

$$\det\left(\psi'(x)\right) \cdot b\!\left(\psi(x)\right) = \lambda a(x),$$

where the constant λ is defined by

$$\int_{D_2} b = \lambda \cdot \int_{D_1} a.$$

If, in particular, vol D_1 = vol D_2 we choose $a = 1$ and $b = 1$ and find $\lambda = 1$ and $\det \psi'(x) = 1$, so that ψ is volume-preserving. We conclude that the total volume is the only invariant of volume preserving diffeomorphisms. For a proof of Theorem 6 with precise regularity conditions for the boundaries and the volume forms, we refer to Dacorogna and Moser [59]. In the special case $\mathbb{R}^2$, the volume form $dx_1 \wedge dx_2$ is a symplectic form and we see that the total area alone distinguishes diffeomorphic symplectic manifolds from each other. In higher dimensions the situation is very different.

In the symplectic space $(\mathbb{R}^{2n}, \omega_0)$ one can ask the same questions for symplectic diffeomorphisms instead of volume preserving diffeomorphisms. Consider, for example, a symplectic diffeomorphism φ of $\mathbb{R}^{2n}$ which maps D_1 onto D_2. Then it maps the boundary ∂D_1 of D_1 onto ∂D_2. The canonical line bundles of the two hypersurfaces are, therefore, isomorphic and the corresponding Hamiltonian flows are conjugated up to reparametrization. In particular, the actions of the corresponding closed characteristics, if there are any, have to agree. We can, moreover, thicken the boundaries into a one parameter family of hypersurfaces. As we shall prove in Chapter 4, the leaves of such a foliation of a neighborhood of a hypersurface carry an abundance of closed characteristics. Consequently, one expects many symplectic invariants and we see that the problem of distinguishing diffeomorphic domains from the symplectic point of view is of quite a different nature than the volume preserving situation. Turning, therefore, to a seemingly simpler but related problem, we consider two open domains U and V in $\mathbb{R}^{2n}$ and look for conditions which allow a symplectic embedding $\varphi : U \to V$. This is an embedding $\varphi : U \to \mathbb{R}^{2n}$ satisfying, in addition, $\varphi^* \omega_0 = \omega_0$ and $\varphi(U) \subset V$.

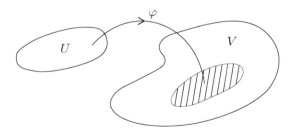

Fig. 1.3

Since such a φ is also volume preserving we clearly have the restriction vol$(U) \le$ vol(V). The condition vol $(U) <$ vol (V) is already sufficient to guarantee the

1.6 THE PROBLEM OF SYMPLECTIC EMBEDDINGS

existence of a volume preserving embedding. The question arises whether there are obstacles apart from the volume for symplectic embeddings. Consider, for example, an open ball $B(R)$ of radius $R > 0$:

$$B(R) = \left\{ (x,y) \in \mathbb{R}^{2n} \mid |x|^2 + |y|^2 < R^2 \right\}$$

which, of course, has finite volume. It can easily be embedded symplectically into the special open cylinders of infinite volume which are defined by

$$\hat{Z}(r) = \left\{ (x,y) \in \mathbb{R}^{2n} \mid x_1^2 + x_2^2 < r^2 \right\}$$

for every $r > 0$. Indeed, simply take the linear symplectic map

$$\varphi(x,y) = \left(\varepsilon x_1, \varepsilon x_2, x_3, \ldots, x_n, \frac{1}{\varepsilon} y_1, \frac{1}{\varepsilon} y_2, y_3, \ldots, y_n \right).$$

Then $\varphi(B(R)) \subset \hat{Z}(r)$ if $\varepsilon > 0$ is chosen sufficiently small. Note that the 2-plane span $\{e_1, e_2\}$ of $\mathbb{R}^{2n}$ defining this cylinder is isotropic and hence inherits no symplectic structure. The situation changes drastically if we replace $\hat{Z}(r)$ by the cylinder

$$Z(r) = \left\{ (x,y) \in \mathbb{R}^{2n} \mid x_1^2 + y_1^2 < r^2 \right\}.$$

This cylinder is defined by the 2-plane span $\{e_1, f_1\}$ which is a symplectic subspace. Trying similarly with the map

$$\psi(x,y) = \left(\varepsilon x_1, \frac{1}{\varepsilon} x_2, x_3, \ldots, x_n, \varepsilon y_1, \frac{1}{\varepsilon} y_2, y_3, \ldots, y_n \right)$$

one finds $\psi(B(R)) \subset Z(r)$ if ε is small. The map ψ is volume preserving. But, ψ is symplectic only if $\varepsilon = 1$ in which case we can have $\psi(B(R)) \subset Z(r)$ only if $r \geq R$. So far we experimented using linear symplectic mappings and one might hope to do better with nonlinear symplectic mappings. This is not the case. In 1985, M. Gromov [107] discovered the following surprising result.

Theorem 7. (Gromov's squeezing theorem) Assume there is a symplectic embedding φ defined on $B(R)$ and satisfying

$$\varphi(B(R)) \subset Z(r)$$

then $r \geq R$ (Fig. 1.4).

This phenomenon demonstrates clearly that the symplectic structure of a map is much more rigid than the volume preserving structure. Gromov's discovery motivated the search for symplectic invariants and a class of such symplectic invariants was constructed in 1987 by I. Ekeland and H. Hofer [68, 69] by means of a variational principle for periodic solutions of Hamiltonian equations; the construction was motivated by earlier results in [67]. These invariants, described in the next

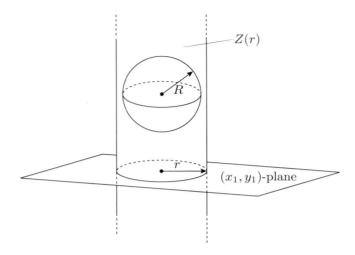

Fig. 1.4

chapter, will give an explanation of rigidity phenomena for many symplectic embeddings, among them the result above.

One is tempted to give an explanation of the rigidity phenomenon above from a dynamical point of view. Observe that the hypersurface $\partial B(R)$ is the standard sphere on which all characteristics are closed. It is an energy surface for the quadratic Hamiltonian function

$$H(x,y) = \frac{1}{2}\left(|x|^2 + |y|^2\right).$$

The Hamiltonian equations $\dot{z} = J\nabla H(z)$ with $z \in \mathbb{R}^{2n}$ are linear and the solutions

$$z(t) = e^{Jt}z(0)$$

satisfy $|z(1)|^2 = |z(0)|^2$. They are periodic with period 2π. Moreover, on $\partial B(R)$ they all have the same action

$$A(z) = \pi R^2, \quad z(t) \in \partial B(R).$$

Similarly, one verifies that the characteristics on $\partial Z(r)$ are all closed, wind around the cylinder, and have the actions $A(z) = \pi r^2$ for $z(t) \subset \partial Z(r)$, quite in contrast to the characteristics on $\partial \hat{Z}(r)$ which are straight lines. If now φ is a symplectic embedding of $\overline{B(R)}$, then on $\partial \varphi(B(R))$ the characteristics are also closed and have the actions πR^2. Imagine now that there is an "optimal embedding" of $B(R)$ into $Z(r)$ in which case one is inclined to think that the boundary of the embedded ball touches the boundary of the cylinder and that this is happening along a closed characteristic. We then conclude in view of the invariance of the actions that $r = R$. A proof of Gromov's theorem along these lines does not exist and we shall proceed differently later on.

1.7 Symplectic classification of positive definite quadratic forms

Before we turn to the subtle problem of embedding an open set in another one using a symplectic map, it is useful to study the special case in which the open sets are ellipsoids and the symplectic maps under consideration are linear. We start with an algebraic observation well-known in mechanics. Recall that a function $q: V \to \mathbb{R}$ on a vector space V is called a quadratic form if there is a symmetric bilinear form $\hat{q}: V \times V \to \mathbb{R}$ such that

$$q(x) = \frac{1}{2}\hat{q}(x,x), \quad x \in V.$$

The quadratic form q is called positive definite if $\hat{q}$ is an inner product in V. Choosing a basis in V, we have the representation

$$\hat{q}(x,y) = \langle Sx, y \rangle$$

$$q(x) = \frac{1}{2}\langle Sx, x \rangle, \quad S = S^T > 0$$

for $x, y \in V$. From now on $(V, \omega) = (\mathbb{R}^{2n}, \omega)$ is the standard symplectic vector space with the symplectic form (previously denoted by ω_0) given by

$$\omega(x,y) = \langle Jx, y \rangle, \quad x, y \in \mathbb{R}^{2n}$$

and

$$J = \begin{pmatrix} 0 & 1 \\ -1 & 0 \end{pmatrix}.$$

Denoting by $\mathcal{P}$ the set of positive definite quadratic forms on V, and by $G = Sp(n)$ the group of linear symplectic maps φ of V, i.e., $\varphi^*\omega = \omega$, we have the G-action on $\mathcal{P}$ defined by the composition

$$\mathcal{P} \times G \to \mathcal{P}: \quad [q, \varphi] \mapsto q \circ \varphi.$$

It turns out that the orbits of $\mathcal{P}$ under this action are characterized by n positive numbers. The form q can be diagonalized by means of a linear map which leaves the form ω invariant. This algebraic fact was already known to K. Weierstrass [220].

Theorem 8. If $q \in \mathcal{P}$ then there is a linear map φ satisfying $\varphi^*\omega = \omega$ and

$$q \circ \varphi(x) = \frac{1}{2}\sum_{j=1}^{n} \lambda_j(x_j^2 + x_{n+j}^2),$$

with $0 < \lambda_n \le \lambda_{n-1} \le \ldots \le \lambda_1$. The numbers λ_j are uniquely determined by q and ω. Indeed $(\pm i\lambda_j)$ for $1 \le j \le n$ are the eigenvalues of the linear Hamiltonian vector field X_q associated with q and ω by

$$\omega(X_q(x), y) = -dq(x)(y)$$

for all $x, y \in \mathbb{R}^{2n}$.

Postponing the proof we observe that if $q(x) = \frac{1}{2}\langle Sx, x\rangle$ with a positive definite symmetric matrix S, then X_q is the linear vector field $X_q(x) = JSx$, and if $\tau \in Sp(n)$ is any symplectic matrix, we conclude by the transformation law of Hamiltonian vector fields

$$X_{q\circ\tau} = \tau^{-1} \cdot X_q \cdot \tau.$$

Thus the two linear vector fields X_q and $X_{q\circ\tau}$ are similar to each other and have, therefore, the same eigenvalues. Associating with q the n positive real numbers $\lambda = \lambda(q) \in \mathbb{R}^n$ guaranteed by the theorem, we conclude that

$$\lambda(q\circ\tau) = \lambda(q)$$

for all $q \in \mathcal{P}$ and $\tau \in Sp(n)$, so that these numbers characterize the orbits of $\mathcal{P}$ under $Sp(n)$.

This is, of course, in contrast to the action of the orthogonal group of matrices satisfying $A^T A = 1$. Since $q \circ A(x) = \frac{1}{2}\langle SAx, Ax\rangle = \frac{1}{2}\langle A^T SAx, x\rangle = \frac{1}{2}\langle A^{-1}SAx, x\rangle$ we see that the orbits of $\mathcal{P}$ under $O(2n)$ are determined by the $2n$ positive eigenvalues of the symmetric matrix S in contrast to the n imaginary eigenvalue pairs $(\pm i\lambda_j)$ of JS. On $\mathbb{R}^2$, for example, a positive quadratic form is, after an orthogonal transformation, of the form $q(x) = \frac{1}{2}\langle Sx, x\rangle$ with a diagonal matrix $S = \text{Diag}(\lambda_1, \lambda_2)$ and $\lambda_j > 0$. The theorem guarantees a linear map τ satisfying $\det \tau = 1$ which transforms q into the symplectic normal form $q \circ \tau(x) = \frac{1}{2}\langle Qx, x\rangle$ with $Q = \text{Diag}(\lambda, \lambda)$. Geometrically the ellipsoid $\{q(x) < 1\}$ is symplectically mapped onto the disc $\{q \circ \tau(x) < 1\}$ having the same area. We see that in $\mathbb{R}^2$ the normal forms correspond to discs distinguished by the area.

Proof. On the product space $(x, y) \in V \times V$, the functions $K(x, y) = q(x) + q(y)$ and $\omega(x, y)$ are quadratic forms. Since K is positive definite we can apply a well-known variational argument. On the compact manifold $M = M(q)$ defined by

$$M = \{(x, y) \in V \times V \mid q(x) + q(y) = 1\},$$

the continuous function $(x, y) \mapsto \omega(x, y)$ attains a maximum, say at $(a, b) \in M$, such that

$$\omega(a, b) = \max_{(x,y) \in M} \omega(x, y).$$

We point out that a critical point $(a, b) \in M$ of $\omega | M$ lies in a whole circle consisting of critical points belonging to the same critical level. Indeed, if we define the circle action on $V \times V$ by

$$e^{it} \cdot (x, y) = (x(t), y(t)), \quad t \in \mathbb{R},$$

where

$$x(t) = (\cos t)x + (\sin t)y, \qquad y(t) = (\cos t)y - (\sin t)x,$$

1.7 SYMPLECTIC CLASSIFICATION OF ...

then one verifies readily that the quadratic forms are invariant under the action, i.e.,
$$K(x(t), y(t)) = K(x, y) \quad \text{and} \quad \omega(x(t), y(t)) = \omega(x, y)$$
for all $(x, y) \in V \times V$.

If $\lambda \in \mathbb{R}$ denotes the Lagrange multiplier of the critical point $(a, b) \in V \times V$, we have for the derivatives $dK(a, b) = \lambda \, d\omega(a, b)$ or explicitly

(1.73)
$$\hat{q}(a, x) = dq(a)(x) = \lambda \omega(x, b)$$
$$\hat{q}(y, b) = dq(b)(y) = \lambda \omega(a, y)$$

for every $x, y \in V$. We have used the representation $dq(x)(y) = \hat{q}(x, y)$ of the derivative of q. In view of $q(x) = \frac{1}{2}\hat{q}(x, x)$ one concludes from (1.73) that $2 = 2(q(a) + q(b)) = 2\lambda \, \omega(a, b)$. Since the maximum $\omega(a, b)$ is positive,

(1.74)
$$\omega(a, b) = \frac{1}{\lambda} > 0.$$

In particular, a, b are linearly independent in V since ω is antisymmetric. As $\omega(a, b) \neq 0$ the 2-dimensional subspace $V_1 = \text{span } \{a, b\} \subset V$ is symplectic. It is the eigenspace of X_q belonging to the pair $\pm i\lambda$ of purely imaginary eigenvalues. Indeed, the equations (1.73) are equivalent to
$$X_q(a) = +\lambda b \text{ and } X_q(b) = -\lambda a$$
so that $X_q(a \pm ib) = \mp(i\lambda)(a \pm ib)$. Normalizing, we introduce $e_1 = \alpha a$, $f_1 = -\alpha b$ with $\alpha^2 = \lambda$ and find $\omega(f_1, e_1) = 1$. Hence the pair e_1, f_1 constitutes a symplectic basis of V_1 for which

(1.75)
$$q(e_1) = q(f_1) = \frac{\lambda}{2} \quad \text{and} \quad \hat{q}(e_1, f_1) = 0.$$

Recalling the identity $\hat{q}(x, y) = q(x + y) - q(x) - q(y)$ for $x, y \in V$, we, therefore, have

(1.76)
$$q(v) = \frac{\lambda}{2}(\xi^2 + \eta^2)$$

if $v = \xi e_1 + \eta f_1 \in V_1$, which is the desired normal form in V_1. In order to iterate this argument we denote by
$$V_1^\perp = \{v \in V \mid \omega(v_1, v) = 0 \text{ for all } v_1 \in V_1\}$$
the symplectic complement in V. Then $V_1^\perp$ is a symplectic subspace and
$$V = V_1 \oplus V_1^\perp.$$

Indeed, $V_1 \cap V_1^\perp = \{0\}$ since ω is nondegenerate on V_1; moreover $v - \omega(v, e_1)f_1 + \omega(v, f_1)e_1 \in V_1^\perp$ for every $v \in V$ so that $V = V_1 + V_1^\perp$. If $v \in V_1$ and $w \in V_1^\perp$ then $\omega(v, w) = 0$ and we claim that v and w are also orthogonal with respect to $\hat{q}$:

$$\hat{q}(v, w) = 0.$$

This is true, since V_1 is an eigenspace. Indeed, by inserting x and $y \in V_1^\perp$ into the equations (1.73), the right hand side vanishes, hence also the left hand side, and since $V_1 = \text{span}\,\{a, b\}$, the claim is proved. Equivalently we have

(1.77) $\qquad q(v + w) = q(v) + q(w) \quad \text{if} \quad v \in V_1 \text{ and } w \in V_1^\perp.$

We can apply the same arguments as above to the restriction of q onto the symplectic space $V_1^\perp$. Iteratively we find a splitting $V = V_1 \oplus V_2 \oplus \cdots \oplus V_n$ into n distinguished 2-dimensional symplectic spaces V_j, which are the eigenspaces of X_q belonging to the pairs $(\mp i\lambda_j)$ of eigenvalues, with $\lambda_j > 0$. In view of (1.77)

$$q(v) = q(v_1) + q(v_2) + \cdots + q(v_n)$$

if $v = v_1 + v_2 + \cdots + v_n$ with $v_j \in V_j$. Moreover, there are symplectic bases (e_j, f_j) in V_j such that

$$q(v) = \frac{1}{2} \sum_{j=1}^n \lambda_j (x_j^2 + x_{n+j}^2)$$

if $v = \sum_{j=1}^n (x_j e_j + x_{n+j} f_j)$. This is the required normal form of q and the proof of Theorem 8 is finished. ∎

Changing the notation and introducing $\lambda_i = \frac{2}{r_i^2}$, $i = 1, 2, \ldots, n$, we shall write for the normal form

$$q \circ \varphi(x) = \sum_{j=1}^n \frac{1}{r_j^2} (x_j^2 + x_{j+n}^2),$$

hence associating with every $q \in \mathcal{P}$, the positive numbers

$$0 < r_1(q) \leq r_2(q) \leq \cdots \leq r_n(q)$$

which characterize the orbits of the $Sp(n)$-action on $\mathcal{P}$. Abbreviating $r = (r_1, r_2, \ldots, r_n)$ we have proved:

Proposition 7.
$$r(q) = r(q \circ \varphi)$$

for all $q \in \mathcal{P}$ and all $\varphi \in Sp(n)$. Moreover, if

$$r(q) = r(Q)$$

for $q, Q \in \mathcal{P}$ then there is a $\varphi \in Sp(n)$ satisfying $Q = q \circ \varphi$. ∎

1.7 Symplectic classification of ...

Proposition 8. Assume q and $Q \in \mathcal{P}$.
$$Q \leq q \implies r_j(q) \leq r_j(Q) \quad \text{for} \quad 1 \leq j \leq n.$$

Proof. From our proof of Theorem 8 we know, recalling the notation above,
$$\frac{1}{2} r_n(q)^2 = \frac{1}{\lambda_n(q)} = \max_{M(q)} \omega = \omega(a, b),$$
where $M(q) = \{(x, y) \in V \times V \mid q(x) + q(y) = 1\}$. Since $Q(x) \leq q(x)$ for all $x \in V$ there exists $\alpha \geq 1$ such that $1 = Q(\alpha a) + Q(\alpha b)$. Hence $(\alpha a, \alpha b) \in M(Q) = \{(x, y) \mid Q(x) + Q(y) = 1\}$ and $\omega(\alpha a, \alpha b) = \alpha^2 \omega(a, b) \geq \omega(a, b)$. Taking the maximum of ω on $M(Q)$, we conclude $r_n(Q) \geq r_n(q)$. More generally, one verifies the estimates $r_j(Q) \geq r_j(q)$ for $1 \leq j \leq n$ readily, using the following minimax characterization of the critical levels:
$$\frac{1}{2} r_j(q)^2 = \min_{W \in F_j} \left\{ \max_{W \cap M(q)} \omega \right\},$$
where F_j denotes the family of all linear subspaces $W \subset V \times V$ of dimension $2n + 2j$ and $1 \leq j \leq n$. Indeed, recalling that ω and K are, on the product space $V \times V$, quadratic forms and that K is positive definite, this is simply the Courant-Hilbert minimax principle for the eigenvalues of the symmetric matrix associated to the form ω with respect to the inner product defined by the positive form K. We took, however, into account that the problem is invariant under the circle action defined above. In particular, if $(a, b) \in V \times V$ is a critical point then $(-b, a)$ is also a critical point on the same critical level. Hence each eigenvalue in the Courant-Hilbert principle has multiplicity at least 2. The general principle will give the positive characteristic levels $c_{4n} \geq c_{4n-1} \cdots \geq c_{2n+1}$, and, in order to find all the positive critical levels, we have chosen above
$$\frac{1}{2} r_j(q)^2 = c_{2n+2j}$$
for $1 \leq j \leq n$. ∎

In order to interpret the above results geometrically, we associate with the form $q \in \mathcal{P}$ the ellipsoid $E(q)$, defined as the open set

(1.78) $$E(q) = \{x \mid q(x) < 1\} \subset \mathbb{R}^{2n}.$$

The group action of $Sp(n)$ is then visualized by

(1.79) $$\varphi^{-1}(E(q)) = E(q \circ \varphi).$$

Observe that

(1.80) $$E(q) \subset E(Q) \iff q \geq Q,$$

so that $E(q) = E(Q)$ if and only if $q = Q$. We can, therefore, associate with the open set E, the numbers $r(E)$, defined by

$$(1.81) \quad r(E) = r(q) \quad \text{if } E = E(q).$$

In view of (1.79), we conclude from Proposition 7 that the numbers r are invariant under the group of linear symplectic mappings. Summarizing, we have proved:

Proposition 9. If E, F are ellipsoids and $\varphi \in Sp(n)$ then

$$r(\varphi(E)) = r(E).$$

Conversely, if $r(E) = r(F)$ then there exists a map $\varphi \in Sp(n)$ satisfying $\varphi(E) = F$. ∎

The crucial observation now is that these invariants have a monotonicity property. Indeed, in view of (1.80) the Proposition 8 can be restated as follows.

Proposition 10. If E, F are ellipsoids, then

$$E \subset F \quad \Rightarrow \quad r_j(E) \leq r_j(F) \quad \text{for } 1 \leq j \leq n .$$

Conversely, if $r_j(E) \leq r_j(F)$ for $1 \leq j \leq n$, then there exists a $\varphi \in Sp(n)$ satisfying $\varphi(E) \subset F$.

Summarizing our considerations, we have defined for the open sets E of ellipsoids in $\mathbb{R}^{2n}$ the invariants $r(E)$ which allow us to answer the question of linear symplectic embeddings of ellipsoids as follows:

Theorem 9. If E and F are two ellipsoids, then there exists $\varphi \in Sp(n)$ such that

$$\varphi(E) \subset F$$

if and only if $r_j(E) \leq r_j(F)$ for all $1 \leq j \leq n$. ∎

We illustrate the statement with an example. If $B = B(R) = \{x|\ |x|^2 < R^2\}$ is the open ball of radius R, then $r_j(B) = R$ for all $1 \leq j \leq n$ and we can find a map $\varphi \in Sp(n)$ embedding B into the ellipsoid E if and only if E is restricted by the condition $r_1(E) \geq R$. All this is, of course, not surprising. After all we consider only linear symplectic mappings and one might hope that the use of nonlinear symplectic embeddings is more flexible. It turns out, however, and this is a very surprising phenomenon discovered only recently, that if $\varphi : B(R) \to E$ is any symplectic embedding then, $r_1(E) \geq R$ still. We shall prove this rigidity statement later on using a global symplectic invariant $c(U)$ defined for every open set $U \subset \mathbb{R}^{2n}$ and satisfying $c(\varphi(U)) = c(U)$ for symplectic diffeomorphisms. This invariant extends r_1 so that $c(E) = \pi r_1(E)^2$ for ellipsoids E. It has, in addition, the monotonicity property in the sense that $c(U) \leq c(V)$ provided there is a symplectic embedding $\varphi : U \to V$.

Since the construction of this global invariant c will be related to special periodic solutions of Hamiltonian equations, it is useful to interpret the symplectic

1.7 SYMPLECTIC CLASSIFICATION OF ...

invariants $r(E)$ from a dynamical point of view. There is a relation between $r(E)$ and certain closed characteristics on the boundary ∂E of E, which is the compact hypersurface
$$\partial E = \{z \mid q(z) = 1\}, \quad \text{if } E = E(q).$$
We may assume q to be in normal form
$$q(z) = \sum_{j=1}^{n} \frac{x_j^2 + y_j^2}{r_j^2}.$$
Correspondingly, V splits into the 2-dimensional symplectic planes $V = V_1 \ominus V_2 \oplus \cdots \oplus V_n$ and $z_j = (x_j, y_j) \in \mathbb{R}^2$ are the symplectic coordinates in V_j. The solutions $z(t) = (z_1(t), \ldots, z_n(t))$ of the associated Hamiltonian vector field X_q are easily computed. Setting $\lambda_j = \frac{2}{r_j^2}$, we have
$$z_j(t) = e^{t\lambda_j J} z_j(0), \quad 1 \le j \le n,$$
where J is the standard 2×2 matrix of the symplectic structure. Consequently, $\partial E \cap V_j$ carries the distinguished periodic solution $w_j(t) = (0, \ldots, 0, z_j(t), 0, \ldots, 0)$ given by
$$z_j(t) = e^{t\lambda_j J} z_j(0) \quad \text{and} \quad |z_j(0)|^2 = r_j^2$$
and having period $T = \pi r_j^2$. Its action, a global symplectic invariant, is
$$A(w_j) = \frac{1}{2} \int_0^T \langle -J\dot{w}_j, w_j \rangle = \pi r_j^2.$$
We see that the invariant r_j of the set E is represented by a closed characteristic on ∂E. In particular,
$$\pi r_1(E)^2 = \min\{\,|A(z)| : z \text{ is a closed characteristic on } \partial E\,\}.$$

In the following, the dynamical approach will be extremely useful in the construction of symplectic invariants.

We would like to add that the quadratic forms play an important role if one studies the orbit structure of a Hamiltonian vector field X_H locally near an equilibrium point, i.e., a point at which X_H vanishes. We shall finish this chapter by describing some well-known results illustrating this. We may assume the equilibrium to be the origin, so that $\nabla H(0) = 0$ and
$$X_H(z) = JH_{zz}(0)z + \cdots$$
for z near the origin in $\mathbb{R}^{2n}$. From now on we shall assume the eigenvalues of $JH_{zz}(0)$ to be purely imaginary, so that $(\pm i\lambda_j)$ for $1 \le j \le n$ are all the eigenvalues. If one assumes, in addition, that $JH_{zz}(0)$ can be diagonalized symplectically

one can write, after a linear symplectic transformation,

$$H(z) = \frac{1}{2} \sum_{j=1}^{n} \lambda_j (x_j^2 + y_j^2) + \cdots,$$

with real numbers $\lambda_j \in \mathbb{R}$. In view of Theorem 8, we can always achieve this normal form if $H_{zz}(0)$ is positive definite. Now $\mathbb{R}^{2n}$ is decomposed into the sum of symplectic 2-planes (x_j, y_j) which are filled with periodic solutions of the linear system $\dot{z} = JH_{zz}(0)z$ having periods $\frac{2\pi}{|\lambda_j|}$. The question arises whether these families have a continuation to solutions of the nonlinear system near $z = 0$. The vector field X_H may not admit any periodic solution except $z = 0$, as examples show. By restricting the λ_j, a well-known result due to Lyapunov [145] guarantees a continuation under infinitely many nonresonance conditions. It is, therefore, quite remarkable that in the positive definite case studied above, no such conditions are needed, as the following result due to A. Weinstein [223] demonstrates:

Theorem 10. Assume $H_{zz}(0) > 0$. Then there are at least n closed characteristics on the hypersurfaces $H(z) - H(0) = \varepsilon^2$ for $\varepsilon > 0$ sufficiently small. As parameterized solutions of the system X_H, they have periods close to the ones of the linearized system $\dot{z} = JH_{zz}(0)z$.

This local existence statement has been generalized by J. Moser [163]. The periodic solutions mentioned so far are prompted by the solutions of the linearized equation and we turn now to genuinely nonlinear phenomena. First we recall a nonlinear normal form of H at $z = 0$ which goes back to G. Birkhoff ([26] 1927), and which establishes nonlinear local symplectic invariants associated with H at the equilibrium point.

1.8 The orbit structure near an equilibrium, Birkhoff normal form

If H is a smooth function of the form

$$H(z) = \sum_{j=1}^{n} \frac{\alpha_j}{2}(x_j^2 + y_j^2) + \ldots = H_2 + \ldots$$

near the equilibrium point $z = 0$, then we consider the group action $H \circ \tau$ where τ belongs to the group G_1 of symplectic and smooth diffeomorphisms defined locally near $z = 0$ and leaving the origin and the quadratic part H_2 of H invariant,

$$G_1 = \{\tau \mid \tau^* \omega_0 = \omega_0 \quad \text{and} \quad \tau(0) = 0, \ d\tau(0) = id\}.$$

The aim is to find a map $\tau \in G_1$ such that $H \circ \tau$ is in a particularly simple form. We emphasize that, in the following, the quadratic part H_2 is fixed and not assumed to be a definite quadratic form.

1.8 THE ORBIT STRUCTURE NEAR AN EQUILIBRIUM ...

Theorem 11. (Birkhoff normal form) Assume the linear invariants $\alpha = (\alpha_1, \alpha_2, \ldots, \alpha_n)$ are nonresonant of order s, i.e.,

$$\langle \alpha, j \rangle \neq 0 \quad \text{for} \quad j \in \mathbb{Z}^n, \ 0 < |j| \leq s,$$

where $|j| = \sum_{\nu=1}^{n} |j_\nu|$. Then there exists an analytic map $\tau \in G_1$ such that

$$H \circ \tau = F(I_1, \ldots, I_n) + O_s,$$

where F is a polynomial of degree $[\frac{s}{2}]$ in the n variables $I = (I_1, I_2, \ldots, I_n)$ of the form

$$F = \langle \alpha, I \rangle + \frac{1}{2} \langle \beta I, I \rangle + \ldots,$$

with a $n \times n$ matrix β. The functions I_j on $\mathbb{R}^{2n}$ are given by

$$I_j(x, y) = \frac{1}{2}(x_j^2 + y_j^2) \quad \text{for} \quad 1 \leq j \leq n.$$

In addition, the polynomial F is uniquely determined by H and does not depend on the choice of $\tau \in G_1$.

The coefficients of F are, therefore, local symplectic invariants of H called the Birkhoff invariants. Postponing the discussion about the significance of this normal form, we first give the proof, which is of an algebraic nature.

Proof. We proceed iteratively and assume that

$$H = H_2 + H_3 + \ldots + H_{m-1} + H_m + \ldots$$

is already in normal form up to order $m-1$, where H_j are homogeneous polynomials of degree j. We look for a symplectic map $\tau \in G_1$ which does not change the normal form already achieved and which puts $H \circ \tau$ into normal form up to order m. We take

$$\tau = \exp X_P,$$

i.e., the time-1 map of the Hamiltonian vector field X_P, with a homogeneous polynomial P of degree m. Since τ belongs to the flow of a Hamiltonian vector field it is symplectic, and of the form

$$\tau(z) = z + J\nabla P(z) + \ldots,$$

where the dots stand for the higher order terms, and one computes readily

$$H \circ \tau = H_2 + H_3 + \ldots + H_{m-1} + (H_m - \{H_2, P\}) + \ldots.$$

The polynomial in the parenthesis is homogeneous of degree m and the bracket $\{\cdot, \cdot\}$ stands for the Poisson bracket defined for two functions F and G as follows:

$$\{F, G\} = \omega(X_G, X_F).$$

We see that we can modify H by functions of the form $\{H_2, P\}$.

Denote by V_m the vector space of homogeneous polynomials of degree m and let $L : V_m \longrightarrow V_m$ be the linear operator defined by $L(P) = \{H_2, P\}$ for $P \in V_m$.

Lemma 1. The kernel $K = K(L)$ and the range $R = R(L)$ of the linear operator L are complementary:
$$V_m = K + R \quad \text{and} \quad K \cap R = \{0\}.$$
If, in addition, $\langle \alpha, j \rangle \neq 0$ for $0 < |j| \leq m$, then $K = \{0\}$ if m is odd. If $m = 2k$ is even, then
$$K = \operatorname{span} \{I_1^{k_1} \ldots I_n^{k_n} \mid \sum_{j=1}^n k_j = k\}.$$

The normal form, therefore, consists of elements in the kernel of L, and we shall abbreviate the set of normal forms by
$$\mathcal{N} = \{P \mid \{H_2, P\} = 0 \text{ and } P \text{ is a polynomial}\}.$$

Proof of the Lemma. In order to diagonalize L in V_m, we go to complex variables and choose the symplectic coordinates
$$\begin{aligned} \xi_\nu &= \tfrac{1}{\sqrt{2}} \, (x_\nu + i y_\nu) \\ \eta_\nu &= \tfrac{1}{\sqrt{2}} \, (y_\nu + i x_\nu) \end{aligned}$$
so that $\xi_\nu \eta_\nu = \tfrac{i}{2} (x_\nu^2 + y_\nu^2)$ and $H_2 = \tfrac{1}{i} \sum_{\nu=1}^n \alpha_\nu (\xi_\nu \eta_\nu)$. Consequently,
$$L(\xi^k \eta^\ell) = \{H_2, \xi^k \eta^\ell\} = \frac{1}{i} \langle \alpha, \ell - k \rangle \xi^k \eta^\ell.$$
We abbreviated $\langle \alpha, \ell - k \rangle = \sum_{\nu=1}^n \alpha_\nu (\ell_\nu - k_\nu)$. If $P \in V_m$ then
$$P = \sum_{|k| + |\ell| = m} p_{k\ell} \xi^k \eta^\ell$$
is decomposed into $P = P^K + P^N$, where $P^K \in K(L)$ contains the coefficients $p_{k\ell}$ for $\langle \alpha, k - \ell \rangle = 0$ and P^N the coefficients for $\langle \alpha, k - \ell \rangle \neq 0$. In case of nonresonance of order m, we have $\langle \alpha, \ell - k \rangle = 0$ if and only if $k = \ell$. In this case $K(L)$ consists of the monomials of the form $\xi^k \eta^k = (\xi \eta)^k$ which are, in the original coordinates, monomials of the form $I^k = I_1^{k_1} \ldots I_n^{k_n}$, as desired. This finishes the proof of the lemma. ∎

In view of the lemma, it is easy to determine the desired map τ. Splitting $H_m = H_m^R + H_m^K$ into $H_m^R \in R(L)$ and $H_m^K \in K(L)$, so that
$$H_m - \{H_2, P\} = H_m^R + H_m^K - \{H_2, P\},$$
we choose the polynomial P as a solution of $\{H_2, P\} - H_m^R = 0$. The function $H \circ \tau$ is now in normal form up to order m. Denoting the map just constructed by

1.8 THE ORBIT STRUCTURE NEAR AN EQUILIBRIUM ...

$\tau = \tau_m$, we see that the desired map which puts H into normal form up to order s is iteratively found as a composition of mappings:

$$\tau_s \circ \tau_{s-1} \circ \ldots \circ \tau_3,$$

where $\tau_j \in G_1$. It remains to prove the uniqueness of the normal form. This follows from the next lemma.

Lemma 2. Assume that α is nonresonant of order s. If $\tau_1, \tau_2 \in G_1$ satisfy

$$H \circ \tau_1 = F + o_s$$
$$H \circ \tau_2 = \hat{F} + o_s,$$

where F and $\hat{F}$ are polynomials of degree s which are in normal form, i.e., $F, \hat{F} \in \mathcal{N}$, then

(i) $\quad F = \hat{F}$

(ii) $\quad \tau = \tau_1^{-1} \circ \tau_2 = \exp X_P + o_s,$

where $P = P_3 + \ldots + P_s$ is a polynomial of degree s and $P \in \mathcal{N}$.

Proof. It clearly suffices to take $H = F$ already in normal form, so that

$$\hat{F} \circ \tau = F + o_s.$$

We have to show that $\hat{F} = F$ and that $\tau \in G_1$ satisfies (ii). Proceeding again by induction, we assume that

$$P = P_3 + \ldots + P_{m-1} + P_m + \ldots + P_s$$

with $P_j \in \mathcal{N}$ for $3 \leq j \leq m-1$, and show that also $P_m \in \mathcal{N}$. In order to determine the terms of order m, we use the formula

$$H \circ \exp X_G = H + \{H, G\} + \{\{H, G\}, G\} + \ldots$$

for two functions H, G. This follows from the Taylor expansion of $H \circ \exp tX_G$ in t at $t = 0$. If now $\hat{F} = H_2 + \hat{F}_3 + \ldots + \hat{F}_s$ and similarly, $F = H_2 + F_3 + \ldots + F_s$ we find

$$\hat{F}_m + \{H_2, P_m\} + A_m = F_m,$$

where, by the induction assumption, A_m is a sum of Poisson brackets of elements in $\mathcal{N}$. Since $\mathcal{N}$ is a subalgebra we have $A_m \subset \mathcal{N}$, and therefore,

$$\{H_2, P_m\} \in \mathcal{N}.$$

In view of the previous lemma we conclude $\{H_2, P_m\} = 0$ and hence $P_m \in \mathcal{N}$. The induction step is finished and we can, therefore, assume that $P \in \mathcal{N}$. The algebra $\mathcal{N}$ is abelian up to order s, since we assume the nonresonance condition

for α to hold up to order s. This follows from Lemma 1. Therefore, $\{\hat{F}, P\} = 0$ so that
$$\hat{F} \circ \tau = \hat{F} \circ \exp X_P = \hat{F}.$$
Since $\hat{F} \circ \tau = F + o_s$ we conclude $\hat{F} = F$ as claimed in the lemma. This finishes the proof of the lemma and hence that of Birkhoff's theorem. ∎

From the dynamical point of view, the significance of Birkhoff's normal form lies in the fact that the symplectic transformation τ makes the Hamiltonian system integrable in a small neighborhood of the equilibrium point, of course, only up to the terms o_s of higher order. Indeed, dropping for the moment the higher order terms, the corresponding integrable system is defined by the Hamiltonian
$$F(I_1, I_2, \ldots, I_n)$$
on $\mathbb{R}^{2n}$, which is a polynomial in the n functions $I_j(x, y) = \frac{1}{2}(x_j^2 + y_j^2)$. These functions are n integrals of F which are, in addition, in involution, i.e., $\{I_i, I_j\} = 0$. The solutions of the Hamiltonian equations X_F can, therefore, be given explicitly and for all times in terms of the integrals. Removing the coordinate axis from $\mathbb{R}^{2n}$ and introducing the symplectic polar coordinates by
$$y_\nu + i x_\nu = \sqrt{2 r_\nu} e^{i \vartheta_\nu},$$
$\nu = 1, 2, \ldots n$, $r_\nu > 0$ and $\vartheta_\nu \pmod{2\pi}$, then
$$\sum_{j=1}^{n} dy_j \wedge dx_j = \sum_{j=1}^{n} d\vartheta_j \wedge dr_j$$
and the symplectic phase space becomes
$$T^n \times \Omega, \quad \Omega = (\mathbb{R}^+)^n \subset \mathbb{R}^n.$$
Since $r_\nu = I_\nu$ the transformed Hamiltonian is given by $F(r_1, r_2, \ldots, r_n)$ and does not depend on the angle variables $\vartheta_1, \ldots, \vartheta_n$. The flow $\phi^t(r)$ of the vector field
$$X_F : \begin{cases} \dot{\vartheta} &= \frac{\partial}{\partial r} F(r) \\ \dot{r} &= -\frac{\partial}{\partial \vartheta} F(r) = 0 \end{cases}$$
can be read off immediately:
$$\varphi^t(\vartheta, r) = (\vartheta + t\omega, r).$$
The so-called frequency vector ω is defined by
$$\omega = \omega(r) = \frac{\partial}{\partial r} F(r) \in \mathbb{R}^n.$$

1.8 THE ORBIT STRUCTURE NEAR AN EQUILIBRIUM ...

The flow leaves every torus $T^n \times \{r\} \subset T^n \times \Omega$ invariant and, moreover, on the torus the flow is a linear Kronecker flow defined by the frequencies $\omega(r)$. In contrast to the linear case these frequencies depend, in general, on the integrals r. This is the case, for example, if $\det \beta \neq 0$ with β as defined in the normal form. In the original coordinates the solutions are quasi-periodic and represented by

$$y_\nu(t) + ix_\nu(t) = \sqrt{2r_\nu} e^{i(\vartheta_\nu + t\omega_\nu(r))},$$

where $y_\nu(0) + ix_\nu(0) = \sqrt{2r_\nu} e^{i\vartheta_\nu}$.

As a side remark we point out that if the α_j are nonresonant of every order, i.e., rationally independent, then we find inductively by proceeding, as in the proof of the normal form, a symplectic map τ in the form of a formal power series map, which puts the Hamiltonian H into the integrable form $H \circ \tau = P_\infty(I)$. It is given in terms of a formal power series in the integrals $I = (I_1, I_2, \ldots I_n)$ representing infinitely many symplectic invariants of H at 0. This is a purely algebraic statement. If H is analytic and if τ can be chosen to be convergent, then the Hamiltonian system can be solved explicitly near the equilibrium point. It is known, however, that divergence of τ is the typical case. This was proved by C.L. Siegel 1954 [195]. At first sight it seems, therefore, that the normal forms are useless for the investigation of the orbit structure of a Hamiltonian system near the equilibrium point. This, however, is not the case. The normal form in Birkhoff's theorem can be used very effectively if one views X_H as a perturbation of the integrable system X_F near the equilibrium point. Indeed, perturbation methods allow us to continue some dynamical phenomena of the integrable system X_F to the full system X_H near the origin as we shall illustrate, although without giving a proof.

Assume that α is nonresonant of order $s = 4$. In view of Birkhoff's theorem, we may assume after a nonlinear symplectic transformation $\tau \in G_1$, that the Hamiltonian is, near $z = 0$, of the form

$$H(z) = F(I) + o_4(z)$$
$$F(I) = \langle \alpha, I \rangle + \frac{1}{2} \langle \beta I, I \rangle.$$

Assuming the system X_H to be nonlinear and nondegenerate, by requiring that

$$\det \beta \neq 0$$

for the symplectic invariant β, one verifies readily that every punctured neighborhood $0 < |z| < \varepsilon$ of the equilibrium point contains a periodic solution of the integrable system X_F, so that $z = 0$ is a cluster point of periodic solutions of X_F. The periods of these solutions tend to infinity. The same statement holds true also for the full system X_H instead of X_F, as was shown by G. Birkhoff and Lewis in 1933 [27]. For a proof we refer to J. Moser [164], who gave a proof of a more general result.

If, in addition, X_H is sufficiently smooth (e.g., if $H \in C^\ell$ for $\ell = 2n+3$) then the celebrated K.A.M.-theory is applicable and allows a deeper insight into the orbit structure of X_H near the equilibrium point. Although the integrability breaks down under the perturbation one can still conclude that many tori $T^n \times \{r\}$ with r small which are invariant under the integrable system X_F, have a continuation as embedded Lagrangian tori $T_r = \varphi(T^n \times \{r\})$ invariant under the flow X_H. On the tori T_r the flow is the Kronecker flow given by $\dot\vartheta = \omega(r)$, so that T_r is covered by quasi-periodic solutions of X_H having the frequencies $\omega(r)$. In addition, these invariant tori fill out a Cantor set in the phase space of relatively large Lebesgue measure.

The proofs are analytical in nature and very subtle due to the difficulty of small denominators. This difficulty was overcome in the sixties by Kolmogorov [128], Arnold [7, 8], and Moser [159]. For a recent presentation and for references we refer to J. Pöschel [178]; see also [189].

It turns out that the invariant tori constructed by the K.A.M. theory lie in the closure of the set of periodic orbits. This observation leads to the following local existence statement for periodic orbits near an equilibrium point, a proof of which can be found in [56].

Theorem 12. Assume $H \in C^\ell$ for $\ell \geq 2n+3$ and satisfies the nonresonance condition for α of order $s = 4$. Assume the system is nonlinear requiring $\det \beta \neq 0$. If S_ε denotes the closure of the set of periodic solutions contained in the open ball $B(\varepsilon)$ of radius ε centered at the equilibrium point, then

$$m\Big(B(\varepsilon)\setminus S_\varepsilon\Big) = O\left(\sqrt{\varepsilon}\right) m\Big(B(\varepsilon)\Big),$$

where m denotes the Lebesque measure.

These periodic solutions, as well as the quasi-periodic solutions near the equilibrium point, are established supposing only finitely many inequalities for finitely many local symplectic invariants of H at $z = 0$. Higher order resonances of the linear invariants play no role because of the postulated nonlinearity. This is a typical phenomenon in Hamiltonian mechanics. On the other hand, it should be recalled that the analytical methods applied require a lot of derivatives for the system.

The orbits near the equilibrium point described so far are concluded by means of perturbation methods from corresponding orbits of the integrable part of the Birkhoff normal form. There exist, in general, other orbits of very different nature which are not compatible with the integrable part and whose existence explain the divergence of the normal form. This is best illustrated by the celebrated sketch due to V. Arnold and A. Avez in their book [12], 1968. It demonstrates the complexity of the orbit structure of a nonlinear symplectic map φ near a stable fixed point in the plane $\mathbb{R}^2$. Such a map arises, for example, as a transversal section map of a Hamiltonian system on a three-dimensional energy surface near a stable periodic orbit. Instead of studying nearby orbits for all times, one can just as well study the iterates of its section map.

1.8 THE ORBIT STRUCTURE NEAR AN EQUILIBRIUM ...

The fixed point 0 in the middle is surrounded by smooth curves close to circles which are invariant under the map φ. They were discovered by J. Moser 1962 [159]. These curves fill out a Cantor set of relatively large measure, reflecting the integrable approximation of the map near the fixed point. Between these curves one discovers generically orbits of elliptic and hyperbolic periodic points. Moreover, the stable and unstable invariant manifolds issuing from the hyperbolic periodic points do intersect transversally in so-called homoclinic points. It is well-known that these points complicate the orbit structure of φ considerably and destroy the integrable pattern. They give, in particular, rise to invariant hyperbolic sets on which the iterates of the map can be described statistically by means of topological Bernoulli shifts. One can even show that the elliptic fixed point is a cluster point of such homoclinic points. In addition, this picture is repeated near the elliptic periodic orbits for the higher iterates of the map and so on. The dotted lines in the picture indicate the recently discovered Mather sets, for which we refer to J. Mather [151]. The generic existence of the homoclinic orbits in the picture was established by E. Zehnder in [230], 1973. For an improved and modern version we point out C. Genecand [104], 1991; further references are A. Chenciner [49] and J. Moser [162].

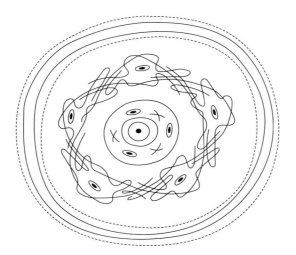

Fig. 1.5

The above picture illustrates the complexity of the orbit structure of Hamiltonian systems and shows, in particular, that orbits of very different longtime behaviour do, in general, coexist side by side. This phenomenon makes the investigation of the long time behaviour by means of solving the Cauchy initial value problem a hopeless task. In contrast, the existence question for periodic solutions with which we shall deal later on, leads to a boundary value problem distinguished by a variational principle.

The local existence results described above are of analytical nature. In contrast, the aim in Chapter 4 is to find global periodic solutions. Similarly as in the situation near an equilibrium point, the orbit structure near a periodic solution can also be studied by means of a Birkhoff normal form. This normal form represents an integrable approximation of the system locally near the periodic solution where the linear invariants are now the Floquet multipliers of the periodic orbit. We should mention that although the qualitative methods employed in finding these global periodic solutions are powerful, they are not sufficient to gain information about the Birkhoff invariants of these periodic solutions. Nevertheless, it is important to keep in mind that a single periodic orbit of a Hamiltonian system carries potential information about a nearby very rich orbit structure including many other periodic orbits having larger periods, quasi-periodic solutions and also hyperbolic phenomena.

Chapter 2
Symplectic capacities

2.1 Definition and application to embeddings

In the following we introduce a special class of symplectic invariants discovered by I. Ekeland and H. Hofer in [68, 69] for subsets of $\mathbb{R}^{2n}$. They were led to these invariants in their search for periodic solutions on convex energy surfaces and called them symplectic capacities. The concept of a symplectic capacity was extended to general symplectic manifolds by H. Hofer and E. Zehnder in [123]. The existence proof of these invariants is based on a variational principle; it is not intuitive, and will be postponed to the next chapter. Taking their existence for granted, the aim of this chapter is rather to deduce the rigidity of some symplectic embeddings and, in addition, the rigidity of the symplectic nature of mappings under limits in the supremum norm, which will give rise to the notion of a "symplectic homeomorphism".

Definition of symplectic capacity. We consider the class of all symplectic manifolds (M, ω) possibly with boundary and of fixed dimension $2n$. A symplectic capacity is a map $(M, \omega) \mapsto c(M, \omega)$ which associates with every symplectic manifold (M, ω) a nonnegative number or ∞, satisfying the following properties A1–A3.

A1. **Monotonicity:** $\qquad c(M, \omega) \leq c(N, \tau)$

if there exists a symplectic embedding $\varphi : (M, \omega) \to (N, \tau)$.

A2. **Conformality:** $\qquad c(M, \alpha\omega) = |\alpha| c(M, \omega)$

for all $\alpha \in \mathbb{R}, \alpha \neq 0$.

A3. **Nontriviality:** $\qquad c(B(1), \omega_0) = \pi = c(Z(1), \omega_0)$

for the open unit ball $B(1)$ and the open symplectic cylinder $Z(1)$ in the standard space $(\mathbb{R}^{2n}, \omega_0)$. For convenience, we recall that with the symplectic coordinates $(x, y) \in \mathbb{R}^{2n}$,

$$B(r) = \left\{ (x, y) \in \mathbb{R}^{2n} \ \big| \ |x|^2 + |y|^2 < r^2 \right\}$$

and

$$Z(r) = \left\{ (x, y) \in \mathbb{R}^{2n} \ \big| \ x_1^2 + y_1^2 < r^2 \right\}$$

for $r > 0$. It is often convenient to replace (A3) by the following weaker axiom (A3′):

A3′. **Weak nontriviality:** $\qquad 0 < c(B(1), \omega_0)$ and $c(Z(1), \omega_0) < \infty$.

It should be pointed out that the axioms (A1)–(A3) do not determine a unique capacity function. There are indeed many ways to construct capacity functions as we shall see later on. We first illustrate the concept and deduce some simple consequences of the axioms (A1)–(A3). In the special case of 2-dimensional symplectic manifolds, $n = 1$, the modulus of the total area

$$c(M, \omega) := \left| \int_M \omega \right|$$

is an example of a symplectic capacity function. It agrees with the Lebesgue measure in $(\mathbb{R}^2, \omega_0)$. In contrast, if $n > 1$, then the symplectic invariant $(\text{vol})^{\frac{1}{n}}$ is excluded by axiom (A3), since the cylinder has infinite volume. If $\varphi : (M, \omega) \to (N, \sigma)$ is a symplectic diffeomorphism between the two manifolds M and N, one applies the monotonicity axiom to φ and φ^{-1} and concludes

$$c(M, \omega) = c(N, \sigma).$$

Therefore, the capacity is indeed a symplectic invariant. Observe also that, by means of the inclusion mapping, we have for open subsets of (M, ω) the monotonicity property

$$U \subset V \implies c(U) \leq c(V).$$

In order to describe some simple examples in $(\mathbb{R}^{2n}, \omega_0)$ we start with

Lemma 1. For $U \subset (\mathbb{R}^{2n}, \omega_0)$ open and $\lambda \neq 0$,

$$c(\lambda U) = \lambda^2 c(U).$$

Proof. This is a consequence of the conformality axiom. The diffeomorphism

$$\varphi : \lambda U \to U, \quad x \mapsto \frac{1}{\lambda} x$$

satisfies $\varphi^*(\lambda^2 \omega_0) = \lambda^2 \varphi^* \omega_0 = \omega_0$. Therefore, $\varphi : (\lambda U, \omega_0) \to (U, \lambda^2 \omega_0)$ is symplectic, so that $c(\lambda U, \omega_0) = c(U, \lambda^2 \omega_0) = \lambda^2 c(U, \omega_0)$ as claimed. ∎

For the open ball of radius $r > 0$ in $\mathbb{R}^{2n}$ we find, in particular

(2.1) $$c(B(r)) = r^2 c\big(B(1)\big) = \pi r^2.$$

Since $B(r) \subset \overline{B(r)} \subset B(r + \varepsilon)$ for every $\varepsilon > 0$ we conclude by monotonicity that $c(\overline{B(r)}) = \pi r^2$. We see that in the special case of $(\mathbb{R}^2, \omega_0)$

$$c\big(B(r)\big) = c\big(\overline{B(r)}\big) = \text{area}\big(B(r)\big)$$

agrees with the Lebesgue measure of the disc. This can be used to show that the capacity agrees with the Lebesgue measure for a large class of sets in $\mathbb{R}^2$, as has been observed by K.F. Siburg [194].

2.1 Definition and application to embeddings

Proposition 1. If $D \subset \mathbb{R}^2$ is a compact and connected domain with smooth boundary, then
$$c(D, \omega_0) = \text{area}(D).$$

Proof. By removing finitely many compact curves from D, we find a simply connected domain $D_0 \subset D$ satisfying $m(D_0) = m(D)$, which in view of the uniformization theorem is diffeomorphic to the unit disc $B(1) \subset \mathbb{R}^2$. Therefore, there exists a $\rho > 0$ and a diffeomorphism $\varphi : B(\rho) \to D_0$ satisfying, in addition, $m(B(\rho)) = m(D_0)$. Given $\varepsilon > 0$ we find $r < \rho$ such that $D_1 := \varphi(\overline{B(r)}) \subset D_0$ satisfies $m(D_1) \geq m(D) - \varepsilon$. By the theorem of Dacorogna-Moser there is, therefore, a measure preserving diffeomorphism $\psi : \overline{B(r)} \to D_1$. Since this ψ is symplectic we can estimate using the monotonicity, the invariance under symplectic diffeomorphisms and the normalization

$$m(D) - \varepsilon \leq m(D_1) = m\left(\overline{B(r)}\right) = c\left(\overline{B(r)}\right) = c(D_1) \leq c(D).$$

On the other hand, there exists a diffeomorphism $\varphi : D \to \overline{B(R)} \setminus \{ \text{finitely many open discs of total measure} \leq \varepsilon \}$. Choosing R appropriately we can assume, again by Dacorogna and Moser's theorem, that φ is symplectic so that

$$c(D) \leq c\left(\overline{B(R)}\right) = \pi R^2 \leq m(D) + \varepsilon.$$

To sum up: $m(D) - \varepsilon \leq c(D) \leq m(D) + \varepsilon$ for every $\varepsilon > 0$ and the result follows. ∎

Clearly,
$$0 < c(U) < \infty$$
for every open and bounded set in $(\mathbb{R}^{2n}, \omega_0)$, since U contains a small ball and is contained in a large ball. A similar argument as for $B(r)$ above shows for symplectic cylinders that

(2.2) $$c(Z(r)) = \pi r^2.$$

Therefore, if an open set U satisfies
$$B(r) \subset U \subset Z(r)$$
for some $r > 0$, we find by the monotonicity that $\pi r^2 = c(B(r)) \leq c(U) \leq c(Z(r)) = \pi r^2$ and hence

$$c(U) = \pi r^2. \quad (\text{Fig. 2.1})$$

This demonstrates that very different (in shape, size, measure and topology) open sets can have the same capacity, if $n > 1$. Recall that the ellipsoids $E \subset \mathbb{R}^{2n}$ introduced in the previous chapter are characterized by the (linear) symplectic

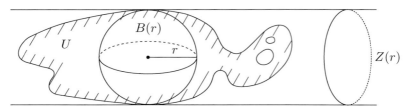

Fig. 2.1

invariants $r_1(E) \leq r_2(E) \leq \ldots \leq r_n(E)$. Applying a linear symplectic map which preserves the capacities we may assume, in view of Theorem 9 of Chapter 1, that
$$B(r_1) \subset E \subset Z(r_1),$$
where $r_1 = r_1(E)$. We conclude that the capacity of an ellipsoid E is then determined by the smallest linear symplectic invariant $r_1(E)$.

Proposition 2. The capacity of an ellipsoid E in $(\mathbb{R}^{2n}, \omega_0)$ is given by
$$c(E) = \pi r_1(E)^2 .$$

We see that every capacity c extends the smallest linear invariants $\pi r_1(E)^2$ of ellipsoids E and the question arises, whether the linear invariants $\pi r_j(E)^2$ for $j > 1$ also have extensions to invariants in the nonlinear case. We shall come back to this question later on. Symplectic cylinders are "based" on symplectic 2-planes. They are very different from cylinders "based" on isotropic 2-planes on which the 2-form ω_0 vanishes, as for example $Z_1(r) = \{(x, y) \in \mathbb{R}^{2n} \mid x_1^2 + x_2^2 < r^2\}$. We claim that
$$c(Z_1(r)) = +\infty, \quad \text{for all } r > 0.$$

This is easily seen as follows. For every ball $B(N)$ there is a linear symplectic embedding $\varphi : B(N) \to Z_1(r)$. Therefore, $\pi N^2 = c(B(N)) \leq c(Z_1(r))$. This holds true for every N and the claim follows.

In order to generalize this example we recall some definitions. If $V \subset \mathbb{R}^{2n}$ is a linear subspace, its symplectic complement $V^\perp$ is defined by
$$V^\perp = \{x \in \mathbb{R}^{2n} \mid \omega_0(v, x) = 0 \text{ for all } v \in V\}.$$
Clearly $(V^\perp)^\perp = V$ and $\dim V^\perp = \dim \mathbb{R}^{2n} - \dim V$, since ω_0 is nondegenerate. A linear subspace V is called isotropic if $V \subset V^\perp$, that is $\omega(v_1, v_2) = 0$ for all v_1 and $v_2 \in V$.

Proposition 3. Assume $\Omega \subset \mathbb{R}^{2n}$ is an open bounded nonempty set and assume $W \subset \mathbb{R}^{2n}$ is a linear subspace with $\operatorname{codim} W = 2$. Consider the cylinder $\Omega + W$, then
$$c(\Omega + W) = +\infty \quad \text{if } W^\perp \text{ is isotropic}$$
$$0 < c(\Omega + W) < \infty \quad \text{if } W^\perp \text{ is not isotropic}.$$

2.1 Definition and application to embeddings

Proof. We may assume that Ω contains the origin. Observe that $\dim W^\perp = 2$. Therefore, in the second case, $W^\perp$ is a symplectic subspace and $\mathbb{R}^{2n} = W^\perp \oplus W$. Choosing a symplectic basis (e_1, f_1) in $W^\perp$ we can, therefore, assume by a linear symplectic change of coordinates that

$$W = \{(x, y) \mid x_1 = y_1 = 0\}.$$

Since Ω is bounded, we have for $z \in \Omega + W$, that $x_1^2 + y_1^2 < N^2$ for some N; consequently $\Omega + W \subset B^2(N) \times \mathbb{R}^{2n-2}$, and hence $c(\Omega + W) \leq c(B^2(N) \times \mathbb{R}^{2n-2}) = \pi N^2 < \infty$ by (2.2). This proves the second statement. To prove the first statement we can assume that

$$W = \{(x, y) \mid x_1 = x_2 = 0\}.$$

There exists $\alpha > 0$, so that the point $(x, y) \in \Omega + W$ if $x_1^2 + x_2^2 < \alpha^2$. Hence every ball $B(R)$ can be symplectically embedded in $\Omega + W$: simply define the linear symplectic map φ by $\varphi(x, y) = (\varepsilon x, \frac{1}{\varepsilon} y)$; then $\varphi(B(R)) \subset \Omega + W$ provided $\varepsilon > 0$ is sufficiently small. Consequently, by the monotonicity of a capacity, $c(\Omega \times W) \geq c(B(R)) = \pi R^2$. This holds true for every $R > 0$ so that $c(\Omega + W) = +\infty$ as claimed. ∎

In view of the monotonicity property, the symplectic invariants $c(M, \omega)$ represent, in particular, obstructions of symplectic embeddings. An immediate consequence of the axioms is the celebrated squeezing theorem of Gromov [107] which gave rise to the concept of a capacity.

Theorem 1. (Gromov's squeezing theorem) *There is a symplectic embedding $\varphi : B(r) \to Z(R)$ if and only if $R \geq r$.*

Proof. If φ is a symplectic embedding, then using the monotonicity property of the capacity, together with (2.1) and (2.2), we have

$$\pi r^2 = c(B(r)) \leq c(Z(R)) = \pi R^2,$$

and the theorem follows. ∎

The next result also illustrates the difference between volume preserving and symplectic diffeomorphisms. We consider in $(\mathbb{R}^4, \omega_0)$ with symplectic coordinates (x_1, y_1, x_2, y_2) the product of symplectic open 2-balls $B(r_1) \times B(r_2)$. By a linear symplectic map we can assume that $r_1 \leq r_2$.

Proposition 4. *There is a symplectic diffeomorphism $\varphi : B(r_1) \times B(r_2) \to B(s_1) \times B(s_2)$ if and only if $r_1 = s_1$ and $r_2 = s_2$.*

Note that, in contrast, there is a linear volume preserving diffeomorphism $\psi : B(1) \times B(1) \to B(r) \times B(\frac{1}{r})$ for every $r > 0$. As $r \to 0$, we evidently have

$$\begin{cases} c\left(B(r) \times B\left(\frac{1}{r}\right)\right) & \to \quad 0 \\ \mathrm{vol}\left(B(r) \times B\left(\frac{1}{r}\right)\right) & = \quad \mathrm{const.} \end{cases}$$

Proof. Since $r_1 \leq r_2$ we can use the diffeomorphism φ to define the symplectic embedding $B^4(r_1) \to B(r_1) \times B(r_2) \xrightarrow{\varphi} B(s_1) \times B(s_2) \to B(s_1) \times \mathbb{R}^2 = Z(s_1)$, where the first and last mappings are the inclusion mappings. By the monotonicity of c, we conclude $s_1 \geq r_1$. Applying the same argument to φ^{-1}, we find $r_1 \geq s_1$, so that $r_1 = s_1$. Now φ is volume preserving; hence $r_1 r_2 = s_1 s_2$ and the result follows. ∎

Clearly, if one assumes that φ is smooth up to the boundary, then the conditions on the radii follow simply from the invariance of the actions $|A(\partial B(r_j))| = \pi r_j^2$ under symplectic diffeomorphism. One might expect the same rigidity as in Proposition 4 to hold also in the general case of a product of n open symplectic 2-balls in $\mathbb{R}^{2n}$. This is indeed the case, but does not follow from the capacity function c alone. Actually, the proof given in [52] is rather subtle and uses the symplectic homology theory, as developed by A. Floer and H. Hofer in [90], see also K. Cieliebak, A. Floer, H. Hofer and K. Wysocki [52]. Finally the restrictions for symplectic embeddings of ellipsoids mentioned in the previous chapter follow immediately from Proposition 2.

Proposition 5. *Assume E and F are two ellipsoids in $(\mathbb{R}^{2n}, \omega_0)$. If $\varphi : E \to F$ is a symplectic embedding, then*

$$r_1(E) \leq r_1(F) \ .$$

The existence of one capacity function permits the construction of many other capacity functions.

As an illustration we shall prove that the Gromov-width $D(M, \omega)$ which appears in Gromov's work [107] and which was explained in the introduction is a symplectic capacity satisfying (A1)–(A3). Recall that there is always a symplectic embedding $\varphi : (B(\varepsilon), \omega_0) \to (M, \omega)$ for ε small by Darboux's theorem and define

$$D(M, \omega) = \sup \left\{ \pi r^2 \ \middle| \ \text{there is a symplectic embedding } \varphi : \Big(B(r), \omega_0\Big) \to (M, \omega) \right\}.$$

Theorem 2. *The Gromov-width $D(M, \omega)$ is a symplectic capacity. Moreover*

$$D(M, \omega) \leq c(M, \omega)$$

for every capacity function c.

Because a compact symplectic manifold (M, ω) has a finite volume we conclude $D(M, \omega) < \infty$ for compact manifolds. This is in contrast to the special capacity function c_0 constructed in the next chapter which can take on the value ∞ for certain compact manifolds.

Proof. We have already verified the monotonicity axiom (A1) in the introduction. In order to verify the conformality axiom (A2), that $D(M, \alpha\omega) = |\alpha| D(M, \omega)$ for $\alpha \neq 0$, it is sufficient to show that to every symplectic embedding

$$\varphi : \Big(B(r), \omega_0\Big) \to (M, \alpha\omega),$$

2.1 Definition and application to embeddings

there corresponds a symplectic embedding

$$\hat{\varphi} : \left(B\left(\frac{r}{\sqrt{|\alpha|}}\right), \omega_0\right) \to (M, \omega),$$

and conversely, so that by definition of D, we conclude that $D(M, \alpha\omega) = |\alpha|D(M, \omega)$. If $\varphi : (B(r), \omega_0) \to (M, \alpha\omega)$ is a symplectic embedding, then $\varphi^*(\alpha\omega) = \omega_0$ so that

$$\varphi^*\omega = \frac{1}{\alpha}\omega_0.$$

Abbreviating $\rho = \frac{r}{\sqrt{|\alpha|}}$ we define the diffeomorphism $\psi : B(\rho) \to B(r)$ by setting $\psi(x) = \sqrt{|\alpha|} \cdot x$ and find

$$\psi^*\left(\frac{1}{\alpha}\omega_0\right) = \frac{|\alpha|}{\alpha}\omega_0.$$

Thus, if $\alpha > 0$, the map $\hat{\varphi} = \varphi \circ \psi : (B(\rho), \omega_0) \to (M, \omega)$ is the desired symplectic embedding. If $\alpha < 0$ we first introduce the symplectic diffeomorphism $\psi_0 : (B(\rho), \omega_0) \to (B(\rho), -\omega_0)$ by setting $\psi_0(u, v) = (-u, v)$ for all $(u, v) \in \mathbb{R}^{2n}$, and find the desired embedding $\hat{\varphi} = \varphi \circ \psi \circ \psi_0 : (B(\rho), \omega_0) \to (M, \omega)$.

The verification of $D(B(r), \omega_0) = \pi r^2$ is easy. If $\varphi : B(R) \to B(r)$ is a symplectic embedding, we conclude $R \leq r$ since φ is volume preserving. On the other hand the identity map induces a symplectic embedding $B(r) \to B(r)$ so that the claim follows. Since there exists a symplectic embedding $\varphi : B(R) \to Z(r)$ if and only if $r \geq R$ by Gromov's squeezing theorem, we conclude that $D(Z(r), \omega_0) = \pi r^2$, hence the Gromov-width satisfies also the nontriviality axiom (A3). In order to prove the last statement of the theorem, we assume $c(M, \omega)$ to be any capacity. If $\varphi : B(r) \to M$ is a symplectic embedding we conclude by monotonicity that $\pi r^2 = c(B(r), \omega_0) \leq c(M, \omega)$. Taking the supremum, we find $D(M, \omega) \leq c(M, \omega)$ as claimed in the theorem. ∎

To a given capacity c one can associate its inner capacity $\check{c}$, defined as follows:

$$\check{c}(M, \omega) = \sup\{c(U, \omega) \mid U \subset M \text{ open and } \overline{U} \subset M \setminus \partial M\}.$$

Correspondingly, we introduce the following

Definition. A capacity c has inner regularity at M if

$$\check{c}(M, \omega) = c(M, \omega).$$

Proposition 6. *The function $\check{c}$ is a capacity having inner regularity and it satisfies $\check{c} \leq c$. In addition, if d is any capacity having inner regularity and satisfying $d \leq c$, then $d \leq \check{c}$.*

Proof. The proof follows readily from the definitions and the axioms for capacity. Assume, for example, that d is a symplectic capacity satisfying $d \leq c$ and having inner regularity.
Then

$$\begin{aligned}
\check{d}(M) &= \sup\{d(U)|\ U \subset M, \overline{U} \subset M\backslash \partial M\} \\
&\leq \sup\{c(U)|\ U \subset M, \overline{U} \subset M\backslash \partial M\} \\
&= \check{c}(M),
\end{aligned}$$

as claimed. ∎

Because it is the smallest capacity, the Gromov-width $D(M,\omega)$ has inner regularity; another example having this property is the capacity c_0 introduced in Chapter 3. If we consider subsets of a given manifold we can also define the concept of outer regularity (relative to the manifold). The outer capacity of a set is defined as the infimum taken over the capacities of open neighborhoods of the closure of the given set. We shall return to this concept in the next section.

2.2 Rigidity of symplectic diffeomorphisms

We consider a sequence $\psi_j \colon \mathbb{R}^{2n} \to \mathbb{R}^{2n}$ of symplectic diffeomorphisms in $(\mathbb{R}^{2n}, \omega_0)$. By definition, the first derivatives satisfy the identity

$$\psi_j'(x)^T J\, \psi_j'(x) = J, \quad x \in \mathbb{R}^{2n}.$$

Therefore, if the sequence ψ_j converges in C^1 then the limit $\psi(x) = \lim \psi_j(x)$ is also a symplectic map. By contrast, we shall now assume that the sequence ψ_j only converges locally uniformly to a map

$$\psi(x) = \lim_{j \to \infty} \psi_j(x),$$

which is, therefore, a continuous map. Since $\det \psi_j'(x) = 1$ for every x, we find, in view of the transformation formula for integrals, that

(2.3) $$\int f\bigl(\psi(x)\bigr)\, dx = \int f(x)\, dx$$

for all $f \in C_c^\infty(\mathbb{R}^{2n})$, so that ψ is measure preserving. Assume now that ψ is differentiable, then evidently $\det \psi'(x) = \pm 1$. However it is a striking phenomenon that ψ is even symplectic,

$$\psi'(x)^T J \psi'(x) = J,$$

if it is assumed to be differentiable. Hence the symplectic nature survives under topological limits.

2.2 RIGIDITY OF SYMPLECTIC DIFFEOMORPHISMS

Theorem 3. Let $\varphi_j : B(1) \to (\mathbb{R}^{2n}, \omega_0)$ be a sequence of symplectic embeddings converging locally uniformly to a map $\varphi : B(1) \to \mathbb{R}^{2n}$. If φ is differentiable at $x = 0$, then $\varphi'(0) = A$ is a symplectic map, i.e., $A^*\omega_0 = \omega_0$.

We see that, in general, a volume preserving diffeomorphism cannot be approximated by symplectic diffeomorphisms in the C^0-topology. By using, locally, Darboux charts we deduce immediately from Theorem 3

Theorem 4. (Eliashberg, Gromov) The group of symplectic diffeomorphisms of a compact symplectic manifold (M, ω) is C^0-closed in the group of all diffeomorphisms of M.

In the early seventies M. Gromov proved the alternative that the group of symplectic diffeomorphisms either is C^0-closed in the group of all diffeomorphisms or its C^0-closure is the group of volume preserving diffeomorphisms. That symplectic diffeomorphisms can be distinguished from volume preserving diffeomorphisms by global properties which are stable under C^0-limits was announced in the early eighties by Y. Eliashberg in his preprint "Rigidity of symplectic and contact structure", (1981) [78], which in full form has not been published. Proofs are partially contained in Eliashberg [71], 1987. Gromov gave a proof of Theorem 4 in [107] using the techniques of pseudoholomorphic curves. Both Eliashberg und Gromov deduced the C^0-stability from non embedding results. In his book [108] Gromov uses so-called Nash-Moser techniques of hard implicit function theorems, while Eliashberg [71], 1987, uses an analogue of Theorem 3. Following the strategy of I. Ekeland and H. Hofer in [68], we shall show next, that Theorem 3 is an easy consequence of the existence of any capacity function c.

It is convenient in the following to extend the capacity to all subsets of $\mathbb{R}^{2n}$. To do so we take a capacity function c given on the open subsets $U \subset \mathbb{R}^{2n}$ and define for an arbitrary subset $A \subset \mathbb{R}^{2n}$:

$$c(A) = \inf \{c(U) | A \subset U \text{ and } U \text{ open}\}.$$

Then the monotonicity property

$$A \subset B \implies c(A) \leq c(B)$$

holds true for all subsets of $\mathbb{R}^{2n}$. From the symplectic invariance of the capacity on open sets, one deduces the invariance

$$c(\varphi(A)) = c(A)$$

under every symplectic embedding φ defined on an open neighborhood of A.

Proof of Theorem 3. Without loss of generality we shall assume in the following that $\varphi(0) = 0$. We first claim that the linear map $\varphi'(0) = A$ is an isomorphism. Indeed, because φ is differentiable at 0, we have $\varphi(x) = Ax + O(|x|)$, so that for the open balls B_ε of radius $\varepsilon > 0$ and centered at 0,

$$\frac{m(\varphi(B_\varepsilon))}{m(B_\varepsilon)} \longrightarrow |\det A| \text{ as } \varepsilon \to 0.$$

On the other hand, because the symplectic diffeomorphisms φ_j are volume preserving and $\varphi_j \to \varphi$ uniformly, we have $m(\varphi(B_\varepsilon)) = m(B_\varepsilon)$ and hence $|\det A| = 1$, so that A is an isomorphism, as claimed. We shall see later on (Lemma 3) that A is an isomorphism under weaker assumption: instead of requiring φ_j to be symplectic, we shall merely require these mappings to preserve a given capacity function.

Next we claim that to prove Theorem 3, it is sufficient to show that

$$(2.4) \qquad A^* \omega_0 = \lambda \omega_0 \text{ for some } \lambda \neq 0.$$

Indeed, with φ_j we can also consider the symplectic embeddings $(\varphi_j, id) : B(1) \times \mathbb{R}^{2n} \to (\mathbb{R}^{2n} \times \mathbb{R}^{2n}, \omega_0 \oplus \omega_0)$ and hence conclude for the derivative at $(0,0)$, given by $\bar{A} = (A, 1)$, that also $\bar{A}^*(\omega_0 \oplus \omega_0) = \mu(\omega_0 \oplus \omega_0)$ for some $\mu \neq 0$. On the other hand, in view of (2.4), $\bar{A}^*(\omega_0 \oplus \omega_0) = (\lambda \omega_0) \oplus \omega_0$ and consequently $\mu = 1 = \lambda$, as required in Theorem 4, proving our claim. In order to prove (2.4) we make use of the following algebraic lemma due to Y. Eliashberg [71].

Lemma 2. Assume A is a linear isomorphism satisfying $A^* \omega_0 \neq \lambda \omega_0$. Then for every $a > 0$ there are symplectic matrices U and V such that $U^{-1}AV$ has the form

$$U^{-1}AV = \left(\begin{array}{cc|c} a & 0 & 0 \\ 0 & a & \\ \hline * & & * \end{array} \right),$$

with respect to the splitting of $\mathbb{R}^{2n} = \mathbb{R}^2 \oplus \mathbb{R}^{2n-2}$ into symplectic subspaces.

Postponing the proof of the lemma, we first show that $A^* \omega_0 = \lambda \omega_0$ for some $\lambda \neq 0$. Arguing by contradiction, we assume that $A^* \omega_0 \neq \lambda \omega_0$ and apply the lemma. Defining the symplectic maps $\psi_j := U^{-1} \varphi_j V$ in the neighborhood of the origin, we conclude that $\psi_j \to \psi := U^{-1} \varphi V$ locally uniformly, and $\psi'(0) = U^{-1} A V$. Choosing a suitable constant a in the lemma, we have $U^{-1}AV(B(1)) \subset Z(\frac{1}{8})$ and hence $U^{-1}\psi V(B(\varepsilon)) \subset Z(\frac{\varepsilon}{4})$ provided ε is sufficiently small. Because $\psi_j \to \psi$ locally uniformly, $U^{-1}\psi_j V(B(\varepsilon)) \subset Z(\frac{\varepsilon}{2})$ if j is sufficiently large and ε sufficiently small. Since $U^{-1}\psi_j V$ is symplectic, we conclude by the invariance and monotonicity property of a capacity that $c(U^{-1}\psi_j V(B(\varepsilon))) = c(B(\varepsilon)) \leq c(Z(\frac{\varepsilon}{2}))$, which contradicts the nontriviality Axiom (A3). We have proved the statement in (2.4) and it remains to prove Lemma 2.

Proof of Lemma 2. Let B be the symplectic adjoint of A satisfying

$$\omega_0(Ax, y) = \omega_0(x, By)$$

for all x, y, and abbreviate $\omega = B^* \omega_0$. Then $\omega \neq \lambda \omega_0$, as is easily verified using the fact that A is an isomorphism. We claim that there is an x such that $\omega(x, \cdot) \neq \lambda \omega_0(x, \cdot)$ for every λ. Arguing by contradiction, we assume that for every x, there exists a $\lambda(x) \in \mathbb{R}$ satisfying $\omega(x, \cdot) = \lambda(x) \omega_0(x, \cdot)$. If $x \neq 0$ there exists ξ such

2.2 Rigidity of symplectic diffeomorphisms

that $\omega_0(\xi, x) \neq 0$, since ω_0 is nondegenerate. This remains true for all y in a neighborhood $U(x)$ of x. Hence

$$\begin{aligned}\lambda(\xi)\omega_0(\xi, y) &= \omega(\xi, y) = -\omega(y, \xi) \\ &= -\lambda(y)\omega_0(y, \xi) \\ &= \lambda(y)\omega_0(\xi, y),\end{aligned}$$

which implies that $\lambda(\xi) = \lambda(y)$ for y in a neighborhood of x. Since $\mathbb{R}^{2n}\setminus\{0\}$ is connected and since the function $\lambda(x)$ on $\mathbb{R}^{2n}\setminus\{0\}$ is locally constant, it is constant. Therefore $\omega(x, \cdot) = \lambda\omega_0(x, \cdot)$ for $x \neq 0$ and hence for every x. This contradicts the assumption that $\omega \neq \lambda\omega_0$ and proves our claim. Consequently there exists an x such that the linear map $(\omega_0(x, \cdot), \omega(x, \cdot)) : \mathbb{R}^{2n} \to \mathbb{R}^2$ is surjective. For a given $a > 0$, we therefore find an y satisfying

$$\omega_0(x, y) = 1 \text{ and } \omega(x, y) = a^2.$$

Recalling $\omega(x, y) = \omega_0(Bx, By)$, we can choose two symplectic bases $(e_1, f_1, \ldots)$ and $(e_1', f_1', \ldots)$ such that $e_1 = x, f_1 = y$ and $e_1' = \frac{1}{a}Bx, f_1' = \frac{1}{a}By$. In these bases

$$Be_1 = ae_1', \quad Bf_1 = af_1'.$$

From $\langle JAx, y\rangle = \langle Jx, By\rangle = \langle B^T Jx, y\rangle$ we read off $A = -JB^T J$. Representing now A in the new bases as a map from $\mathbb{R}^{2n}$ with basis $(e_1, f_1, \ldots)$ onto $\mathbb{R}^{2n}$ with basis $(e_1', f_1', \ldots)$, we find the representation $U^{-1}AV$ of the desired form. The symplectic matrices are defined by their column vectors as $U = [e_1, f_1, \ldots]$ and $V = [e_1', f_1', \ldots]$. ∎

We know that symplectic diffeomorphisms preserve the capacities. Theorem 3 can, therefore, be deduced from the following, even more surprising statement for continuous mappings due to I. Ekeland and H. Hofer [68].

Theorem 5. Let c be a capacity. Assume $\psi_j : B(1) \to \mathbb{R}^{2n}$ is a sequence of continuous mappings satisfying

$$c(\psi_j(E)) = c(E)$$

for all (small) ellipsoids $E \subset B(1)$ and converging locally uniformly to

$$\psi(x) = \lim \psi_j(x).$$

If ψ is differentiable at 0, then $\psi'(0) = A$ is either symplectic or antisymplectic:

$$A^*\omega_0 = \omega_0 \text{ or } A^*\omega_0 = -\omega_0.$$

Note that the mappings are not required to be invertible.

In order to prove Theorem 5, we start with

Lemma 3. Let c be a capacity. Consider a sequence φ_j of continuous mappings in $\mathbb{R}^{2n}$ converging locally uniformly to the map φ. Assume that $c(\varphi_j(E)) = c(E)$ for the open ellipsoids for all j. If $\varphi'(0)$ exists it is an isomorphism.

Proof. Arguing by contradiction, we assume that A is not surjective, so that $A(\mathbb{R}^{2n})$ is contained in a hyperplane H. Composing, if necessary, with a linear symplectic map we may assume that

$$(2.5) \qquad A(\mathbb{R}^{2n}) \subset H = \{(x,y) | \, x_1 = 0\} .$$

Defining the linear symplectic map ψ by

$$\psi(x,y) = \left(\frac{1}{\alpha} x_1, x_2, \ldots, x_n, \alpha\, y_1, y_2, \ldots, y_n\right)$$

we can choose $\alpha > 0$ so small that

$$\psi A\bigl(B(1)\bigr) \subset B^2(\tfrac{1}{16}) \times \mathbb{R}^{2n-2} = Z(\tfrac{1}{16}) ,$$

where the open 2-disc B^2 on the right hand side is contained in the symplectic plane with the coordinates $\{x_1, y_1\}$. Be definition of a derivative, we have $|\psi\varphi(x) - \psi A(x)| \leq a(|x|)\, |x|$ where $a(s) \to 0$ as $s \to 0$. Consequently

$$\psi\varphi\bigl(B(\varepsilon)\bigr) \subset Z(\tfrac{\varepsilon}{4})$$

if ε is sufficiently small. Since $\psi\varphi_j$ converges locally uniformly to $\psi\varphi$,

$$\psi\varphi_j\bigl(B(\varepsilon)\bigr) \subset Z(\tfrac{\varepsilon}{2}) ,$$

provided j is sufficiently large. By assumption $\psi\varphi_j$ preserves the capacity and so by monotonicity

$$c\bigl(B(\varepsilon)\bigr) = c\bigl(\psi\varphi_j(B(\varepsilon))\bigr) \leq c\bigl(Z(\tfrac{\varepsilon}{2})\bigr) = \tfrac{1}{4} c\bigl(B(\varepsilon)\bigr) .$$

This contradiction shows that A is surjective. ∎

Proof of Theorem 5. We may assume that $\psi(0) = 0$. By Lemma 3, $A = \psi'(0)$ is an isomorphism and we shall prove first that $A^*\omega_0 = \lambda\omega_0$ for some $\lambda \neq 0$. Arguing by contradiction, we assume $A^*\omega_0 \neq \lambda\omega_0$ and find (by Lemma 2) symplectic maps U and V satisfying $U^{-1}AV(B(1)) \subset Z(\tfrac{1}{8})$. Proceeding now as in Theorem 4, we define the sequence $\varphi_j := U^{-1}\psi_j V$. Then $\varphi_j \to \varphi := U^{-1}\psi V$ locally uniformly, and $\varphi'(0) = U^{-1}AV$. Hence $\varphi(B(\varepsilon)) \subset Z(\tfrac{\varepsilon}{4})$ and consequently $\varphi_j(B(\varepsilon)) \subset Z(\tfrac{\varepsilon}{2})$ for j sufficiently large and $\varepsilon > 0$ sufficiently small. Since, by assumption on ψ_j, we have $c(\varphi_j(B(\varepsilon))) = c(B(\varepsilon))$, we infer by the monotonicity of a capacity that $c(B(\varepsilon)) \leq c(Z(\tfrac{\varepsilon}{2}))$, contradicting Axiom (A3) for a capacity.

2.2 Rigidity of symplectic diffeomorphisms

We have demonstrated that $A^*\omega_0 = \lambda \omega_0$. By conformality a linear antisymplectic map preserves the capacities. Composing the maps ψ_j and ψ with the symplectic map $B = \left(\frac{1}{\sqrt{\lambda}} A\right)^{-1}$ if $\lambda > 0$ and with the antisymplectic map $B = \left(\frac{1}{\sqrt{-\lambda}} A\right)^{-1}$ if $\lambda < 0$, we are therefore reduced to the case

$$A = \alpha \mathbf{1}, \text{ with } \alpha > 0$$

and we have to show that $\alpha = 1$. If $\alpha < 1$ there is a small ball and an $\alpha < r < 1$ such that $\psi_j(B(\varepsilon)) \subset B(r\varepsilon)$ for j large. Since ψ_j preserves the capacities we find $c(B(\varepsilon)) = c(\psi_j(B(\varepsilon))) \leq c(B(r\varepsilon)) = r^2 c(B(\varepsilon))$ which is a contradiction. In the case $\alpha > 1$ we shall show that

(2.6) $$\psi_j(B(\varepsilon)) \supset B(r\varepsilon)$$

for some $\alpha > r > 1$, ε small and j large, which leads to the contradiction $r^2 c(B(\varepsilon)) = c(B(r\varepsilon)) \leq c(B(\varepsilon))$. Consequently, $\alpha = 1$ as claimed in the theorem.

To prove (2.6) we have to show that for every $y \in B(r\varepsilon)$ there exists an $x \in B(\varepsilon)$ solving $\psi_j(x) = y$. We use an index argument based on the Brouwer mapping degree. Fix $1 < r < \alpha$. Then, in view of $A(B(\varepsilon)) = B(\alpha\varepsilon)$,

$$\deg(B(\varepsilon), A, y) = 1,$$

for $y \in B(r\varepsilon)$, and the proof follows if we can show that

$$\deg(B(\varepsilon), \psi_j, y) = \deg(B(\varepsilon), A, y).$$

In view of the homotopy invariance of the degree, it is sufficient to verify that the homotopy $h(t, x) = t\psi_j(x) + (1-t)Ax$ satisfies $h(t, x) \neq y$ if $x \in \partial B(\varepsilon)$ and $0 \leq t \leq 1$.

Recall $\psi(0) = 0$ and hence $|\psi(x) - Ax| \leq a(|x|)|x|$, with $a(s) \to 0$ as $s \to 0$. Since $\psi_j \to \psi$ uniformly on compact sets we have, for every $\sigma > 0$ and $j \geq j_0(\sigma)$

(2.7) $$|\psi_j(x) - Ax| \leq a(|x|)|x| + \sigma.$$

Arguing by contradiction, we assume $t\psi_j(x) + (1-t)Ax = y$ for $x \in \partial B(\varepsilon)$ and $y \in B(r\varepsilon)$. Then

$$t(\psi_j(x) - Ax) = y - Ax.$$

The right hand side is larger than $|Ax| - |y| = \alpha\varepsilon - |y| \geq (\alpha - r)\varepsilon$. The left hand side, however is smaller than $a(|\varepsilon|)\varepsilon + \sigma$ according to (2.7). Therefore, choosing $\sigma = a(|\varepsilon|)\varepsilon$ and ε sufficiently small we get a contradiction, hence proving that $h(t, x) \neq y$ for all $0 \leq t \leq 1$ and all $x \in \partial B(\varepsilon)$. This finishes the proof of Theorem 5. ∎

As an interesting special case we conclude from Theorem 5 the following

Corollary. A linear map A in $\mathbb{R}^{2n}$ preserving the capacities of ellipsoids, $c(A(E)) = c(E)$ is either symplectic or antisymplectic.

$$A^*\omega_0 = \omega_0 \quad \text{or} \quad A^*\omega_0 = -\omega_0.$$

Let c be any capacity function and consider a homeomorphism h of $\mathbb{R}^{2n}$ satisfying

$$c(h(E)) = c(E) \text{ for all (small) ellipsoids } E.$$

Then, if h is in addition differentiable, we conclude from Theorem 5 that h is a diffeomorphism which is either symplectic or antisymplectic, $h'(x)^*\omega_0 = \pm\omega_0$. This is analogous to a measure preserving homeomorphism, i.e., a homeomorphism satisfying (2.3). Here we conclude that $\det h'(x) = \pm 1$ in case h is differentiable. We see that every capacity function c singles out the distinguished group of homeomorphisms preserving the capacity of all open sets. The elements of this group of homeomorphisms have the additional property that they are symplectic or antisymplectic in case they are differentiable. This group can, therefore, be viewed as a topological version of the group of symplectic diffeomorphisms. It is not known whether this group is closed under locally uniform limits. But the following weaker results hold true.

Theorem 6. Assume $h_j : \mathbb{R}^{2n} \to \mathbb{R}^{2n}$ is a sequence of homeomorphisms satisfying

$$c(h_j(E)) = c(E)$$

for all (resp. all small) ellipsoids E. Assume h_j converges locally uniformly to a homeomorphism h of $\mathbb{R}^{2n}$. Then

$$c(h(E)) = c(E)$$

for all (resp. all small) ellipsoids E.

Proof. Since $h^{-1} \circ h_j \to \text{id}$ locally uniformly, we conclude for every ellipsoid E and every $0 < \varepsilon < 1$

$$h^{-1} \circ h_j((1-\varepsilon)E) \subset E \subset h^{-1} \circ h_j((1+\varepsilon)E),$$

if only j is sufficiently large. This topological fact is easily verified using the same degree argument as above. Hence

$$h_j((1-\varepsilon)E) \subset h(E) \subset h_j((1+\varepsilon)E).$$

By monotonicity and conformality, $(1-\varepsilon)^2 c(E) = c((1-\varepsilon)E) = c(h_j((1-\varepsilon)E)) \leq c(h(E)) \leq c(h_j((1+\varepsilon)E)) = c((1+\varepsilon)E) = (1+\varepsilon)^2 c(E)$. In short

$$(1-\varepsilon)^2 c(E) \leq c(h(E)) \leq (1+\varepsilon)^2 c(E).$$

This holds true for every $\varepsilon > 0$ so that $c(h(E)) = c(E)$ as desired. ∎

2.2 Rigidity of symplectic diffeomorphisms

In order to generalize this statement, we denote by $\mathcal{O}$ the family of open and bounded sets of $\mathbb{R}^{2n}$ and associate with $\Omega \in \mathcal{O}$

$$\check{c}(\Omega) = \sup\{c(U)\,|\,U \text{ open and } \overline{U} \subset \Omega\}$$
$$\hat{c}(\Omega) = \inf\{c(U)\,|\,U \text{ open and } U \supset \overline{\Omega}\}.$$

The distinguished family $\mathcal{O}_c$ of open sets is defined by the condition

$$\mathcal{O}_c = \{\Omega \in \mathcal{O}\,|\,\hat{c}(\Omega) = \check{c}(\Omega)\}.$$

Proposition 7. Assume h_j is a sequence of homeomorphisms of $\mathbb{R}^{2n}$ converging locally uniformly to a homeomorphism h of $\mathbb{R}^{2n}$. If $c(h_j(U)) = c(U)$ for all $U \in \mathcal{O}$ and all j, then $c(h(\Omega)) = c(\Omega)$ for all $\Omega \in \mathcal{O}_c$.

Proof. If $\Omega \in \mathcal{O}_c$ and $\varepsilon > 0$ we find $\hat{U}, \check{U} \in \mathcal{O}$ with $\hat{U} \supset \overline{\Omega}$ and $\overline{\check{U}} \subset \Omega$ satisfying

$$c(\hat{U}) \leq c(\Omega) + \varepsilon \quad \text{and} \quad c(\check{U}) \geq c(\Omega) - \varepsilon.$$

For j large we have by the above degree argument $h_j(\hat{U}) \supset h(\Omega) \supset h_j(\check{U})$, and hence using the monotonicity property of the capacity

$$c(\hat{U}) = c(h_j(\hat{U})) \geq c(h(\Omega)) \geq c(h_j(\check{U})) = c(\check{U}),$$

so that $c(\Omega) + \varepsilon \geq c(h(\Omega)) \geq c(\Omega) - \varepsilon$. This holds true for every $\varepsilon > 0$ and the theorem is proved. ∎

Definition. A capacity c is called inner regular, respectively, outer regular if

$$c(U) = \check{c}(U) \quad \text{resp. if} \quad c(U) = \hat{c}(U)$$

for all $U \in \mathcal{O}$.

Proposition 8. Assume the capacity c is inner regular or outer regular. Assume φ_j and φ are homeomorphisms of $\mathbb{R}^{2n}$ and

$$\varphi_j \to \varphi \quad \text{and} \quad \varphi_j^{-1} \to \varphi^{-1}$$

locally uniformly. If $c(\varphi_j(U)) = c(U)$ for all $U \in \mathcal{O}$ and all j, then also

$$c\Big(\varphi(U)\Big) = c(U) \quad \text{for all} \quad U \in \mathcal{O}.$$

Proof a) Assume c is inner regular and let Ω be open and bounded. Then if U is open and $\overline{U} \subset \Omega$ we have $\varphi_j(\overline{U}) \subset \varphi(\Omega)$ if j is large and thus $c(\varphi_j(\overline{U})) = c(\overline{U}) \leq c(\varphi(\Omega))$ so that $c(U) \leq c(\varphi(\Omega))$. Hence, taking the supremum we find $c(\Omega) \leq c(\varphi(\Omega))$. Similarly one shows that $c(\Omega) \leq c(\varphi^{-1}(\Omega))$ for every open and bounded set Ω. Consequently, since φ is a homeomorphism $c(\Omega) = c(\varphi^{-1} \circ \varphi(\Omega)) \geq c(\varphi(\Omega)) \geq c(\Omega)$ and thus $c(\varphi(\Omega)) = c(\Omega)$ as desired.

b) If c is outer regular and Ω is open and bounded we conclude for $\varepsilon > 0$ that $\varphi_j(\Omega) \subset U_\varepsilon(\varphi(\Omega))$ if j is sufficiently large; here $U_\varepsilon = \{x|\ \text{dist}\,(x,U) < \varepsilon\}$. Therefore, $c(\Omega) = c(\varphi_j(\Omega)) \leq c(U_\varepsilon(\varphi(\Omega)))$ and taking the infimum on the right hand side we find $c(\Omega) \leq c(\varphi(\Omega))$. Arguing as in the part a) we conclude that $c(\varphi(\Omega)) = c(\Omega)$ for every $\Omega \in \mathcal{O}$. ∎

So far we have deduced from the existence of a symplectic capacity some surprising phenomena about symplectic mappings. Nevertheless, the notion itself is still rather mysterious and raises many questions. It is, for example, not known whether the knowledge of the capacities of small sets is sufficient to understand the capacity of larger sets. To be more precise, does, for example, a homeomorphism preserving the capacity of small sets preserve the capacity of large sets?

We have no example of a homeomorphism which preserves one capacity but not another. Neither do we know whether a homeomorphism preserving the capacity of open sets, also preserves the Lebesgue measure of open sets. But it is easy to see that a homeomorphism preserving all capacities of open sets is necessarily measure preserving. In order to prove this we first define a special embedding capacity γ. Introducing for $r > 0$ the open cube

$$Q(r) = (0,r)^{2n} \subset \mathbb{R}^{2n},$$

having edges parallel to the coordinate axis, we define

$$\gamma(M,\omega) := \sup\left\{r^2|\ \text{there is a symplectic embedding}\ \varphi: Q(r) \to M\right\}.$$

Clearly γ is a capacity satisfying the Axioms (A1), (A2) and the weak nontriviality condition (A3'). It is not normalized. One can prove, by using the n-th order capacity function c_n of Ekeland and Hofer in [69], that $\gamma(B(r)) = \frac{1}{n}\pi r^2$, moreover $\gamma(Z(r)) = \pi r^2$.

Proposition 9. Assume h is a homeomorphism of $\mathbb{R}^{2n}$ satisfying

$$\gamma\Big(h(\Omega)\Big) = \gamma(\Omega)$$

for all $\Omega \in \mathcal{O}$. Then h preserves the Lebesgue measure μ of open sets, i.e.,

$$\mu\Big(h(\Omega)\Big) = \mu(\Omega)$$

for all $\Omega \in \mathcal{O}$.

2.2 Rigidity of symplectic diffeomorphisms

Proof. From the definition of γ we infer
$$\mu\big(Q(r)\big) = r^{2n} = \gamma\big(Q(r)\big)^n.$$

Since every symplectic embedding is volume preserving we find for the capacity $\gamma(\Omega)$ of the open and bounded set $\Omega \subset \mathbb{R}^{2n}$ that
$$\mu(\Omega) \geq \sup_r \mu\big(Q(r)\big) = \gamma(\Omega)^n,$$

where the supremum is taken over those r, for which there is a symplectic embedding $Q(r) \to \Omega$. If Q is any open cube having its edges parallel to the coordinate axes we conclude that
$$\mu(Q) = \gamma(Q)^n = \gamma\big(h(Q)\big)^n \leq \mu\big(h(Q)\big).$$

It follows that
$$\mu(\Omega) \leq \mu\big(h(\Omega)\big),$$

for every $\Omega \in \mathcal{O}$. Indeed assume $\Omega \in \mathcal{O}$; then given $\varepsilon > 0$ we find, in view of the regularity of the Lebesgue measure, finitely many disjoint open cubes Q_j contained in Ω such that $\mu(\Omega) - \varepsilon \leq \sum \mu(Q_j)$. Hence by the estimate above
$$\begin{aligned}\mu(\Omega) - \varepsilon &\leq \sum \mu\big(h(Q_j)\big) \\ &= \mu\big(h(\bigcup_j Q_j)\big) \\ &\leq \mu\big(h(\Omega)\big).\end{aligned}$$

This holds true for every $\varepsilon > 0$ and the claim follows. By the same argument, $\mu(\Omega) \leq \mu(h^{-1}(\Omega))$ and hence $\mu(\Omega) \leq \mu(h(\Omega)) \leq \mu(h^{-1} \circ h(\Omega)) = \mu(\Omega)$, proving the proposition. ∎

Corollary. *If a homeomorphism of $\mathbb{R}^{2n}$ preserves all the capacities of open sets in $\mathbb{R}^{2n}$, then it also preserves the Lebesgue measure.*

All our considerations so far are based on the existence of a capacity not yet established. In the next chapter we shall construct a very special capacity function defined dynamically by means of Hamiltonian systems.

Chapter 3
Existence of a capacity

This chapter is devoted to the existence proof of a distinguished capacity function, denoted by c_0, and introduced by H. Hofer and E. Zehnder in [123]. It is intimately related to periodic orbits of Hamiltonian systems. The capacity $c_0(M,\omega)$ measures the minimal C^0-oscillation of special Hamiltonian functions $H : M \to \mathbb{R}$, needed in order to conclude the existence of a distinguished periodic orbit having small period and solving the associated Hamiltonian system X_H on M. For 2-dimensional manifolds, c_0 agrees with the total area, and in the special case of convex, open and smooth domains $U \subset (\mathbb{R}^{2n}, \omega_0)$ it is represented by a distinguished closed characteristic of the boundary ∂U having minimal (reduced) action equal to $c_0(U, \omega_0)$. This can be used, for example, to exhibit a class of compact hypersurfaces in $\mathbb{R}^{2n}$, which are not symplectically diffeomorphic to convex ones. The proof will be based on the action principle. The variational approach will be explained in detail. In the analytical framework of the Sobolev space $H^{1/2}$, we shall introduce a special minimax argument which originated in the work of P. Rabinowitz. The construction of the capacity c_0 completes the proof of the symplectic rigidity phenomena described in Chapter 2. The special features of c_0, however, will also be useful later on for the dynamics of Hamiltonian systems.

3.1 Definition of the capacity c_0

If (M, ω) is a symplectic manifold we recall that to a smooth function $H : M \to \mathbb{R}$ there belongs a unique Hamiltonian vector field X_H on M defined by

$$\omega(X_H(x), v) = -dH(x)(v), \quad v \in T_x M$$

and $x \in M$. A T-periodic solution $x(t)$ of the Hamiltonian equation

$$\dot{x} = X_H(x) \quad \text{on} \quad M$$

is a solution satisfying the boundary conditions $x(T) = x(0)$ for some $T > 0$. We now single out a distinguished class of Hamiltonian functions H which look qualitatively as depicted in the following figure 3.1.

When M has a boundary ∂M, the Hamiltonian vector field X_H has compact support contained in the interior of M. Our guiding principle is the hope that if the oscillation of the function H is sufficiently large, then the corresponding Hamiltonian vector field possesses, independently of the size of its support, a fast periodic solution, i.e., a periodic solution having a small period T, say $0 < T \leq 1$.

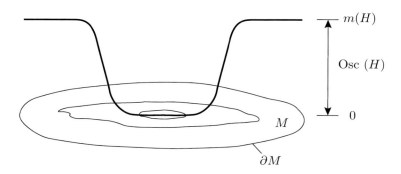

Fig. 3.1

To be precise we denote by $\mathcal{H}(M,\omega)$ the set of smooth functions H on M satisfying the following three properties:

P.1. There is a compact set $K \subset M$ (depending on H) such that $K \subset (M\backslash \partial M)$ and
$$H(M\backslash K) \equiv m(H) \text{ (a constant)}$$

P.2. There is an open set $U \subset M$ (depending on H) on which
$$H(U) \equiv 0$$

P.3. $0 \leq H(x) \leq m(H)$ for all $x \in M$.

The constant $m(H) = \max H - \min H$ is the oscillation of H. A function $H \in \mathcal{H}(M,\omega)$ will be called admissible if all the periodic solutions of $\dot{x} = X_H(x)$ on M are either constant, i.e., $x(t) \equiv x(0)$ for all t or have a period $T > 1$. Abbreviating the set of admissible functions by $\mathcal{H}_a(M,\omega) \subset \mathcal{H}(M,\omega)$, we define
$$c_0(M,\omega) = \sup \{m(H) \mid H \in \mathcal{H}_a(M,\omega)\}.$$

Therefore, if $c_0(M,\omega) < \infty$ then this number is characterized by the property that for every function H in $\mathcal{H}(M,\omega)$ whose oscillation satisfies $m(H) > c_0(M,\omega)$, the vector field X_H possesses a nonconstant T-periodic solution for some $0 < T \leq 1$, and $c_0(M,\omega)$ is the infimum of the real numbers having this property. The main result of this chapter is

Theorem 1. The function c_0 is a symplectic capacity. Moreover, $c_0 = \check{c}_0$, i.e., c_0 has the property of inner regularity.

As an illustration, we consider the ellipsoid E in $(\mathbb{R}^{2n},\omega_0)$ determined by the invariants $r_1(E) \leq r_2(E) \leq \ldots r_n(E)$. We know that $c(E,\omega_0) = \pi r_1(E)^2$ for every capacity c. This linear invariant can, therefore, dynamically be defined as follows:

3.1 Definition of the capacity c_0

Proposition 1. If $H \in \mathcal{H}(E, \omega_0)$ satisfies $m(H) > \pi r_1(E)^2$, then the Hamiltonian vector field X_H on E possesses a nonconstant periodic solution of period $0 < T \leq 1$. Moreover, $\pi r_1(E)^2$ is the infinimum of the real numbers with this property.

Corollary. If $H \in \mathcal{H}(B(r), \omega_0)$ satisfies
$$m(H) > \pi r^2,$$
then the Hamiltonian system $\dot{x} = X_H(x)$ on $B(r)$ has a nonconstant periodic solution of period $0 < T \leq 1$.

Observe first that the statement about inner regularity $c_0(M, \omega) = \check{c}_0(M, \omega)$ follows immediately from the definitions. We only consider the case $c_0(M, \omega) < \infty$, the case $c_0(M, \omega) = \infty$ if treated similarly. If $\varepsilon > 0$ then there exists $H \in \mathcal{H}_a(M, \omega)$ with $m(H) > c_0(M, \omega) - \varepsilon$. Let $K = \text{supp}(X_H) \subset (M \backslash \partial M)$ and pick an open set U with $K \subset U \subset \bar{U} \subset (M \backslash \partial M)$. Clearly $H \in \mathcal{H}_a(U, \omega)$ and hence $c_0(U, \omega) \geq c_0(M, \omega) - \varepsilon$ so that $c_0(M, \omega) = \sup\{c_0(U, \omega) \mid U \subset M \text{ open and } \bar{U} \subset M \backslash \partial M\}$ which by definition is equal to $\check{c}_0(M, \omega)$.

We have to prove that $c_0(M, \omega)$ meets the axioms of a capacity c which we recall for the convenience of the reader. All the symplectic manifolds considered are of fixed dimension $2n$.

A.1. $c(M, \omega) \leq c(N, \sigma)$ provided there exists a symplectic embedding $\varphi : M \to N$.
A.2. $c(M, \alpha \omega) = |\alpha| c(M, \omega)$, $\alpha \neq 0$.
A.3. $c(Z(1), \omega_0) = \pi = c(B(1), \omega_0)$.

The first two properties follow very easily from the definition of c_0.

Lemma 1. c_0 satisfies the monotonicity axiom A.1.

Proof. If $\varphi : M \to N$ is a symplectic embedding we can define the map $\varphi_* : \mathcal{H}(M, \omega) \to \mathcal{H}(N, \sigma)$ as follows:
$$\varphi_*(H) = \begin{cases} H \circ \varphi^{-1}(x) & \text{if } x \in \varphi(M) \\ m(H) & \text{if } x \notin \varphi(M). \end{cases}$$

Note that if $K \subset M \backslash \partial M$ for a compact set $K \subset M$, then also $\varphi(K) \subset N \backslash \partial N$, so that indeed $\varphi_*(H) \in \mathcal{H}(N, \sigma)$. Clearly $m(\varphi_*(H)) = m(H)$, so that the lemma follows if we show that $\varphi_*(\mathcal{H}_a(M, \omega)) \subset \mathcal{H}_a(N, \sigma)$. Since φ is symplectic $\varphi^*(X_H) = X_{\varphi_*(H)}$ on $\varphi(M)$ so that the flows are conjugated. In particular, the nonconstant periodic solutions together with their periods correspond to each other. ∎

Lemma 2. c_0 satisfies the conformality axiom A.2.

Proof. Assume $\alpha \neq 0$ and define the obvious bijection $\psi : \mathcal{H}(M, \omega) \to \mathcal{H}(M, \alpha \omega)$ by $\psi : H \mapsto |\alpha| \cdot H = H_\alpha$. Clearly $m(H_\alpha) = |\alpha| \cdot m(H)$ so that the lemma follows if we show that ψ is also a bijection between $\mathcal{H}_a(M, \omega)$ and $\mathcal{H}_a(M, \alpha \omega)$. By the definition of Hamiltonian vector fields $(\alpha \omega)(X_{H_\alpha}, \cdot) = -dH_\alpha = -|\alpha|dH = |\alpha|\,\omega(X_H, \cdot)$. Hence, since ω is nondegenerate

$$\frac{\alpha}{|\alpha|} X_{H_\alpha} = X_H \text{ on } M.$$

Therefore, $X_{H_\alpha} = X_H$ or $X_{H_\alpha} = -X_H$ and the two vector fields have the same periodic solutions with the same periods. ∎

Lemma 3. $c_0(B(1), \omega_0) \geq \pi$.

Proof. Pick $0 < \varepsilon < \pi$, we shall construct a function $H \in \mathcal{H}_a(B(1))$ satisfying $m(H) = \pi - \varepsilon$, hence proving $c_0(B(1), \omega_0) \geq \pi - \varepsilon$. Choose a smooth function $f : [0, 1] \to [0, \infty)$ satisfying

$$0 \leq f'(t) < \pi$$
$$f(t) = 0 \quad \text{for } t \text{ near } 0$$
$$f(t) = \pi - \varepsilon \quad \text{for } t \text{ near } 1.$$

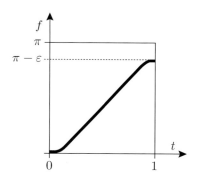

Fig. 3.2

Define $H(x) = f(|x|^2)$ for $x \in B(1)$. Then $m(H) = \pi - \varepsilon$ and we claim that $H \in \mathcal{H}_a(M, \omega)$. Indeed, the Hamiltonian system

$$-J\dot{x} = \nabla H(x) = 2f'(|x|^2)x$$

has the function $G(x) = \frac{1}{2}|x|^2$ as integral, since $\langle \nabla G, J \nabla H \rangle = 2\,f'(|x|^2)\,\langle x, Jx \rangle = 0$. Therefore, if $x(t)$ is a solution, then

$$2\,f'\,(\,|x(t)|^2\,) = a$$

is constant and the solution satisfies

$$-J\dot{x} = ax.$$

3.1 Definition of the capacity c_0

Consequently, all the solutions are periodic and given by $x(t) = e^{aJt}x(0) = (\cos at)x(0) + (\sin at)Jx(0)$. If $a = 0$ then the solution is constant, while if $a > 0$ the period is given by $T = \frac{2\pi}{a} > 1$ since, by construction, $0 \leq a < 2\pi$. Therefore, H is admissible as we wanted to prove. ∎

The inclusion map $B(1) \to Z(1)$ is a symplectic embedding and we deduce from Lemma 1 and Lemma 3

$$c_0(Z(1)) \geq c_0(B(1)) \geq \pi.$$

In order to show that c_0 meets the normalization A.3 we have to prove that $c_0(Z(1)) \leq \pi$. This is the subtle part of the proof. It requires an existence proof and our next aim is to prove

Theorem 2. If $H \in \mathcal{H}(Z(1))$ satisfies $m(H) > \pi$, then the Hamiltonian system $\dot{x} = J\nabla H(x)$ on $Z(1)$ possesses a periodic solution in $Z(1)$ having period $T = 1$ and which is not a constant solution.

The theorem is qualitative in nature. We observe that, in proving Theorem 2, we may replace H by $H \circ \psi$, where ψ is a compactly supported symplectic diffeomorphism of $Z^{2n}(1)$, i.e., the closure of the set $\{x|\psi(x) \neq x\}$ is a compact subset of $Z^{2n}(1)$. Indeed, in view of the transformation law of Hamiltonian vector fields (Chapter 1), the flows of X_H and $X_{H\circ\psi}$ are conjugated. Using this observation, we may assume without loss of generality that H vanishes in a neighborhood of the origin. This is proved as follows. By assumption $H|U \equiv 0$. We pick a point $z_0 \in U$ and choose a smooth function $\rho : \mathbb{R}^{2n} \to \mathbb{R}$ with compact support in $Z^{2n}(1)$ and which is equal to 1 on a neighborhood V of the line $\{tz_0 | 0 \leq t \leq 1\}$ connecting the origin with z_0. Define the Hamiltonian $K : Z^{2n}(1) \to \mathbb{R}$ by

$$K(z) = \rho(z)\langle z, -Jz_0\rangle .$$

Then the time-1 map ψ of X_K is symplectic and has compact support in $Z^{2n}(1)$. Since $X_K(z) = z_0$ if $z \in V$, the line $\psi^t(0) = tz_0$ is a solution for $0 \leq t \leq 1$ moving the origin to the point z_0. Hence

$$\psi(0) = z_0 ,$$

and the Hamiltonian $H \circ \psi$ vanishes in a neighborhood of 0.

From now on we therefore assume that the Hamiltonian H vanishes in an open neighborhood of the origin and first extend H defined on $Z(1)$ to a function defined on the whole space $\mathbb{R}^{2n}$. This is, of course, easy since the function is constant near the boundary of $Z(1)$. However, in view of the proof later on, we shall choose a very special continuation. Observe first that if $H \in \mathcal{H}(Z(1), \omega_0)$ then there exists an ellipsoid $E = E_N$ such that $H \in \mathcal{H}(E, \omega_0)$, where

$$E_N = \left\{ z \in \mathbb{R}^{2n} \;\middle|\; q(z) < 1 \right\},$$

is, in the coordinates $z = (x, y)$ given by,

$$q(z) = (x_1^2 + y_1^2) + \frac{1}{N^2} \sum_{j=2}^{n} (x_j^2 + y_j^2)$$

and $N \in \mathbb{Z}^+$ is sufficiently large (Fig. 3.3).

Since $H \in \mathcal{H}(Z(1), \omega_0)$ satisfies $m(H) > \pi$ there is an $\varepsilon > 0$ such that $m(H) > \pi + \varepsilon$. We can, therefore, pick a smooth function $f : \mathbb{R} \to \mathbb{R}$ satisfying

$$\begin{aligned} f(s) &= m(H) & \text{for } s \leq 1 \\ f(s) &\geq (\pi + \varepsilon)s & \text{for all } s \in \mathbb{R} \\ f(s) &= (\pi + \varepsilon)s & \text{for } s \text{ large} \\ 0 < f'(s) &\leq \pi + \varepsilon & \text{for } s > 1. \quad \text{(Fig. 3.4)} \end{aligned}$$

The extension of H is now defined by

$$\overline{H}(z) = \begin{cases} H(z) & z \in E \\ f(q(z)) & z \notin E. \end{cases}$$

Clearly $\overline{H} \in C^\infty(\mathbb{R}^{2n})$ and $\overline{H}$ is quadratic at infinity,

$$\overline{H}(z) = (\pi + \varepsilon)q(z), \quad \text{if} \quad |z| \geq R$$

for some large R. The following crucial proposition describes the distinguished periodic solutions we are looking for.

Proposition 2. Assume $x(t)$ is a periodic solution of $\dot{x} = X_{\overline{H}}(x)$ having period 1. If it satisfies

$$\Phi(x) := \int_0^1 \left\{ \frac{1}{2} \langle -J\dot{x}, x \rangle - \overline{H}(x(t)) \right\} dt > 0,$$

then $x(t)$ is nonconstant and $x(t) \subset E$. Hence $x(t)$ is a nonconstant 1-periodic solution of the original system $\dot{x} = X_H(x)$ on $Z(1)$.

3.1 Definition of the capacity c_0

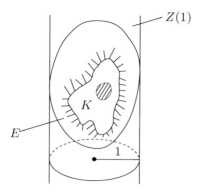

Fig. 3.3

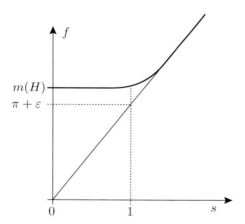

Fig. 3.4

Proof. If $x(t) = x^*$ is a constant solution, then $\Phi(x^*) \leq 0$ since $\overline{H} \geq 0$. The Hamiltonian vector field $X_{\overline{H}}$ vanishes on ∂E. Therefore, if for a solution $x(t_0) \notin E$ for some t_0, then $x(t) \notin E$ for all $t \in \mathbb{R}$ and hence it solves the equation

$$-J\dot{x} = \nabla \overline{H}(x) = f'(q(x)) \cdot \nabla q(x).$$

Note that, outside of E, the function q is an integral of this equation since $\langle \nabla q, J f'(q) \nabla q \rangle = f'(q) \langle \nabla q, J \nabla q \rangle = 0$. Consequently, if $x(t)$ is a solution then

$$q(x(t)) = q(x(t_0)) = \tau$$

is constant in t, and in view of $\langle \nabla q(z), z \rangle = 2q(z)$, for $z \in \mathbb{R}^{2n}$, we can compute

$$\begin{aligned} \Phi(x) &= \int_0^1 \{\tfrac{1}{2}\langle -J\dot{x}, x\rangle - \overline{H}(x)\}\, dt \\ &= \int_0^1 \tfrac{1}{2} f'(\tau) \langle \nabla q(x), x\rangle dt - \int_0^1 f(q(x)) dt \\ &= \tfrac{1}{2} \int_0^1 f'(\tau)\, 2q(x(t)) dt - f(\tau) \\ &= f'(\tau) \cdot \tau - f(\tau). \end{aligned}$$

By definition of f, this is smaller than or equal to $(\pi + \varepsilon)\tau - (\pi + \varepsilon)\tau = 0$. Hence $\Phi(x) \leq 0$ for all solutions outside of E and the proposition is proved. ∎

In view of this remark we have to find a 1-periodic solution $x = x(t)$ of the Hamiltonian system $\dot{x} = X_{\overline{H}}(x)$ which satisfies $\Phi(x) > 0$. In the following we shall change the notation and replace $\overline{H}$ by H. In order to establish the existence of a periodic solution, we make use of a well-known variational principle for which the critical points are the required periodic solutions. To introduce this principle, we proceed at first on a informal level and consider the loop space

$$\Omega = C^\infty(S^1, \mathbb{R}^{2n}) \text{ where } S^1 = \mathbb{R}/\mathbb{Z},$$

of smooth loops in $\mathbb{R}^{2n}$. We define a function $\Phi : \Omega \to \mathbb{R}$ by setting

$$\Phi(x) = \int_0^1 \frac{1}{2} \langle -J\dot{x}, x\rangle\, dt - \int_0^1 H(x(t)) dt, \quad x \in \Omega,$$

and claim that the critical points of Φ are the periodic solutions of $\dot{x} = X_H(x)$. Computing the derivative at $x \in \Omega$ in the direction of $y \in \Omega$, we find

$$\begin{aligned} \Phi'(x)(y) &= \tfrac{d}{d\varepsilon} \Phi(x + \varepsilon y)\Big|_{\varepsilon=0} \\ &= \int_0^1 \langle -J\dot{x} - \nabla H(x), y\rangle\, dt. \end{aligned}$$

Consequently, $\Phi'(x)(y) = 0$ for all $y \in \Omega$ if and only if the loop x satisfies the equation

$$-J\dot{x}(t) - \nabla H(x(t)) = 0,$$

i.e., $x(t)$ is a solution of the Hamiltonian equation which also satisfies $x(0) = x(1)$. Hence it is periodic of period 1. Therefore, the principle picks out precisely the 1-periodic solutions among the intricate set of all solutions of a Hamiltonian vector field. This principle is the well-known action principle in mechanics, where it is

3.2 The minimax idea

often used to derive the transformation law of Hamiltonian vector fields. It might seem rather awkward to use this principle for existence proofs, since it is highly degenerate. Indeed, taking the special loops x_j,

$$x_j(t) = e^{j2\pi Jt}\xi = (\cos j2\pi t)\xi + (\sin j2\pi t)J\xi$$

where $|\xi| = 1$, one computes readily for the first part of Φ

$$\int_0^1 \frac{1}{2}\langle -J\dot{x}_j, x_j\rangle\, dt = \pi j, \quad \|x_j\|_{L^2} = 1,$$

while the second part of Φ stays bounded. We see that the functional is bounded neither below nor above. In particular, the variational techniques based on minimizing sequences do not apply. This is in sharp contrast to the variational principle for closed geodesics on Riemannian manifolds. This geometric problem gave rise to two powerful variational methods, the so-called Morse theory and the Ljusternik-Schnirelman theory. We should also mention that the Morse theory is not applicable directly to our functional. Indeed, a priori the Morse indices of critical points are infinite and hence topologically not visible. It is this difficulty which forced Andreas Floer to develop powerful new techniques extending the Morse theory, which led him to the so-called Floer complex described in more detail later on.

It was only relatively recently that P. Rabinowitz [180, 181] demonstrated that the degenerate variational principle can be used very effectively for existence proofs. For this purpose he introduced special minimax principles adapted to the structure of the functional Φ. Before we describe the technical details, it might be helpful to recall in an abstract setting the minimax idea.

3.2 The minimax idea

We consider a differentiable function

$$f: E \to \mathbb{R}, \quad f \in C^1(E, \mathbb{R})$$

defined on a Hilbert space E whose inner product is denoted by $\langle \cdot, \cdot \rangle$ and whose norm is $\|x\|^2 = \langle x, x \rangle$ for $x \in E$. We search for critical points of f, i.e., points $x^* \in E$ at which the derivative of f vanishes:

$$df(x^*) = 0 \iff df(x^*)(y) = 0 \text{ for all } y \in E.$$

The derivative $df(x)$ is, by definition, an element of the dual space $\mathcal{L}(E, \mathbb{R}) = E^*$. Recall that by a well-known theorem due to F. Riesz, there is a distinguished linear isometry I from E^* onto E such that $e^*(x) = \langle I(e^*), x \rangle$ for every $x \in E$, where $e^* \in E^*$. Therefore,

$$df(x)(y) = \langle v(x), y \rangle \quad \text{and} \quad \|df(x)\|_* = \|v(x)\|$$

where $v(x) \in E$ is uniquely determined by $df(x)$. It is called the gradient of f at x (with respect to the inner product $\langle \cdot, \cdot \rangle$) and denoted by

$$v(x) = \nabla f(x) \in E.$$

The critical points of f are the zeros of the gradient function $x \mapsto \nabla f(x)$ in E. Using a dynamical approach, we interpret the critical points of f as the equilibrium points of the gradient equation

$$\dot{x} = -\nabla f(x), \quad x \in E.$$

This is an ordinary differential equation on E, and we shall assume now that it has a global flow $\varphi^t(x)$, i.e., that we can solve the Cauchy initial value problem uniquely and for all times $t \in \mathbb{R}$:

$$\frac{d}{dt}\varphi^t(x) = -\nabla f(\varphi^t(x))$$
$$\varphi^0(x) = x,$$

for all initial conditions $x \in E$. Such a global flow does exist, for example, if there is a constant $M \geq 0$ such that $\|\nabla f(x) - \nabla f(y)\| \leq M\|x - y\|$ for all $x, y \in E$, as is, of course, well-known. The crucial property of the gradient flow is that $f(\varphi^t(x))$ decreases along nonconstant solutions $\varphi^t(x)$. This follows immediately from the definitions; indeed

$$\frac{d}{ds} f(\varphi^s(x)) = df(\varphi^s(x)) \left(\frac{d}{ds}\varphi^s(x) \right)$$
$$= df(\varphi^s(x)) (-\nabla f(\varphi^s(x)))$$
$$= -\langle \nabla f(\varphi^s(x)), \nabla f(\varphi^s(x)) \rangle.$$

In particular,

(3.1) $\quad f(\varphi^t(x)) - f(x) = \displaystyle\int_0^t \frac{d}{ds} f(\varphi^s(x))\, ds = -\int_0^t \|\nabla f(\varphi^s(x))\|^2 ds.$

Since E is not compact we cannot expect f to have any critical points unless we impose additional conditions on f, as the example $f(x) = e^{-x}$ on $x \in \mathbb{R}$ already shows. The following strong compactness condition goes back to Palais and Smale [175] and guarantees a critical point in many variational problems.

Definition. f is said to satisfy P.S., if every sequence $x_j \in E$ satisfying

$$\nabla f(x_j) \to 0 \text{ in } E \text{ and } |f(x_j)| \leq c < \infty,$$

for some $c \geq 0$, possesses a convergent subsequence. Since f is assumed to be C^1, the limit of this subsequence is a critical point of f.

3.2 THE MINIMAX IDEA

If $\mathcal{F}$ is a family of subsets $F \subset E$ one defines the minimax $c(f, \mathcal{F})$ belonging to f and $\mathcal{F}$ by

$$c(f, \mathcal{F}) = \inf_{F \in \mathcal{F}} \sup_{x \in F} f(x) \quad \in \mathbb{R} \cup \{\infty\} \cup \{-\infty\}$$

Minimax Lemma. Assume $f \in C^1(E, \mathbb{R})$ and $\mathcal{F}$ meet the following conditions:

(i) f satisfies P.S.

(ii) $\dot{x} = -\nabla f(x)$ defines a global flow $\varphi^t(x)$

(iii) The family $\mathcal{F}$ is positively invariant under the flow, i.e., if $F \in \mathcal{F}$ then $\varphi^t(F) \in \mathcal{F}$ for every $t \geq 0$

(iv) $-\infty < c(f, \mathcal{F}) < \infty$.

Then the real number, $c(f, \mathcal{F})$ is a critical value of f, i.e., there exists $x^* \in E$ satisfying

$$\nabla f(x^*) = 0 \quad \text{and} \quad f(x^*) = c(f, \mathcal{F}).$$

Proof. Abbreviating $c = c(f, \mathcal{F})$ which, by assumption, is a real number, we shall show that for every $\varepsilon > 0$ there exists an $x \in E$ satisfying

$$c - \varepsilon \leq f(x) \leq c + \varepsilon \quad \text{and} \quad \| \nabla f(x) \| \leq \varepsilon.$$

Then choosing $\varepsilon_j = 1/j$ we find a sequence x_j which has, by the P.S. condition, a convergent subsequence whose limit satisfies $\nabla f(x^*) = 0$ and $f(x^*) = c$, hence proving the Lemma. Arguing by contradiction we find an $\varepsilon > 0$ such that

(3.2) $$\| \nabla f(x) \| > \varepsilon$$

for all x satisfying $c - \varepsilon \leq f(x) \leq c + \varepsilon$. By the definition of c, there exists an $F \in \mathcal{F}$ satisfying

(3.3) $$\sup_{x \in F} f(x) \leq c + \varepsilon.$$

Pick a point $x \in F$; then $f(x) \leq c + \varepsilon$ and we claim that the solution $\varphi^t(x)$ satisfies $f(\varphi^{t^*}(x)) \leq c - \varepsilon$ if $t^* = 2/\varepsilon$. Indeed, if $f(\varphi^t(x)) \leq c - \varepsilon$ for some $0 \leq t \leq t^*$ there is nothing to prove since $f(\varphi^t(x))$ is decreasing. Assume now, by contradiction, that $f(\varphi^t(x)) > c - \varepsilon$ for all $0 \leq t \leq t^*$. Then by, (3.2), we have $\|\nabla f(\varphi^t(x))\| \geq \varepsilon$, for $0 \leq t \leq t^*$, and hence, by (3.1), $f(\varphi^t(x)) \leq f(x) - \varepsilon^2 t$, so that $f(\varphi^{t^*}(x)) \leq c + \varepsilon - \varepsilon^2 t^* = c - \varepsilon$, a contradiction. Setting $F^* = \varphi^{t^*}(F)$ we have shown that

$$\sup_{x \in F^*} f(x) \leq c - \varepsilon.$$

Since, by assumption, $F^* \in \mathcal{F}$, this estimate contradicts the definition of c. ∎

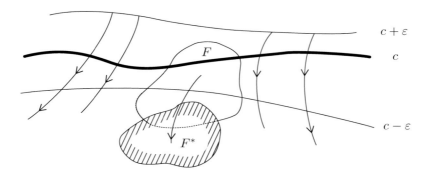

Fig. 3.5

For example, if f is, in addition, bounded from below we can choose $\mathcal{F}$ to be the family of subsets $\{x\}$ consisting of the elements of E so that

$$c(f, \mathcal{F}) = \inf_{x \in E} f(x).$$

The lemma then guarantees a minimum, i.e., we find $x^* \in E$ satisfying $\nabla f(x^*) = 0$ and

$$f(x^*) = \inf_{x \in E} f(x).$$

As another illustration we describe the celebrated Mountain Pass Lemma discovered by A. Ambrosetti and P. Rabinowitz [6]. We reproduce the visually attractive version found in R. Palais and C. Terng [174]. A subset $R \subset E$ is called a mountain range relative to f if it separates E, if

$$\alpha := \inf_R f > -\infty$$

and if on every component of $E \backslash R$, the function f assumes a value strictly less than α. Consider an $f \in C^1(E, \mathbb{R})$ satisfying the P.S. condition and assume that the gradient equation generates a global flow.

Mountain Pass Lemma. If $R \subset E$ is a mountain range relative to f, then f has a critical value c satisfying

$$c \geq \inf_R f.$$

Proof. We choose two different components E^0 and E^1 of $E \backslash R$ and define $E_\alpha^j = \{x \in E^j \mid f(x) < \alpha\}$ for $j = 0, 1$. By assumption, these sets are not empty. Let Γ be the set of all continuous paths $\gamma : [0, 1] \to E$ satisfying $\gamma(0) \in E_\alpha^0$ and $\gamma(1) \in E_\alpha^1$ and define the family $\mathcal{F}$ of compact subsets of E by

$$\mathcal{F} = \Big\{ \text{image } (\gamma) \; \Big| \; \gamma \in \Gamma \Big\}.$$

3.2 THE MINIMAX IDEA

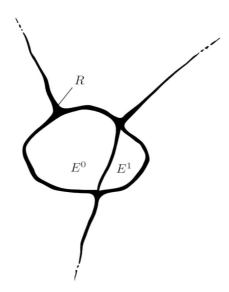

Fig. 3.6

Since $\gamma(0)$ and $\gamma(1)$ belong to different components of $E \backslash R$, we conclude that $\gamma(t_0) \in R$ for some $0 < t_0 < 1$ so that $f(\gamma(t_0)) \geq \alpha$ and hence

$$\alpha \leq c(f, \mathcal{F}) < \infty.$$

In view of the minimax lemma, it suffices to verify that $\mathcal{F}$ is positively invariant under the flow φ^t of $-\nabla f$. For this it is sufficient to show that if $x \in E_\alpha^j$ then also $\varphi^t(x) \in E_\alpha^j$ for all $t \geq 0$. But this follows in view of $f(\varphi^t(x)) \leq f(\varphi^0(x)) = f(x) < \alpha$ for all $t \geq 0$. ∎

For other versions of the Mountain Pass Lemma as well as for many applications, we refer to the books *Variational methods* by M. Struwe [208] and *Minimax methods in critical point theory with applications to differential equations* by P. Rabinowitz [182]. It should be said that if f is bounded neither from above nor from below, it is often a subtle task to analyze the qualitative behaviour of f and to find an appropriate family $\mathcal{F}$ meeting the estimate (iv) above. This will be done for our special functional Φ in the next two sections. First, however, we have to define the appropriate functional analytic setting.

As for more sophisticated classical techniques in critical point theory, we would like to mention the classical paper by R.S. Palais [173] (1970) and the monographs by P. Blanchard, E. Brüning [28] (1992), by J. Mawhin, M. Willem [152] (1989) and, quite recently, by Kung-Ching Chang [39] (1993). New methods designed for the action functional on the loop space of a compact symplectic manifold will be described in Chapter 6.

3.3 The analytical setting

Let $H \in C^\infty(\mathbb{R}^{2n}, \mathbb{R})$ denote the special Hamiltonian function introduced in Section 3.1. It vanishes on an open neighborhood of the origin and satisfies

(3.4) $$H(z) = Q(z), \quad |z| \geq R,$$

with the quadratic form $Q(z) = (\pi + \varepsilon)q(z), z \in \mathbb{R}^{2n}$. We shall translate these properties of the Hamiltonian into properties of the corresponding functional Φ on an appropriate Hilbert space. Recall that on the space $\Omega = C^\infty(S^1, \mathbb{R}^{2n})$ of smooth loops Φ is defined by

(3.5) $$\Phi(x) = \int_0^1 \tfrac{1}{2}\langle -J\dot{x}, x\rangle dt - \int_0^1 H(x(t))\, dt, \quad x \in \Omega.$$

In order to find the convenient Hilbert space, we represent the periodic loops $x \in C^\infty(S^1, \mathbb{R}^{2n})$ by their Fourier-series

(3.6) $$x(t) = \sum_{k \in \mathbb{Z}} e^{k 2\pi J t} x_k, \quad x_k \in \mathbb{R}^{2n}$$

which converge, together with all their derivatives, in the supremum norm. Considering first the dominant part of Φ which reflects the symplectic structure, we abbreviate

(3.7) $$a(x, y) = \int_0^1 \tfrac{1}{2}\langle -J\dot{x}, y\rangle dt, \quad x, y \in \Omega.$$

Then, inserting the Fourier expansions of $x, y \in \Omega$ into $a(x, y)$, and observing that

$$\int_0^1 \langle e^{j 2\pi J t} x_j, e^{k 2\pi J t} y_k\rangle dt = \delta_{jk}\langle x_j, y_k\rangle,$$

one obtains

(3.8) $$2a(x,y) = 2\pi \sum_{j \in \mathbb{Z}} j \langle x_j, y_j\rangle$$

$$= 2\pi \sum_{j>0} |j|\langle x_j, y_j\rangle - 2\pi \sum_{j<0} |j|\langle x_j, y_j\rangle.$$

Consequently, $a(x, y)$ can be defined as a continuous bilinear form on the larger space $H^{1/2} = H^{1/2}(S^1)$ which is a special Sobolev space. Recall that the spaces

3.3 THE ANALYTICAL SETTING

$H^s = H^s(S^1)$ for $s \geq 0$ are defined by

$$H^s = \left\{ x \in L^2(S^1) \mid \sum_{j \in \mathbb{Z}} |j|^{2s} |x_j|^2 < \infty \right\}$$

where

$$x = \sum_{j \in \mathbb{Z}} e^{j 2\pi J t} x_j, \quad x_j \in \mathbb{R}^{2n}$$

is the Fourier series of x which converges in $L^2(S^1)$. The space H^s is a Hilbert space with inner product and associated norm defined by

$$\langle x, y \rangle_s = \langle x_0, y_0 \rangle + 2\pi \sum_{k \in \mathbb{Z}} |k|^{2s} \langle x_k, y_k \rangle$$
$$\|x\|_s^2 = \langle x, x \rangle_s,$$

for $x, y \in H^s$. Note that the norm $\|x\|_0$ is equivalent to the L^2-norm. We shall denote the distinguished Hilbert space $H^{1/2}$ which will be the underlying space of our functional by

$$E \equiv H^{1/2}$$
$$\langle , \rangle \equiv \langle , \rangle_{\frac{1}{2}} \text{ and } \| \| = \| \|_{\frac{1}{2}}.$$

There is an orthogonal splitting of E

$$E = E^- \oplus E^0 \oplus E^+$$

into the spaces of $x \in E$ having only Fourier coefficients for $j < 0, j = 0, j > 0$. The corresponding orthogonal projections are denoted by P^-, P^0, P^+. Therefore, every $x \in E$ has a unique decomposition

$$x = x^- + x^0 + x^+.$$

In view of (3.8) we define for $x, y \in E$

(3.9) $$a(x, y) = \tfrac{1}{2} \langle x^+, y^+ \rangle - \tfrac{1}{2} \langle x^-, y^- \rangle$$
$$= \tfrac{1}{2} \langle (P^+ - P^-) x, y \rangle,$$

which is a continuous bilinear form on E and which agrees for $x, y \in C^\infty(S^1, \mathbb{R}^{2n})$ with (3.7). The function $a : E \to \mathbb{R}$, defined by

(3.10) $$a(x) = a(x, x) = \tfrac{1}{2} \|x^+\|^2 - \tfrac{1}{2} \|x^-\|^2,$$

is differentiable with derivative

$$da(x)(y) = \langle (P^+ - P^-) x, y \rangle,$$

so that the gradient of a is

(3.11) $$\nabla a(x) = (P^+ - P^-)x = x^+ - x^- \in E,$$

at every point $x \in E$.

Next we point out some useful properties of the spaces H^s needed later on. Clearly the spaces decrease,
$$H^t \subset H^s \subset H^0 \quad \text{for } t \geq s \geq 0,$$
while the norms increase:
$$\|x\|_t \geq \|x\|_s \geq \|x\|_0 \quad \text{for } x \in H^t.$$
In particular, the inclusion maps $I : H^t \to H^s$ for $t \geq s$ are continuous.

Proposition 3. Assume $t > s \geq 0$. Then the inclusion map $I : H^t \to H^s$ is compact, i.e., it maps bounded sets of H^t onto relatively compact subsets of H^s.

Proof. The continuous linear operator $P_N : H^t \to H^s$ defined by
$$P_N x = \sum_{|k| \leq N} e^{k 2\pi J t} x_k$$
has finite dimensional range and hence is a compact operator. The estimate
$$\|(P_N - I)x\|_s^2 = \|\sum_{|k|>N} e^{k 2\pi J t} x_k\|_s^2$$
$$= 2\pi \sum_{|k|>N} |k|^{2s} |x_k|^2 = 2\pi \sum_{|k|>N} |k|^{2(s-t)} |k|^{2t} |x_k|^2$$
$$\leq N^{2(s-t)} 2\pi \sum_{|k|>N} |k|^{2t} |x_k|^2 \leq N^{2(s-t)} \|x\|_t^2$$

shows that $P_N \to I$ in the operator norm, i.e., the norm of $\mathcal{L}(H^s, H^t)$. Consequently I is also compact, which is well-known and easily seen as follows. Let $\varepsilon > 0$ and choose N so large that $\|P_N - I\| < \varepsilon/2$ in the operator-norm. If B is the unit ball in H^t, then $P_N(B) \subset H^s$ is covered by finitely many balls of radius $\varepsilon/2$ which are centered at the points $y_1, y_2, \ldots, y_m \in H^s$. Hence, if $x \in B$ we find an y_j satisfying $\|P_N(x) - y_j\|_s < \varepsilon/2$ and together with $\|P_N(x) - I(x)\|_s < \varepsilon/2$ this implies that $\|I(x) - y_j\|_s < \varepsilon$. Thus $I(B)$ is covered by finitely many balls of radius $\varepsilon > 0$. This holds true for every $\varepsilon > 0$ and we conclude that $I(B)$ is relatively compact in H^s. ∎

The set $C^\infty(S^1, \mathbb{R}^{2n})$ is dense in H^s for every $s \geq 0$. However, not all elements of $H^{1/2}$ are represented by continuous functions. In order to find continuous functions the following proposition will be helpful.

3.3 THE ANALYTICAL SETTING

Proposition 4. Assume $s > \frac{1}{2}$. If $x \in H^s$ then $x \in C(S^1)$. Moreover, there is a constant $c = c_s$ such that

$$\sup_{0 \le t \le 1} |x(t)| \le c \|x\|_s, \quad x \in H^s(S^1).$$

Proof. We show that the Fourier series

$$x = \sum_k e^{k 2\pi J t} x_k,$$

which converges in L^2, also converges in the sup-norm. This follows from the estimates:

$$\sum_{k \ne 0} |e^{k 2\pi J t} x_k| = \sum_{k \ne 0} |x_k| = \sum_{k \ne 0} |k|^{-s} |k|^s |x_k|$$

$$\le \left(\sum_{k \ne 0} \frac{1}{|k|^{2s}}\right)^{\frac{1}{2}} \cdot \left(\sum_{k \ne 0} |k|^{2s} |x_k|^2\right)^{\frac{1}{2}}$$

$$\le c \|x\|_s,$$

where we have used that $2s > 1$. ∎

The same argument shows that, more generally, if $s > \frac{1}{2} + r$ for an integer r, then $x \in H^s$ belongs to $C^r(S^1)$ and

$$\sup_{\substack{0 \le j \le r \\ 0 \le t \le 1}} |D^j x(t)| \le c \|x\|_s, \quad x \in H^s(S^1).$$

This is again a special case of the Sobolev embedding theorem. In view of Proposition 3, the inclusion map

(3.12) $$j : H^{1/2} \longrightarrow L^2$$

is compact. Its adjoint operator

(3.13) $$j^* : L^2 \longrightarrow H^{1/2}$$

is, as usual, defined by

(3.14) $$\big(j(x), y\big)_{L^2} = \langle x, j^*(y)\rangle_{\frac{1}{2}}$$

for all $x \in H^{1/2}$ and $y \in L^2$. The following property of j^* will be useful later on.

Proposition 5.
$$j^*(L^2) \subset H^1 \quad \text{and} \quad \|j^*(y)\|_1 \leq \|y\|_{L^2}.$$

The map j^* factors: $L^2 \to H^1 \to H^{1/2}$, thus we have verified using Proposition 3 that j^* is a compact operator. This follows also from the compactness of j, as is well-known.

Proof. By definition of the adjoint
$$\sum_k \langle x_k, y_k \rangle = \langle x_0, j^*(y)_0 \rangle + 2\pi \sum_k |k| \langle x_k, j^*(y)_k \rangle$$
for $x \in H^{1/2} \subset L^2$ and $y \in L^2$. One reads off the following formula for j^*
$$j^*(y) = y_0 + \sum_{k \neq 0} \frac{1}{2\pi |k|} e^{k 2\pi J t} y_k,$$
if $y \in L^2$. The estimate
$$\|j^*(y)\|_1 \leq \|y\|_{L^2}$$
is now obvious. ■

Coming back to the functional Φ, we next study the function
$$b(x) = \int_0^1 H\big(x(t)\big) \, dt ,$$
recalling that H vanishes in a neighborhood of the origin. In view of $|H(z)| \leq M|z|^2$, for all $z \in \mathbb{R}^{2n}$, this map b is defined for $x \in L^2$ and hence also for $x \in E \subset L^2$. If we consider b as a function on L^2, we shall denote it by $\hat{b}$, so that, with the inclusion map $j : E \to L^2$, we can write

(3.15) $$b(x) = \hat{b}(j(x)), \quad x \in E.$$

In order to prove that $\hat{b} : L^2 \to \mathbb{R}$ is differentiable, we start from the identity

(3.16) $$H(z + \xi) = H(z) + \langle \nabla H(z), \xi \rangle + \int_0^1 \langle \nabla H(z + t\xi) - \nabla H(z), \xi \rangle \, dt$$

for $z, \xi \in \mathbb{R}^{2n}$. Since $|H_{zz}(z)| \leq M$ the last term is $\leq M|\xi|^2$. Assume now that $x \in L^2$ then $\nabla H(x) = \nabla H(x(t)) \in L^2$ in view of $|\nabla H(z)| \leq M|z|$ for all $z \in \mathbb{R}^{2n}$. Therefore, given x and $h \in L^2$, we find by integration

$$\hat{b}(x + h) = \hat{b}(x) + \int_0^1 \langle \nabla H(x), h \rangle \, dt + R(x, h).$$

3.3 THE ANALYTICAL SETTING

Moreover, $|R(x,h)| \leq M \, \|h\|_{L^2}^2$. This estimate shows that $\hat{b}$ is differentiable with derivative at x given by

$$d\hat{b}(x)(h) = \int_0^1 \langle \nabla H(x), h \rangle \, dt = (\nabla H(x), h)_{L^2}$$
$$= (\nabla \hat{b}(x), h)_{L^2}.$$

For the gradient (with respect to L^2) we read off $\nabla \hat{b}(x) = \nabla H(x) \in L^2$. The derivative of $b : E \to \mathbb{R}$ is in view of (3.14) given by

$$\langle \nabla b(x), y \rangle = db(x)(y) = d\hat{b}\,(j(x))\,(j(y))$$
$$= \left(\nabla \hat{b}(j(x)), j(y) \right)_{L^2} = \langle j^* \nabla \hat{b}(j(x)), y \rangle \, .$$

Therefore,

(3.17) $$\nabla b(x) = j^* \nabla \hat{b}\left(j(x)\right) = j^* \nabla H(x) \, .$$

Lemma 4. The map $b : E \to \mathbb{R}$ is differentiable. Its gradient $\nabla b : E \to E$ is continuous and maps bounded sets into relatively compact sets. Moreover,

$$\|\nabla b(x) - \nabla b(y)\| \leq M \|x - y\|$$

and $|b(x)| \leq M \|x\|_{L^2}^2$ for all $x, y \in E$.

Proof. Clearly $x \mapsto \nabla H(x)$ is globally Lipschitz continuous on L^2 and maps, therefore, bounded sets into bounded sets. The first claim follows from (3.17), since $j^* : L^2 \to E$ is compact. Moreover,

$$\|\nabla b(x) - \nabla b(y)\|_{\frac{1}{2}} = \|j^*\left(\nabla H(x) - \nabla H(y)\right)\|_{\frac{1}{2}}$$
$$\leq \|\nabla H(x) - \nabla H(y)\|_{L^2} \leq M \, \|x - y\|_{L^2}$$
$$\leq M \, \|x - y\|_{\frac{1}{2}}.$$

The last estimate follows from $|H(z)| \leq M|z|^2$, for all $z \in \mathbb{R}^{2n}$. ∎

It should be mentioned that $b \in C^\infty(E, \mathbb{R})$ provided $H \in C^\infty(\mathbb{R}^{2n}, \mathbb{R})$. This is readily seen by using the Taylor formula together with the nontrivial fact that E is continuously embedded in L^p for every $p \geq 1$ which is proved in the Appendix. To keep our presentation elementary we shall, however, not use this observation here. Summarizing the discussion so far, we have extended Φ from the space Ω of smooth loops to the Hilbert space $E \supset \Omega$ by

$$\Phi(x) = a(x) - b(x) \, , \quad x \in E.$$

This function $\Phi : E \to \mathbb{R}$ is differentiable and its gradient is given by
$$\nabla \Phi(x) = x^+ - x^- - \nabla b(x).$$
We are interested in classical solutions of differential equations. It is, therefore, important to observe now that a critical point of Φ is not simply an element in E which might not even be a continuous function, but it is actually a smooth periodic solution of the Hamiltonian equation X_H having period 1. Indeed the following "regularity" statement holds true.

Lemma 5. Assume $x \in E$ is a critical point, $\nabla \Phi(x) = 0$. Then $x \in C^\infty(S^1)$. Moreover, it solves the Hamiltonian equation
$$\dot{x}(t) = J \nabla H\left(x(t)\right), \quad 0 \le t \le 1,$$
so that $x \in \Omega$ is a 1-periodic solution.

Proof. Represent x and $\nabla H(x) \in L^2$ by their Fourier expansions in L^2
$$x = \sum e^{k\, 2\pi\, Jt}\, x_k$$
$$\nabla H(x) = \sum e^{k\, 2\pi\, Jt}\, a_k.$$
By assumption $d\Phi(x)(v) = 0$.
Hence, using $\langle \nabla b(x), v \rangle = \langle j^* \nabla H(x), v \rangle = (\nabla H(x), v)_{L^2}$,
$$\langle (P^+ - P^-)x, v \rangle = \int_0^1 \langle \nabla H(x), v \rangle\, dt$$
for all $v \in E$. Choosing the test functions $v(t) = e^{k 2\pi J t}\, v$, we find
(3.18) $\qquad\qquad 2\pi k\, x_k = a_k, \quad k \in \mathbb{Z}$
and $a_0 = 0$. We infer that
$$\sum |k|^2\, |x_k|^2 \le \sum |a_k|^2 < \infty,$$
so that $x \in H^1$ and by Proposition 4, the element $x \in E$ belongs to $C(S^1)$. Consequently $\nabla H(x(t)) \in C(S^1)$, and, therefore,
$$\xi(t) = \int_0^t J \nabla H\left(x(t)\right)\, dt \in C^1(\mathbb{R}).$$
Comparing the Fourier coefficients we find in view of (3.18), that $\xi(t) = x(t) - x(0)$, hence x belongs to $C^1(S^1)$ and also solves the equation $\dot{x}(t) = J \nabla H(x(t))$. The

3.3 THE ANALYTICAL SETTING

right hand side of the equation is in C^1, hence $x \in C^2$, and iterating this argument we conclude that $x \in C^\infty(S^1)$ as claimed. ∎

The careful choice of the asymptotic behaviour of the Hamiltonian function H implies that X_H has, in the region of $\mathbb{R}^{2n}$ where $|z|$ is large, no periodic solutions of period 1. From this dynamical behaviour, we shall conclude that Φ satisfies the P.S. condition.

Lemma 6. Every sequence $x_j \in E$ satisfying $\nabla\Phi(x_j) \to 0$ contains a convergent subsequence. In particular, Φ satisfies the P.S. condition.

Proof. Assume $\nabla\Phi(x_j) \to 0$ so that

(3.19) $$x_j^+ - x_j^- - \nabla b(x_j) \to 0.$$

Recall the splitting $x_j = x_j^- + x_j^0 + x_j^+$. If x_j is bounded in E, then $x_j^0 \in \mathbb{R}^{2n}$ is bounded and we conclude in view of the compactness of ∇b that there exists a convergent subsequence. To prove that x_j is bounded we argue by contradiction and assume $\|x_j\| \to \infty$. Define

$$y_k = \frac{x_k}{\|x_k\|}$$

so that $\|y_k\| = 1$. By assumption, using (3.17),

$$(P^+ - P^-)y_k - j^*\left(\frac{1}{\|x_k\|}\nabla H(x_k)\right) \to 0.$$

Since $|\nabla H(z)| \leq M|z|$ the sequence

$$\frac{\nabla H(x_k)}{\|x_k\|} \in L^2$$

is bounded in L^2. Since $j^* : L^2 \to E$ is compact, $(P^+ - P^-)y_k$ is relatively compact, and since, in addition, y_k^0 is bounded in $\mathbb{R}^{2n}$, the sequence y_k is relatively compact in E. After taking a subsequence we can assume $y_k \to y$ in E and hence $y_k \to y$ in L_2.

$$\left\|\frac{\nabla H(x_k)}{\|x_k\|} - \nabla Q(y)\right\|_{L_2} \leq \frac{1}{\|x_k\|}\|\nabla H(x_k) - \nabla Q(x_k)\|_{L_2} + \|\nabla Q(y_k - y)\|_{L_2}.$$

Since $|\nabla H(z) - \nabla Q(z)| \leq M$ for all $z \in \mathbb{R}^{2n}$ and since ∇Q defines a continuous linear operator of L_2, we conclude

$$\frac{\nabla H(x_k)}{\|x_k\|} \to \nabla Q(y) \quad \text{in} \quad L^2.$$

Consequently,

$$\frac{\nabla b(x_k)}{\|x_k\|} = j^*\left(\frac{\nabla H(x_k)}{\|x_k\|}\right) \to j^*\left(\nabla Q(y)\right) \quad \text{in} \quad E.$$

This implies that $y \in E$ solves the linear equation in E

$$y^+ - y^- - j^* \nabla Q(y) = 0$$

$$\|y\| = 1.$$

As in Lemma 5 one verifies that y belongs to $C^\infty(S^1, \mathbb{R}^{2n})$ and also solves the linear Hamiltonian equation

$$\dot{y}(t) = X_Q\big(y(t)\big)$$
$$y(0) = y(1).$$

Recall now that $Q = (\pi + \varepsilon)q$, and $q(z) = (x_1^2 + y_1^2) + \frac{1}{N^2} \sum_{j=2}^{n} (x_j^2 + y_j^2)$. We see that the symplectic 2-planes $\{x_j, y_j\}$ are filled with periodic solutions of X_Q having periods $T \neq 1$. Since the linear equation does not admit any nontrivial periodic solutions of period 1 we conclude $y(t) = 0$. This contradicts $\|y\| = 1$ and we conclude that the sequence x_k must be bounded. ∎

The gradient equation

$$\dot{x} = -\nabla \Phi(x), \quad x \in E$$

is in view of Lemma 4 globally Lipschitz continuous and, therefore, defines a unique global flow

$$\mathbb{R} \times E \to E \quad : \quad (t, x) \mapsto \varphi^t(x) \equiv x \cdot t,$$

which maps bounded sets into bounded sets. All this is well-known from the theory of ordinary differential equations. Our flow has, in addition, a compactness property which will be crucial for the topological arguments in the next section.

Lemma 7. *The flow of $\dot{x} = -\nabla \Phi(x)$ admits the representation*

(3.20) $$x \cdot t = e^t x^- + x^0 + e^{-t} x^+ + K(t, x),$$

where $K : \mathbb{R} \times E \to E$ is continuous and maps bounded sets into precompact sets.

Proof. Prompted by the variation of constant formula, we define K by

$$K(t, x) = -\int_0^t \left(e^{t-s} P^- + P^0 + e^{-t+s} P^+\right) \nabla b\,(x \cdot s)\, ds.$$

We have to verify that K has the desired properties. Abbreviating the right hand side of (3.20) by $y(t)$, one verifies readily that

$$\dot{y}(t) = (P^- - P^+) y(t) - \nabla b(x \cdot t).$$

Since $y(0) = x$, the function $\xi(t) = y(t) - x \cdot t$ solves the linear equation
$$\dot{\xi}(t) = (P^- - P^+)\xi(t) \quad \text{and} \quad \xi(0) = 0.$$
By the uniqueness of the initial value problem $\xi(t) = 0$ so that $y(t) = x \cdot t$ as required. In view of (3.16) and (3.17) we can write
$$K(t,x) = j^* \left\{ -\int_0^t \left(e^{t-s} P^- + P^0 + e^{-t+s} P^+ \right) \nabla H\left(j(x \cdot s) \right) ds \right\}.$$
Abbreviating the map in the bracket by $k(t,x)$ then $k : \mathbb{R} \times E \to L^2$ is continuous and maps bounded sets into bounded sets. By Proposition 5, the map $j^* : L^2 \to E$ maps bounded sets into precompact sets and, therefore, K has the desired properties. ∎

To sum up this rather technical section we have extended the classical Hamiltonian functional originally given on the loop space of smooth loops, and constructed a functional $\Phi \in C^1(E, \mathbb{R})$ on the Hilbert space E. Its critical points, if there are any, are smooth periodic solutions of the Hamiltonian equations with period 1. The functional Φ has, in addition, crucial compactness properties. It remains to find a special critical point $x^* \in E$ of Φ satisfying $\Phi(x^*) > 0$. This will be done in the next section.

3.4 The existence of a critical point

A special minimax argument will guarantee the existence of a special critical point of the functional Φ introduced in the previous section:
$$\Phi(x) = a(x) - b(x)$$
$$= \tfrac{1}{2}\|x^+\|^2 - \tfrac{1}{2}\|x^-\|^2 - \int_0^1 H\left(x(t)\right) dt,$$
where $x = x^- + x^0 + x^+ \in E = E^- \oplus E^0 \oplus E^+$. Recall that $H \geq 0$ and that H is identically zero in a neighbourhood of $z = 0 \in \mathbb{R}^{2n}$. In addition $H(z) = (\pi + \varepsilon) q(z)$ for $|z|$ sufficiently large. Evidently, every critical point $z \in \mathbb{R}^{2n}$ of the function H, i.e., a point in $\{z | \nabla H(z) = 0\}$, is a critical point of Φ satisfying $\Phi(z) \leq 0$. We are looking for critical points of Φ which are not constant loops.

Proposition 6. *There exists $x^* \in E$ satisfying $\nabla \Phi(x^*) = 0$ and $\Phi(x^*) > 0$.*

In order to prove this proposition we first single out two distinguished subsets Σ and Γ of E. The bounded set $\Sigma = \Sigma_\tau \subset E$ is defined by
$$\Sigma_\tau = \left\{ x \,\middle|\, x = x^- + x^0 + se^+ \,,\, \|x^- + x^0\| \leq \tau \quad \text{and} \quad 0 \leq s \leq \tau \right\}$$

where $\tau > 0$. Here $e^+ \in E^+$ is the element
$$e^+(t) = e^{2\pi Jt} e_1 \quad \text{and} \quad e_1 = (1, 0, \ldots, 0) \in \mathbb{R}^{2n}.$$
Clearly $\|e^+\|^2 = 2\pi$ and $\|e^+\|_{L^2} = 1$. By $\partial\Sigma$ we denote the boundary of Σ in $E^- + E^0 + \mathbb{R}e^+$.

Lemma 8. There exists $\tau^* > 0$ such that for $\tau \geq \tau^*$
$$\Phi \Big| \partial\Sigma \leq 0.$$

Proof. We use the asymptotic behaviour of Φ. From $a|E^- \oplus E^0 \leq 0$ and $b \geq 0$ we infer
$$\Phi \Big| E^- \oplus E^0 \leq 0.$$
It remains to consider those parts of the boundary $\partial\Sigma$ which are defined by $\|x^- + x^0\| = \tau$ or $s = \tau$. By the construction of H there exists a constant $\gamma > 0$ such that
$$H(z) \geq (\pi + \varepsilon) q(z) - \gamma \quad \text{for all} \quad z \in \mathbb{R}^{2n}.$$
Therefore,
$$\Phi(x) \leq a(x) - (\pi + \varepsilon) \int_0^1 q(x) + \gamma, \quad \text{for all} \quad x \in E.$$
Recalling the definition of the quadratic form q, one verifies for the orthogonal splitting $x = x^- + x^0 + se^+ \in E^- \oplus E^0 + E^+$ that
$$\int_0^1 q(x^- + x^0 + se^+) \, dt = \int_0^1 q(x^-) \, dt + \int_0^1 q(x^0) \, dt + \int_0^1 q(se^+) \, dt.$$
Recalling that $\|e^+\|^2 = 2\pi$ we can, therefore, estimate for $x = x^- + x^0 + se^+$
$$\Phi(x) \leq \tfrac{1}{2} s^2 \|e^+\|^2 - \tfrac{1}{2} \|x^-\|^2 - (\pi + \varepsilon) q(x^0) - (\pi + \varepsilon) \int_0^1 q(se^+) + \gamma$$
$$= -\tfrac{1}{2} \|x^-\|^2 - \varepsilon s^2 \|e^+\|^2_{L^2} - (\pi + \varepsilon) q(x^0) + \gamma.$$
We thus find a constant $c > 0$ such that
$$\Phi(x^- + x^0 + se^+) \leq \gamma - c \|x^- + x^0\|^2 - c \|se^+\|^2.$$
The right hand side is ≤ 0 if $\|x^- + x^0\| = \tau$ or $s = \tau$ for τ sufficiently large. This finishes the proof of the lemma. ∎

The subset $\Gamma = \Gamma_\alpha \subset E^+$ is defined by
$$\Gamma = \left\{ x \in E^+ \, \Big| \, \|x\| = \alpha \right\}.$$

3.4 THE EXISTENCE OF A CRITICAL POINT

Lemma 9. *There exists an $\alpha > 0$ and $\beta > 0$ such that*
$$\Phi \big| \Gamma \geq \beta > 0.$$

Proof. We use the behaviour of Φ locally near $x = 0 \in E$. Recall that H vanishes identically near the origin in $\mathbb{R}^{2n}$. We shall prove that
$$b(x) \frac{1}{\|x\|^2} \to 0 \quad \text{as} \quad \|x\| \to 0.$$

It then follows for $x \in E^+$ that
$$\Phi(x) \frac{1}{\|x\|^2} = \frac{1}{2} - \frac{1}{\|x\|^2} b(x) \to \frac{1}{2}.$$

Consequently, $\Phi(x) \geq \frac{1}{4}\|x\|^2$ provided $\|x\|$ is small enough and this implies the desired result. Arguing indirectly, we find a sequence $x_j \in E$ and a constant $d > 0$ satisfying
$$x_j \to 0 \quad \text{and} \quad \frac{1}{\|x_j\|^2} b(x_j) \geq d > 0.$$

Set $y_j = \frac{x_j}{\|x_j\|}$ so that $\|y_j\| = 1$. We claim that there is a subsequence y_j and functions $y, w \in L^2$ satisfying
$$y_j \longrightarrow y \quad \text{in } L^2$$
$$y_j(t) \longrightarrow y(t)$$
$$|y_j(t)| \leq w(t)$$
$$x_j(t) \longrightarrow 0$$

for almost every t.

Indeed, by Proposition 3, E is compactly embedded in L^2. Hence we find a subsequence y_j which is a Cauchy sequence in L^2 and the claim follows by the following well-known argument. Choosing a fast Cauchy subsequence we may assume that
$$\|y_{k+1} - y_k\|_{L^2} \leq \frac{1}{2^k}, \quad k \geq 1.$$

Then the sequence $f_n \in L^2$, defined by
$$f_n(t) := \sum_{k=1}^{n} |y_{k+1}(t) - y_k(t)|,$$

is monotone increasing and satisfies $\|f_n\|_{L^2} \leq 1$. Hence by the monotone convergence theorem $f_n(t) \to f(t)$ a.e and $f \in L^2$. Since, for $m > n$,
$$|y_m(t) - y_n(t)| \leq |y_m(t) - y_{m-1}(t)| + \cdots + |y_{n+1}(t) - y_n(t)|$$
$$\leq f(t) - f_{n-1}(t),$$

we conclude that for a.e t the sequence $y_n(t)$ is a Cauchy sequence in $\mathbb{R}$ and hence defines $y(t) := \lim y_n(t)$ for a.e t. In view of $|y(t) - y_n(t)| \leq f(t)$ we therefore find, by Lebesgue's convergence theorem, that $y_n \to y$ in L^2 and, since $|y_n(t)| \leq f(t) + |y(t)| =: w(t)$ for a.e t and $w \in L^2$, the claim is proved. By definition

$$b(x_j)\frac{1}{\|x_j\|^2} = \int_0^1 H(x_j)\frac{1}{\|x_j\|^2}\,dt.$$

From $H(z) \leq M|z|^2$, for all $z \in \mathbb{R}^{2n}$, we infer

$$H\big(x_j(t)\big)\frac{1}{\|x_j\|^2} \leq Mw(t)^2$$

for almost every t, so that the L^1 function $w(t)^2$ is a majorant. Using that H vanishes at $z = 0$ with its derivatives up to second order we find, using the Taylor formula,

$$H\big(x_j(t)\big)\frac{1}{\|x_j\|^2} = O\big(|x_j(t)|\big)\,|y_j(t)|^2 \to 0$$

for almost every t. Hence by means of Lebesgue's theorem

$$\int_0^1 H\big(x_j(t)\big)\frac{1}{\|x_j\|^2}\,dt \to 0\,,\quad j \to \infty.$$

This contradicts the assumption that $d > 0$ and Lemma 9 is proved. ∎

The rather clumsy argument in the proof can, of course, be avoided if one makes use of the following well-known, but not quite elementary Sobolev estimate. The space $H^{1/2}(S^1)$ is continuously embedded in $L^p(S^1)$ for every $p \geq 1$. Hence there are constants $M = M_p$ such that

$$\|u\|_{L^p} \leq M\|u\|_{1/2}\,,\quad u \in H^{1/2}(S^1)\,.$$

A proof can be found in the Appendix. Taking this estimate for granted and observing that $|H(z)| \leq c|z|^3$ for all $z \in \mathbb{R}^{2n}$, we find a constant $K > 0$ such that

$$\int_0^1 |H\big(x(t)\big)|\,dt \leq c\|x\|_{L^3}^3 \leq K\|x\|_{1/2}^3\,,$$

for all $x \in H^{1/2}$. Now, if $x \in E_+$, then $\Phi(x) \geq \frac{1}{2}\|x\|^2 - K\|x\|^3$ and the lemma is now obvious. The following picture (Fig. 3.7) illustrates our situation more geometrically.

The estimate $\Phi|\partial\Sigma \leq 0$ reflects our assumptions $H \geq 0$ and $m(H) > \pi + \varepsilon$ for our original function $H \in \mathcal{H}(Z(1), \omega_0)$, while $\Phi|\Gamma \geq \beta > 0$ is a consequence of

3.4 THE EXISTENCE OF A CRITICAL POINT

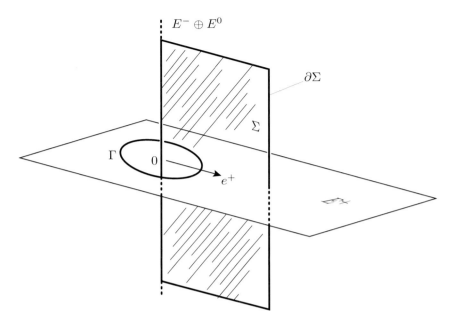

Fig. 3.7

the assumption that H vanishes identically on some open set for which we could choose a neighbourhood of the origin in $\mathbb{R}^{2n}$.

Next we study what happens if we translate the set Σ by means of the flow $\varphi^t(x)$ belonging to the gradient equation $\dot{x} = -\nabla\Phi(x)$ on E. Since $\Phi(\varphi^t(x))$ decreases in t we conclude immediately from Lemma 8 that

(3.21) $$\Phi \Big| \varphi^t(\partial\Sigma) \leq 0 \quad \text{all} \quad t \geq 0.$$

By Lemma 9, on the other hand, $\Phi|\Gamma > 0$ and consequently $\varphi^t(\partial\Sigma) \cap \Gamma = \emptyset$ for all $t \geq 0$; the "frame" $\varphi^t(\partial\Sigma)$ cannot cross the "circle" Γ as t increases. Intuitively it is, therefore, clear that the "surface" $\varphi^t(\Sigma)$ spanned by the frame must intersect Γ for every $t \geq 0$. This requires, however, a proof. It will be based on a topological argument.

Lemma 10.
$$\varphi^t(\Sigma) \cap \Gamma \neq \emptyset, \quad \text{all} \quad t \geq 0.$$

Postponing the proof we shall first finish the proof of Proposition 6. In fact, we are now in business. We can apply the following minimax argument. We take the family $\mathcal{F}$ consisting of the subsets $\varphi^t(\Sigma)$, for every $t \geq 0$ and define

$$c(\Phi, \mathcal{F}) = \inf_{t \geq 0} \sup_{x \in \varphi^t(\Sigma)} \Phi(x).$$

We claim that $c(\Phi, \mathcal{F})$ is finite. Indeed, since $\varphi^t(\Sigma) \cap \Gamma \neq \emptyset$ and $\Phi|\Gamma \geq \beta$ we conclude that
$$\beta \leq \inf_{x \in \Gamma} \Phi(x) \leq \sup_{x \in \varphi^t(\Sigma)} \Phi(x) < \infty.$$
For the second estimate we have used that Φ maps, in view of Lemma 4, bounded sets into bounded sets. Therefore,
$$-\infty < \beta \leq c(\Phi, \mathcal{F}) < \infty.$$
We know already that the functional Φ satisfies the P.S. condition (Lemma 6) and that the gradient equation generates a unique global flow (Lemma 7). Moreover, by the properties of a flow, determined by time-independent equations, the family $\mathcal{F}$ is positively invariant. Consequently the Minimax Lemma guarantees that $c(\Phi, \mathcal{F})$ is a critical value. We deduce a point $x^* \in E$ satisfying $\nabla \Phi(x^*) = 0$ and
$$\Phi(x^*) = c(\Phi, \mathcal{F}) \geq \beta > 0,$$
and the Proposition is proved. It remains to prove Lemma 10.

Proof of Lemma 10. We shall use the Leray-Schauder degree. This degree extends the Brouwer mapping degree to infinite dimensional spaces for the restricted class of continuous mappings which are of the form $id+$ compact. Abbreviating the flow by $\varphi^t(x) \equiv x \cdot t$, we wish to verify that $(\Sigma \cdot t) \cap \Gamma \neq \emptyset$ for all $t \geq 0$. We can rewrite this by requiring

(3.22)
$$(P^- + P^0)(x \cdot t) = 0$$
$$\|x \cdot t\| = \alpha$$
$$x \in \Sigma.$$

Recall that, by Lemma 7, the flow has the representation $x \cdot t = e^t x^- + x^0 + e^{-t} x^+ + K(t, x)$, so that (3.22) becomes

(3.23)
$$0 = e^t x^- + x^0 + (P^- + P^0) K(t, x)$$
$$0 = \alpha - \|x \cdot t\|$$
$$x \in \Sigma.$$

Multiplying the E^- part by e^{-t} one finds the equivalent equations

(3.24)
$$0 = x^- + x^0 + (e^{-t} P^- + P^0) K(t, x)$$
$$0 = \alpha - \|x \cdot t\|$$
$$x \in \Sigma.$$

3.4 THE EXISTENCE OF A CRITICAL POINT

Since $x \in \Sigma$ is represented by $x = x^- + x^0 + se^+$, with $0 \leq s \leq \tau$, we can rewrite (3.24) as follows:

(3.25) $$0 = x + B(t, x) \quad \text{and} \quad x \in \Sigma,$$

where the operator B is defined by

$$B(t, x) = (e^{-t} P^- + P^0) K(t, x) + P^+ \left\{ \left(\|x \cdot t\| - \alpha \right) e^+ - x \right\}.$$

Abbreviating $F = E^- \oplus E^0 + \mathbb{R} e^+$ the map $B : \mathbb{R} \times F \to F$ is continuous and maps bounded sets into relatively compact sets. This was proved in Lemma 7. We therefore can apply the Leray-Schauder index theory. The equation (3.25) possesses, for given $t \geq 0$, a solution $x \in \Sigma$ if we can show that $\deg(\Sigma, id + B(t, \cdot), 0) \neq 0$. In order to compute this degree we first observe, in view of $\varphi^t(\partial\Sigma) \cap \Gamma = \emptyset$ for $t \geq 0$, that there is no solution of (3.25) on $\partial\Sigma$, i.e.,

$$0 \notin \left(id + B(t, \cdot) \right)(\partial\Sigma),$$

if $t \geq 0$. Hence by the homotopy invariance of the degree,

$$\deg\left(\Sigma, id + B(t, \cdot), 0\right) = \deg\left(\Sigma, id + B(0, \cdot), 0\right).$$

Since $K(0, x) = 0$ we find $B(0, x) = P^+\{(\|x\| - \alpha)e^+ - x\}$. Defining the homotopy

$$L_\mu(x) = P^+\left\{(\mu\|x\| - \alpha) e^+ - \mu x\right\} \quad \text{for} \quad 0 \leq \mu \leq 1,$$

we claim $x + L_\mu(x) \neq 0$ for $x \in \partial\Sigma$. Indeed, if $x \in \Sigma$ satisfies $x + L_\mu(x) = 0$ then $x = se^+$ and, therefore, $s\left((1 - \mu) + \mu \|e^+\|\right) = \alpha$. Consequently $0 < s \leq \alpha$ so that $x \notin \partial\Sigma$ if $\tau > \alpha$ as claimed. Therefore, by homotopy again

$$\deg\left(\Sigma, id + B(t, \cdot), 0\right) =$$
$$\deg\left(\Sigma, id + L_0, 0\right) =$$
$$\deg\left(\Sigma, id - \alpha e^+, 0\right) =$$
$$\deg\left(\Sigma, id, \alpha e^+\right) = 1,$$

provided that $\alpha e^+ \in \Sigma$, which holds true for $\tau > \alpha$. This finishes the proof of Lemma 10 and therefore the proof of the proposition is completed. ∎

The critical point found in the proposition is a smooth periodic solution of the Hamiltonian equation $\dot{z} = J\nabla H(z)$ in $\mathbb{R}^{2n}$ having period 1, as we know from the "regularity" Lemma 5. This solution is the distinguished nonconstant 1-periodic solution we are looking for in Theorem 2 whose proof, therefore, is completed. With this existence theorem for periodic solutions, we have also established the existence of the special capacity function $c_0(M, \omega)$. Consequently, the rigidity results deduced in the previous chapter from the existence of a capacity function c are now completely proved. We have demonstrated that these recently discovered rigidity phenomena in symplectic geometry can be traced back to a simple minimax argument applied to the old action principle of classical mechanics.

3.5 Examples and illustrations

In Chapter 2 we deduced from the axioms that open balls and cylinders in $(\mathbb{R}^{2n}, \omega_0)$ have the capacities $c(B(r)) = e(Z(r)) = \pi r^2$ for every capacity function. This agrees with the actions of the closed characteristics of $\partial B(r)$ and $\partial Z(r)$ as we have seen in Chapter 1. For an ellipsoid E, the capacity $c(E)$ also agrees with the action of a particular closed characteristic of its boundary ∂E. Recall that the (reduced) action $A(\gamma)$ of a loop γ in $(\mathbb{R}^{2n}, \omega_0)$ is defined by

$$A(\gamma) = \int_\gamma p\,dq\ .$$

We consider now, more generally, a convex bounded and smooth domain D of $\mathbb{R}^{2n}$. Its boundary ∂D carries at least one closed characteristic as we have demonstrated in Chapter 1. It turns out, for the special capacity function c_0 constructed above, that $c_0(D)$ is also represented by a distinguished closed characteristic of the boundary of D.

Theorem 3. Assume $C \subset \mathbb{R}^{2n}$ is a convex bounded domain with smooth boundary ∂C. Then there exists a distinguished closed characteristic $\gamma^* \subset \partial C$ satisfying

$$c_0(C, \omega_0) = |A(\gamma^*)|$$

and $|A(\gamma^*)| = \inf \Big\{ |A(\gamma)| \ :\ \gamma \subset \partial C \ \ \text{is a closed characteristic} \Big\}$.

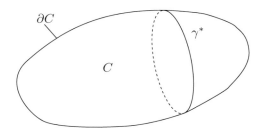

Fig. 3.8

3.5 EXAMPLES AND ILLUSTRATIONS

The proof of this theorem is quite similar to the proof of Theorem 2 and we refer to H. Hofer and E. Zehnder [123]. It is useful to point out that the same representation formula holds true for the different capacity function defined on subsets of $\mathbb{R}^{2n}$ which was originally introduced by I. Ekeland and H. Hofer [68]. Consequently, these two capacities agree on such convex subsets $C \subset \mathbb{R}^{2n}$. As another immediate consequence we conclude that the following estimate for the action of periodic solutions x on convex and compact smooth hypersurfaces ∂C holds true:

$$|A(x)| \geq \sup \left\{ \pi r^2 \mid \text{ there exists} \right.$$
$$\left. \text{a symplectic embedding } \varphi : B(r) \to C \right\} = D(C, \omega_0).$$

Here we have used the monotonicity of c_0. This estimate considerably sharpens an inequality due to C. Croke and A. Weinstein [58] for periodic solutions of convex hypersurfaces. A. Künzle [131] observed that Theorem 3 can also be used to describe a simple class of compact hypersurfaces in $\mathbb{R}^{2n}$ which cannot be mapped onto convex hypersurfaces by means of symplectic diffeomorphisms. Consider an open and bounded domain $M \subset \mathbb{R}^{2n}$ with smooth boundary $S = \partial M$ satisfying

$$B(r) \subset M \subset Z(r)$$

which is a full Bordeaux bottle as illustrated in the figure:

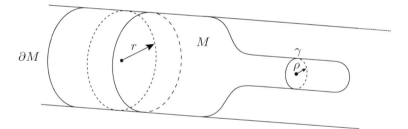

Fig. 3.9

M contains in particular a piece of a cylinder $Z(\rho)$ having radius $\rho < r$. On this piece the characteristics γ of $\partial Z(\rho)$ are closed and have the action $A(\gamma) = \pi \rho^2$. Assume now by contradiction that φ is a symplectic diffeomorphism of $\mathbb{R}^{2n}$ mapping M onto a convex domain C. Then it maps the hypersurface ∂M onto the convex hypersurface ∂C. From the invariance of the capacity function, we deduce

$$\pi r^2 = c_0(M) = c_0(\varphi(M)) = c_0(C).$$

By Theorem 3 we have, therefore, $A(\gamma) \geq \pi r^2$ for all characteristic loops γ of ∂C. But φ leaves the action invariant, so that

$$\pi \rho^2 = A(\gamma) = A\big(\varphi(\gamma)\big) < \pi r^2 \leq A\big(\varphi(\gamma)\big)$$

which is a contradiction. Evidently, the capacity of the above manifold M is also represented by a closed characteristic on ∂M which, however, does not minimize the action. As for general domains with smooth boundaries, it should be recalled that it is not known whether every compact hypersurface in $\mathbb{R}^{2n}$ admits a closed characteristic.

In the special case of 2-dimensional compact symplectic manifolds, the regular energy surfaces are, of course, 1-dimensional and consist of finitely many periodic orbits. This observation allowed K.F. Siburg [194] to compute the capacity c_0 in this case:

Theorem 4. (K.F. Siburg) Assume (M, ω) is a compact and connected symplectic manifold, possibly with boundary, with $\dim M = 2$. Then

$$c_0(M, \omega) = \left| \int_M \omega \right|.$$

We see that c_0 agrees with the total volume so that in dimension 2 this invariant classifies symplectically diffeomorphic manifolds. This was shown in Chapter 1.

Proof. We first show that for every capacity function c

$$c(M, \omega) \geq \left| \int_M \omega \right|.$$

Given $\varepsilon > 0$, then by removing finitely many compact curves on M, we find a simply connected manifold $N \subset M$ which is diffeomorphic to a closed disc $D \subset \mathbb{R}^2$ and which satisfies

$$\left| \int_N \omega \right| \geq \left| \int_M \omega \right| - \varepsilon,$$

and we can choose the radius of the disc D such that its Lebesgue measure, $\mu(D)$, agrees with the integral of ω over N. (Recall that a Riemannian surface is obtained by a polygon in the plane by identifying suitably the edges; hence we might just remove the curves corresponding to these edges to get a disc, see W. Massey [148].) By the theorem of Dacorogna and Moser, see Chapter 1, this diffeomorphism can be chosen to be symplectic. Therefore, by the monotonicity property of the capacity

$$c(M, \omega) \geq c(N, \omega) = c(D) = \mu(D) \geq \left| \int_M \omega \right| - \varepsilon.$$

This holds true for every $\varepsilon > 0$ and the desired estimate is proved. It remains to prove that

$$c_0(M, \omega) \leq \left| \int_M \omega \right|.$$

3.5 Examples and Illustrations

In view of the conformality property we may assume the integral on the right hand side to be positive. Pick $\varepsilon > 0$ and $H \in \mathcal{H}(M, \omega)$ satisfying

$$m(H) \geq \int_M \omega + \varepsilon.$$

We have to establish the existence of a nonconstant T-periodic solution of $\dot{x} = X_H(x)$ on M with period $0 < T \leq 1$. Denote by $R \subset [0, m(H)]$ the set of regular values of H. Since the set of critical levels is compact, and, in view of Sard's theorem, of Lebesgue measure zero, we find finitely many mutually disjoint intervals $I_j = [a_j, b_j] \subset R$ satisfying

$$\sum_j (b_j - a_j) \geq m(H) - \frac{\varepsilon}{2} \geq \int_M \omega + \frac{\varepsilon}{2}.$$

By definition of $\mathcal{H}(M, \omega)$ we have $H \equiv m(H)$ near ∂M. Therefore, if $h \in R$, then $H^{-1}(h)$ consists of finitely many embedded circles $S^1 \subset M$ each of which carries precisely one (nonconstant) periodic solution $\gamma(t, h)$ of X_H. Choose, for every j, one of the components of $H^{-1}[a_j, b_j]$ and call it A_j. It is simply covered by a smooth family $\gamma(t, h)$ of periodic solutions having periods denoted by $T(h) > 0$, when $h \in I_j$. Define the diffeomorphism $\varphi = \varphi_j$

$$\varphi : (t, h) \mapsto \gamma(t, h) \in A_j,$$

where $0 \leq t < T(h)$ and $a_j \leq h \leq b_j$. We claim

$$\varphi^* \omega = dt \wedge dh.$$

Indeed, since $H(\gamma(t, h)) = h$ for all t we conclude, denoting the partial derivatives with subscripts, $\omega(\gamma_t, \gamma_h) = dH(\gamma(t, h))(\gamma_h) = 1$. Therefore, one computes for $\xi, \eta \in \mathbb{R}^2$, that $\varphi^* \omega(\xi, \eta) = \omega(d\varphi\,\xi, d\varphi\,\eta) = \omega(\xi_1 \gamma_t + \xi_2 \gamma_h, \eta_1 \gamma_t + \eta_2 \gamma_h) = \xi_1 \eta_2 - \xi_2 \eta_1 = (dt \wedge dh)(\xi, \eta)$, proving the claim. Consequently,

$$\int_{A_j} \omega = \int_{\varphi^{-1}(A_j)} \varphi^* \omega = \int_{a_j}^{b_j} T(h)\, dh.$$

Arguing now by contradiction we assume $T(h) > 1$ for all $h \in R$. Then

$$\int_M \omega \geq \sum_j \int_{A_j} \omega > \sum_j (b_j - a_j) \geq \int_M \omega + \frac{\varepsilon}{2},$$

by the assumption on $m(H)$. This contradiction shows that there is indeed an $h \in R$ and a periodic solution $\gamma(h)$ lying in some A_j and having period $0 < T(h) \leq 1$.

In view of the definition of c_0 the desired estimate follows and the proof of the theorem is finished. ∎

To illustrate this theorem we consider a compact 2-dimensional manifold without boundary. In this case the proof shows that for every $H \in C^2(M, R)$ satisfying $\max H - \min H > |\int \omega|$ the Hamiltonian vector field X_H admits a "fast" periodic solution of period $0 < T \leq 1$. We would like to add another elementary observation concerning the special case of subsets of $\mathbb{R}^2$. We know from Chapter 2, that for every capacity function, $c(D) = \text{area}(D)$ for a bounded connected subset D of $\mathbb{R}^2$ having smooth boundary. Since the special capacity c_0 has the additional property of inner regularity we can easily extend this statement to open subsets of $\mathbb{R}^2$.

Proposition 7. If $\Omega \subset \mathbb{R}^2$ is an open bounded and path connected set, then $c_0(\Omega, \omega_0) = \text{area}(\Omega)$.

Proof. Exhaust $\Omega = \bigcup K_j, j \geq 1$ by compact sets $K_j \subset \overset{\circ}{K}_{j+1}$ and define smooth functions $\beta_j : \Omega \to \mathbb{R}$ by setting $\beta_j(x) = 0$ if $x \in K_j$ and $\beta_j(x) = 1$ if $x \in \Omega \setminus K_{j+1}$. Define $\beta \in C^\infty(\Omega)$ by

$$\beta(x) = \sum_{j \geq 1} \beta_j(x), \quad x \in \Omega.$$

Then $\beta(x) \to +\infty$ as $x \to \mathbb{R}^2 \setminus \Omega$. Now take a sequence $b_n \to \infty$ of regular values of β and consider the compact symplectic manifolds $D_j = \beta^{-1}(-\infty, b_j]$ having smooth boundaries. Then $\Omega = \cup D_j, j \geq 1$. Since c_0 has the property of inner regularity we conclude $c_0(\Omega) = \lim c_0(D_j)$, as $j \to \infty$. From the regularity of the Lebesgue measure we find, on the other hand, area $(\Omega) = \lim \text{area}(D_j)$ as $j \to \infty$. Since we already know that $c_0(D_j) = \text{area}(D_j)$, the proposition is proved. ∎

Although the invariant c_0 is defined by means of special periodic solutions, it is not at all surprising that in the 2-dimensional case it is intimately related to the area, since the symplectic form is the area form. In higher dimensions the capacity c_0 is, in contrast, not yet understood. It turns out that it is extremely difficult to compute this capacity. But some examples are known which are mentioned next.

Theorem 5. Consider $(\mathbb{C}P^n, \omega)$ with the standard symplectic structure related to the Fubini-Study metric. Then

$$c_0(\mathbb{C}P^n, \omega) = \pi.$$

This result is due to H. Hofer and C. Viterbo [122]. We do not give a proof here and remark that the proof analyzes the structure of the holomorphic spheres in $\mathbb{C}P^n$ and is based on a Fredholm theory for first order elliptic systems. Now assume that (M, ω) is any compact symplectic manifold without boundary, and denote by $\pi_2(M)$ the homotopy classes of maps $u : S^2 \to M$. Then there is a map $\pi_2(M) \to \mathbb{R}$ given by

$$u \mapsto \langle \omega, u \rangle = \int_{S^2} u^* \omega \in \mathbb{R}.$$

3.5 Examples and illustrations

Define $\alpha^*(M) \in \mathbb{R} \cup \{\infty\}$ by

$$\alpha^*(M) = \inf \left\{ \langle \omega, u \rangle \,\Big|\, u \in \pi_2(M) \text{ and } \langle \omega, u \rangle > 0 \right\}.$$

Moreover, if $\pi_2(M) = 0$, we set $\alpha^* = \infty$. Then the following result due to A. Floer – H. Hofer – C. Viterbo[88] and R. Ma [146] holds true.

Theorem 6. Consider the symplectic manifold $(M \times \mathbb{R}^{2n}, \omega \oplus \omega_0)$ with $n \geq 1$ and assume $\alpha^* = \alpha^*(M) > 0$ or $\alpha^* = \infty$. Then

$$c_0\Big(M \times B(r)\Big) = c_0\Big(M \times Z(r)\Big) = \pi r^2,$$

provided $\pi r^2 < \alpha^*$.

The case of the symplectic torus T^{2n} will be discussed in the next chapter.

Incidentally, by restricting the class of Hamiltonian functions admitted in the definition of $c_0(M, \omega)$, one finds different capacity functions. For example, if $\tilde{\mathcal{H}}(M, \omega) = \{H \in \mathcal{H}(M, \omega) |$ support of ∇H is contractible in $M\}$ and correspondingly $\tilde{\mathcal{H}}_a(M, \omega) = \mathcal{H}_a(M, \omega) \cap \tilde{\mathcal{H}}(M, \omega)$, then we can define

$$\tilde{c}_0(M, \omega) = \sup \left\{ m(H) \,\Big|\, H \in \tilde{\mathcal{H}}_a(M, \omega) \right\}.$$

Our proof shows that $\tilde{c}_0$ is a capacity function; the related periodic solutions are all contractible. It clearly satisfies

$$D(M, \omega) \leq \tilde{c}_0(M, \omega) \leq c_0(M, \omega)$$

where $D(M, \omega)$ is the Gromov-width introduced in Chapter 1.

Chapter 4
Existence of closed characteristics

In the previous chapter, the dynamical approach to the symplectic invariants led to the special capacity function c_0. Its construction was based on a variational principle for periodic solutions of certain Hamiltonian systems. The period of these periodic solutions was prescribed. In this chapter we shall deduce from this symplectic invariant the existence of periodic solutions on prescribed energy surfaces. If we neglect the parametrization of such solutions the aim is to find closed characteristics of a distinguished line bundle over a hypersurface in a symplectic manifold. It is determined by the symplectic structure, as explained in the introduction. A very special symplectic structure on the torus will lead us to M. Herman's counterexample to the closing lemma in the smooth category.

4.1 Periodic solutions on energy surfaces

The flow φ^t of a Hamiltonian vector field

(4.1) $$\dot{x} = X_H(x), \quad x \in M$$

on a symplectic manifold (M, ω) leaves the level sets of the smooth function H on M invariant, i.e., $H(\varphi^t(x)) = H(x)$ as long as t is defined. Fixing a value of this energy function which we can assume to be $E = 1$, we shall require the subset

(4.2) $$S = \left\{ x \in M \mid H(x) = 1 \right\}$$

to be compact and a regular, i.e.,

(4.3) $$dH(x) \neq 0 \quad \text{for} \quad x \in S.$$

Thus $S \subset M$ is a smooth and compact submanifold of codimension 1 whose tangent space at $x \in S$ is given by

(4.4) $$T_x S = \left\{ \xi \in T_x M \mid dH(x)\xi = 0 \right\}.$$

By the definition of a Hamiltonian vector field we have $X_H(x) \in T_x S$ so that X_H is a nonvanishing vector field on S whose flow exists for all times since S is compact. Our aim is to find a periodic solution for X_H on the energy surface S. This is a qualitative existence problem described in the introduction.

In order to illustrate the method employed we start with a preliminary result. The trick is to thicken the given energy surface S and to consider a 1-parameter family of surfaces parameterized by the energy. Since S is compact and regular there is an open and bounded neighborhood U of S which is filled with compact and regular energy surfaces having energy values near $E = 1$:

(4.5) $$U = \bigcup_{\lambda \in I} S_\lambda,$$

where I is an open interval around $\lambda = 1$ and where $S_\lambda = \{x \in U \mid H(x) = \lambda\}$ is diffeomorphic to the given surface S, which corresponds to the parameter value $\lambda = 1$. Indeed, take any Riemannian metric on M, then the gradient of H with respect to this metric does not vanish, $\nabla H \neq 0$, in a neighborhood of S, in view of (4.3). We can, therefore, define the modified gradient vector field

$$\dot{x} = \frac{\nabla H(x)}{|\nabla H(x)|^2}$$

near S, which is transversal to S. Its flow ψ^t satisfies

$$H\big(\psi^t(x)\big) = 1 + t$$

for $x \in S$ and, therefore, defines a diffeomorphism $(x, t) \mapsto \psi^t(x)$ from $S \times (-\varepsilon, \varepsilon)$ onto an open neighborhood U of S as claimed.

Theorem 1. (Hofer-Zehnder) Let S be a compact and regular energy surface for the Hamiltonian vector field X_H on (M, ω). Assume there is an open neighborhood U of S having bounded capacity: $c_0(U, \omega) < \infty$. Then there exists a sequence $\lambda_j \to 1$ of energy values, such that X_H possesses a periodic solution on every energy surface S_{λ_j}.

Actually, the proof establishes more solutions:

Corollary. There is a dense subset $\Sigma \subset I$ such that for $\lambda \in \Sigma$ the energy surface S_λ has a periodic solution of X_H, provided $c_0(U, \omega) < \infty$.

It should be emphasized that there are no assumptions required for the given regular and compact energy surface S other than that a bounded open neighborhood must have finite capacity. For example, in the special case of the standard symplectic manifold $(\mathbb{R}^{2n}, \omega_0)$ a bounded set U is contained in a large ball $B(R)$ and, consequently, by the monotonicity of the capacity, always has finite capacity. Hence every compact regular energy surface $S \subset \mathbb{R}^{2n}$ gives rise to an abundance of periodic solutions for the Hamiltonian vector fields X_H having energies near the prescribed energy.

Proof. The statement is an immediate consequence of the definition of the special capacity c_0 if we recall from the introduction that we have the freedom to choose a convenient Hamiltonian function representing the energy surfaces. We shall construct an auxiliary Hamiltonian function F on U which is constant on

4.1 Periodic solutions on energy surfaces

every surface S_λ contained in U and which, in addition, belongs to the set $\mathcal{H}(U, \omega)$ of functions defined in the previous chapter.

If $I = \{1 - \rho < \lambda < 1 + \rho\}$ for some $\rho > 0$ we pick an ε in $0 < \varepsilon < \rho$. Since, by assumption, $c_0(U, \omega) < \infty$ we can choose a smooth function $f : \mathbb{R} \to \mathbb{R}$ satisfying:

$$\begin{aligned}
f(s) &= c_0(U, \omega) + 1 \quad &&\text{for} \quad s \leq 1 - \varepsilon \quad \text{and} \quad s \geq 1 + \varepsilon \\
f(s) &= 0 \quad &&\text{for} \quad 1 - \tfrac{\varepsilon}{2} \leq s \leq 1 + \tfrac{\varepsilon}{2} \\
f'(s) &< 0 \quad &&\text{for} \quad 1 - \varepsilon < s < 1 - \tfrac{\varepsilon}{2} \\
f'(s) &> 0 \quad &&\text{for} \quad 1 + \tfrac{\varepsilon}{2} < s < 1 + \varepsilon.
\end{aligned}$$

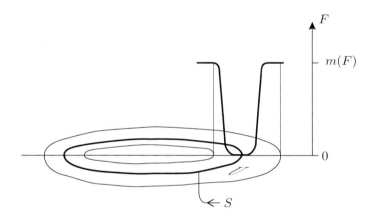

Fig. 4.1

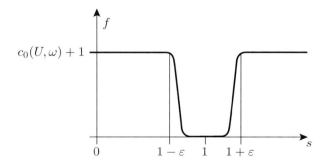

Fig. 4.2

Define $F : U \to \mathbb{R}$ by
$$F(x) = f\big(H(x)\big), \quad x \in U.$$
Then $F \in \mathcal{H}(U,\omega)$ and $m(F) > c_0(U,\omega)$. Consequently, in view of the definition of the capacity $c_0(U,\omega)$ there exists a nonconstant periodic solution $x(t)$ having period $0 < T \leq 1$ of the Hamiltonian system:
$$\dot{x} = X_F(x), \quad x \in U.$$
In view of the definition of a Hamiltonian vector field $\omega(X_F(x), \cdot) = -dF(x) = -f'(H(x))\,dH(x) = \omega(f'(H(x))X_H(x), \cdot)$ and, therefore,
$$X_F(x) = f'\big(H(x)\big) \cdot X_H(x), \quad x \in U.$$
Moreover,
$$H\big(x(t)\big) = \lambda$$
is constant in t, since $\frac{d}{dt} H(x(t)) = dH(x(t)) \cdot \dot{x}(t) = f'(H) \cdot \omega(X_H, X_H) = 0$. Since $x(t)$ is not a constant solution we conclude
$$f'\big(H(x(t))\big) = f'(\lambda) = \tau \neq 0.$$
Thus, in view of the definition of the function f, the value λ belongs to the set $1-\varepsilon < \lambda < 1-\frac{\varepsilon}{2}$ or $1+\frac{\varepsilon}{2} < \lambda < 1+\varepsilon$. In particular $|\lambda - 1| < \varepsilon$. Reparameterizing, we define the closed curve $y : \mathbb{R} \to S_\lambda$ by
$$y(t) = x\left(\frac{t}{\tau}\right)$$
which has period τT and satisfies
$$\dot{y}(t) = \frac{1}{\tau} \dot{x}\left(\frac{t}{\tau}\right) = X_H\big(y(t)\big),$$
hence is a periodic solution of the original Hamiltonian vector field X_H on the energy surface $H(y(t)) = \lambda$. Moreover, $|\lambda - 1| < \varepsilon$. Since ε is arbitrary the theorem is proved. Replacing $\lambda = 1$ by any other value of λ in the interval I we have proved, by the same argument, also the corollary. ∎

One has to keep in mind that the periodic solutions guaranteed by Theorem 1 are very special. They are indeed found indirectly by a special variational principle hidden in the construction of the symplectic invariant c_0 and there may exist many other periodic solutions. Moreover, the periodic solutions do not necessarily lie on the given energy surface S for X_H but only nearby. However, if we know in addition, that the periods T_j of the periodic solutions x_j on S_{λ_j} for X_H are bounded, we can conclude that S too has a periodic solution. This simple observation will be

4.1 Periodic solutions on energy surfaces

very useful later on. To be precise we choose a Riemannian metric g on M and abbreviate $g(x)(\xi, \eta) = \langle \xi, \eta \rangle$ and $g(x)\langle \xi, \xi \rangle = |\xi|^2$. If $x(t)$ is a periodic solution of period T we can introduce its length by

$$l(x) = \int_0^T |\dot{x}(t)|\, dt.$$

Possibly after shrinking the open neighborhood U, we may assume

$$C^{-1} \leq |X_H(x)| \leq C, \quad x \in U$$

for a constant $C > 0$. Since the $x_j(t)$ solve the Hamiltonian equations $\dot{x}_j(t) = X_H(x_j(t))$ we conclude

$$C^{-1} \cdot T_j \leq l(x_j) \leq C \cdot T_j.$$

Proposition 1. Let $\lambda_j \to 1$ and assume that the periods T_j (or equivalently the lengths $l(x_j)$) of the periodic solutions $x_j(t)$ on S_{λ_j} are bounded. Then $S = S_1$ possesses a periodic solution of X_H.

Proof. Normalizing the periods to 1 we define $y_j(t) = x_j(T_j \cdot t)$ for $0 \leq t \leq 1$, so that

$$\dot{y}_j(t) = T_j X_H\big(y_j(t)\big)$$

and $H(y_j(t)) = \lambda_j$. Since, by assumption, the right hand side is bounded, we find by means of the Arzelà-Ascoli theorem a subsequence such that $T_j \to T$ and $y_j \to y$ with convergence, at first in C^0, and, by making use of the equation, in C^∞. The function $y : [0, 1] \to U$ is periodic with period 1, satisfies $H(y(t)) = 1$, and

$$\dot{y}(t) = T X_H\big(y(t)\big),$$

and is thus the desired periodic solution if $T \neq 0$. Arguing by contradiction, we assume $T = 0$. Then $y_j \to \{y^*\}$ shrinks to a point y^* and $X_H(y(t)) \to X_H(y^*) = V$. Since the energy surface S is regular, $V \neq 0$. We can estimate in local coordinates near the point y^*;

$$\langle X_H(y_j), V \rangle \geq (1 - \varepsilon)\, |V|^2$$

for some small $\varepsilon > 0$ and all j sufficiently large, so that

$$T_j^{-1} \langle \dot{y}_j(t), V \rangle \geq (1 - \varepsilon)\, |V|^2.$$

Integrating over the period we conclude $V = 0$ contradicting $V \neq 0$ and proving the proposition. ∎

Having established the existence of periodic solutions of a Hamiltonian vector field X_H, it is useful to remember that, in general, a periodic solution belongs to a whole family of periodic solutions parameterized by the energy. This will be proved by the so-called Poincaré continuation method. This local technique is based on the implicit function theorem and requires some knowledge of the Floquet multipliers of a given reference solution.

Consider a periodic solution $x(t) = x(t, E^*)$ of X_H having energy $E^* = H(x(t))$ and period $T = T^*$. Since the vector field X_H is time independent one of the Floquet multipliers of $x(t)$ must be equal to 1. Indeed, abbreviating $x = x(0)$ we can differentiate the flow $\varphi^t \circ \varphi^s(x) = \varphi^s \circ \varphi^t(x)$ in s at $s = 0$ and obtain

$$d\varphi^t(x) X_H(x) = X_H\left(\varphi^t(x)\right)$$

for all t. If $t = T$ is the period and hence $\varphi^T(x) = x$ we obtain

$$d\varphi^T(x) X_H(x) = X_H(x) \in T_x M.$$

By definition, the eigenvalues of the linear map $d\varphi^T(x) : T_x M \to T_x M$ are the Floquet multipliers of the periodic solution $x(t)$ and we have just verified that 1 is always a Floquet multiplier. But since $d\varphi^T(x)$ is a symplectic linear map, the eigenvalue 1 has necessarily even multiplicity, so that $x(t)$ possesses always at least two-Floquet multipliers equal to 1. Actually we shall verify this during the proof of the following proposition which was known to H. Poincaré.

Proposition 2. Assume a periodic solution $x(t, E^*)$ of X_H on M having energy $E^* = H(x(t, E^*))$ and period T^* has exactly two Floquet multipliers equal to 1. Then there exists a unique and smooth 1-parameter family $x(t, E)$ of periodic solutions with periods $T(E)$ close to T^* and lying on the energy surfaces

$$H\left(x(t, E)\right) = E$$

for $|E - E^*|$ sufficiently small. Moreover, $T(E) \to T(E^*)$ as $E \to E^*$.

Postponing the proof we observe that the reference solution $x(t, E^*)$ is not isolated in M as a periodic solution. The periodic solutions $x(t, E)$ correspond to different values of E and it will turn out in the proof that, on the fixed energy surface E, this solution is isolated among those periodic solutions having periods close to T. Geometrically, the periodic solutions $x(t, E)$ fill out an embedded cylinder in M as illustrated in the following figure 4.3.

Proof of Proposition 2. It is useful to recall the construction of a transversal section map belonging to a periodic solution. Assume $x(t)$ to be a nonconstant periodic solution of a vector field X on M and denote by $T > 0$ its period. Then, we can intersect $x(t)$ at the point $p = x(0)$ with a $(2n - 1)$ dimensional hypersurface, i.e., a local submanifold $\Sigma \subset M$ with codim $\Sigma = 1$ to which the vector field X is nowhere tangential. This means that $T_x \Sigma$ and $X(x)$ span the tangent space $T_x M$ for every $x \in \Sigma$ near p:

$$T_x M = T_x \Sigma \oplus \langle X(x) \rangle, \quad x \in \Sigma.$$

4.1 Periodic solutions on energy surfaces

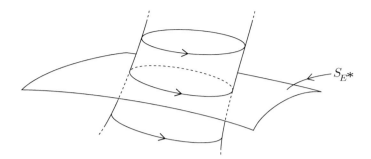

Fig. 4.3

If φ^t denotes the flow of X then $\varphi^T(p) = p \in \Sigma$. Since the flow depends differentiably on x we can define a smooth map $\psi: \Sigma \to \Sigma$ locally near p by following an initial point $x \in \Sigma$ along its solution $\varphi^t(x)$ until it meets Σ again, i.e.,

$$\psi(x) = \varphi^{\tau(x)}(x), \quad x \in \Sigma.$$

Here $\tau = \tau(x)$ is close to $T > 0$ and uniquely determined so that $\varphi^\tau(x) \in \Sigma$. Clearly $\psi(p) = p$. This map ψ is called the Poincaré section map of the periodic solution $x(t)$. It is important for us to know the eigenvalues of the linearized map

$$d\psi(p): T_p\Sigma \to T_p\Sigma$$

at the fixed point $p \in \Sigma$. The relation of these eigenvalues to the Floquet multipliers of $x(t)$, i.e., the eigenvalues of $d\varphi^T(p)$

$$d\varphi^T(p): T_pM \to T_pM$$

is given by the following

Lemma 1. $d\varphi^T(p)$ has 1 as an eigenvalue with eigenvector $X(p)$ and the remaining eigenvalues are those of $d\psi(p)$: i.e., the characteristic polynomials are related by

$$\det\left(\lambda - d\varphi^T(p)\right) = (\lambda - 1)\det\left(\lambda - d\psi(p)\right).$$

Proof. We have already verified the first part of the lemma. Differentiating ψ at p, we obtain for $\xi \in T_p\Sigma$

$$\begin{aligned}
d\psi(p)\xi &= d\varphi^T(p)\xi + \tfrac{d}{dt}\varphi^t\Big|_{t=T} \cdot d\tau(p)\xi \\
&= d\varphi^T(p)\xi - \lambda(\xi)X(p),
\end{aligned}$$

where we have abbreviated the linear form $d\tau(p)\xi$ by $-\lambda(\xi)$. With respect to the splitting $T_pM = \langle X(p)\rangle \oplus T_p\Sigma$, the linear map $d\varphi^T(p)$ has, therefore, the

representation
$$d\varphi^T(p) = \left(\begin{array}{c|c} 1 & \lambda \\ \hline 0 & d\psi(p) \end{array}\right),$$
from which the lemma follows immediately. ∎

Evidently the fixed points of ψ near p are the initial conditions of all the periodic solutions near the reference solutions $x(t)$ which have periods near T. We shall use this observation in order to prove Proposition 2. Let $x(t)$ be the periodic solution of the proposition. Then $H(x(t)) = E^*$ and $dH(x) \neq 0$ for x near $x(t)$. Moreover, $dH(X_H) = 0$. We can, therefore, introduce near $p = x(0)$ convenient local coordinates $(x_1, \ldots, x_{2n}) \in \mathbb{R}^{2n}$ such that p corresponds to $x^* = (E^*, 0, \ldots, 0)$ and such that $H(x_1, \ldots, x_{2n}) = x_1$ and moreover, such that $x_{2n} = 0$ is a local transversal section Σ. In view of $H(\varphi^t(x)) = H(x)$ the section map ψ is, in these coordinates expressed by

$$\psi : \begin{pmatrix} x' \\ x'' \end{pmatrix} \longrightarrow \begin{pmatrix} \psi'(x', x'') = x' \\ \psi''(x', x'') \end{pmatrix}.$$

where the coordinates $(x', x'') \in \Sigma$ stand for $x' = x_1$ and $x'' = (x_2, \ldots, x_{2n-1})$. The aim is to find the fixed points of ψ near x^*. In order to do this we only have to solve the equation
$$\psi''(x', x'') = x'',$$
since the first equation $\psi'(x', x'') = x'$ is automatically satisfied. Since, by assumption, $\psi(x^*) = x^*$ is a reference solution, we can apply the implicit function theorem, observing that in our coordinates the linearized map is expressed by

$$d\psi(x^*) = \left(\begin{array}{c|c} 1 & 0 \\ \hline * & \frac{\partial}{\partial x''}\psi''(x^*) \end{array}\right).$$

Recalling the assumption of Proposition 2 and Lemma 1, the matrix
$$\frac{\partial}{\partial x''}\psi''(x^*) - \mathbb{1}_{2n-2}$$
is nonsingular. Therefore, there exists a unique map $x' \mapsto x'' = a(x')$ solving the equation $\psi''(x', a(x')) = a(x')$. Thus $x' = E$ and $x'' = a(E)$ are the initial conditions on Σ of the desired periodic solutions satisfying Proposition 2. ∎

One is tempted to apply Proposition 2 to the periodic solutions on S_{λ_j} guaranteed by Theorem 1 in order to establish the existence of a periodic solution on the given energy surface S as $\lambda_j \to 0$. But unfortunately nothing can be said about the Floquet multipliers of these solutions. In addition, Proposition 2 is of a local nature: globally the cylinder of periodic solutions may eventually become tangential to an energy surface. This situation has been studied by C. Robinson in [183], 1970.

4.2 The characteristic line bundle of a hypersurface

The search for periodic orbits of a Hamiltonian vector field lying on a prescribed energy surface S is independent of the choice of the Hamiltonian representing S: it only depends on the submanifold S and the symplectic structure ω. Indeed, as explained in Chapter 1, this dynamical problem can be described geometrically. We consider a submanifold $S \subset M$ of codimension 1, i.e., a hypersurface of the symplectic manifold (M, ω). Then the symplectic structure ω and S determine a distinguished line bunde $\mathcal{L}_S \subset TS$, as follows. If $x \in S$, then restriction of the 2-form ω to the odd-dimensional subspace $T_x S \subset T_x M$ is necessarily degenerate. The kernel of this restriction if of dimension 1, because ω is nondegenerate on $T_x M$, and hence defines a line bundle, $\mathcal{L}_S \subset TS$,

$$\mathcal{L}_S = \{(x, \xi) \in T_x S \,|\, \omega_x(\xi, \eta) = 0 \text{ for all } \eta \in T_x S\}.$$

This line bundle gives the direction of every Hamiltonian vector field having S as a regular energy surface.

Proposition 3. Assume the smooth function H represents S as $S = \{x \,|\, H(x) = \text{const}\}$ and satisfies $dH \neq 0$ on S. Then

$$X_H(x) \in \mathcal{L}_S(x) \text{ for all } x \in S.$$

Proof. Since the tangent space is given by $T_x S = \{\xi \in T_x M \,|\, dH(x)\xi = 0\}$ we have $\omega(X_H(x), \xi) = -dH(x)\xi = 0$ for all $\xi \in T_x S$. ∎

Notation. The bundle $\mathcal{L}_S$ is called the characteristic line bundle of the hypersurface S. A closed characteristic of S, or a periodic Hamiltonian orbit of S, is an embedded circle $P \subset S$ satisfying

$$TP = \mathcal{L}_S \,|\, P.$$

In the following we shall denote the set of closed characteristics of S by

$$\mathcal{P}(S).$$

It agrees with the set of unparameterized periodic solutions of every Hamiltonian vector field X_H on S having S as a regular energy surface.

Assume now that $\mathcal{L}_S \to S$ is orientable, i.e., the bundle possesses a nonvanishing section. We shall show that there exists a Hamiltonian function $H : U \to \mathbb{R}$ defined on a neighborhood of S and representing $S = H^{-1}(0)$ as a regular energy surface. We make use of an almost complex structure on M compatible with ω, i.e., a smooth family of linear maps $J(x) \in \mathcal{L}(T_x M), x \in M$, satisfying $J^2 = -\mathbb{1}$ and

$$\omega(x)\Big(\xi, J(x)\eta\Big) = g(x)\,(\xi, \eta)$$

for all $\xi, \eta \in T_x M$, with a Riemannian metric g. If $N_S \to S$ is the normal bundle of S whose fibre at $x \in S$ is defined by

$$N_S(x) = \left\{ \eta \in T_x M \,\middle|\, g(x)(\eta, \xi) = 0 \ \text{ for all } \ \xi \in T_x S \right\},$$

we have the bundle isomorphism

$$\mathcal{L}_S \to N_S : (x, \xi) \mapsto \left(x, J(x)\xi \right).$$

This map is indeed well-defined and injective on each fibre: if $\omega(\xi, \eta) = 0$ for all $\eta \in T_x S$ then $g(\eta, J\xi) = \omega(\eta, J^2 \xi) = -\omega(\eta, \xi) = 0$. The normal bundle N_S is orientable since $\mathcal{L}_S$ is orientable. Therefore, N_S and $S \times \mathbb{R}$ are isomorphic as vector bundles. Taking a nonvanishing section σ of N_S, so that $0 \neq \sigma(x) \in N_S(x) \subset T_x M$ we can define the map

$$\psi : S \times (-\varepsilon, \varepsilon) \to U \subset M$$

by means of the exponential map $(x, t) \mapsto exp_x(t\sigma(x))$. It is a diffeomorphism onto an open neighborhood U of S if S is compact and ε sufficiently small. If

$$F : S \times (-\varepsilon, \varepsilon) \to \mathbb{R}$$

is the smooth function $F(x, t) = t$, then the composition $H = F \circ \psi^{-1} : U \to \mathbb{R}$ is the desired Hamiltonian; it satisfies $S = H^{-1}(0)$ and $dH \neq 0$ as claimed.

Definition. Let S be a compact hypersurface in (M, ω). A parameterized family of hypersurfaces modelled on S is a diffeomorphism

$$\psi : S \times I \to U \subset M,$$

I being an open interval containing 0, onto a bounded neighborhood U of S satisfying

$$\psi(x, 0) = x \quad \text{for} \quad x \in S.$$

We shall abbreviate in the following $S_\varepsilon = \psi(S \times \{\varepsilon\})$ and later on we often simply write (S_ε) instead of ψ.

We remark that the following statements are equivalent: (i) The line bundle $\mathcal{L}_S \to S$ is orientable. (ii) The normal bundle $N_S \to S$ is orientable. (iii) S is orientable. (iv) There exists a parameterized family of hypersurfaces modelled on S. (v) There exists a smooth function $H \in C^\infty(U, \mathbb{R})$ defined on an open neighborhood U of S representing $S = \{x \in U | H(x) = \text{const.}\}$ and satisfying $dH \neq 0$. Using these concepts we can reformulate Theorem 1 of the previous section as follows:

4.2 THE CHARACTERISTIC LINE BUNDLE OF A HYPERSURFACE

Theorem 2. (Hofer-Zehnder) Let S be a compact hypersurface in (M, ω) and let (S_ε) be a parameterized family of hypersurfaces modelled on S. If $c_0(U, \omega) < \infty$, then there exists a dense set $\Sigma \subset I$ such that
$$\mathcal{P}(S_\varepsilon) \neq \emptyset \quad \text{for} \quad \varepsilon \in \Sigma \ .$$

In order to find closed characteristics on the hypersurface S and not merely close by we shall now restrict the class of hypersurfaces under consideration. We assume that S is the boundary of a compact symplectic manifold $(B, \omega) \subset (M, \omega)$, i.e., $\partial B = S$.

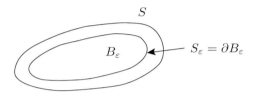

Fig. 4.4

Then if (S_ε) is a parameterized family modelled on S we have $S_\varepsilon = \partial B_\varepsilon$, for symplectic manifolds B_ε and we assume the parametrization to be chosen such that
$$B_\varepsilon \subset B_{\varepsilon'} \quad \text{if} \quad \varepsilon \leq \varepsilon'.$$
In view of the monotonicity property of a capacity we then have
$$c_0(B_\varepsilon, \omega) \leq c_0(B_{\varepsilon'}, \omega) \quad \text{if} \quad \varepsilon \leq \varepsilon',$$
so that the function
$$C(\varepsilon) := c_0(B_\varepsilon, \omega)$$
is monotone increasing.

Definition. The hypersurface S_{ε^*} is called of c_0-Lipschitz type if there are positive constants L and μ such that
$$C(\varepsilon) \leq C(\varepsilon^*) + L(\varepsilon - \varepsilon^*)$$
for every ε in the interval $\varepsilon^* \leq \varepsilon \leq \varepsilon^* + \mu$.

Using the monotonicity property of c_0 one verifies easily that the definition does not depend on the choice of the family S_ε modelled on S_{ε^*}. We illustrate the concept by an example. Let $S = \partial B$ be the boundary of the compact manifold B and assume there exists a vector field X in a neighborhood of B satisfying

(i) $\qquad L_X \omega = \omega$

(ii) $\qquad X(x) \notin T_x S \quad \text{if} \quad x \in S \ .$

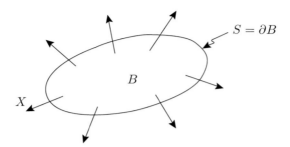

Fig. 4.5

The flow φ^t of X defines a distinguished parameterized family of hypersurfaces modelled on S by
$$\psi(x,t) = \varphi^t(x), \quad x \in S.$$
for $|t|$ sufficiently small. From $L_X \omega = \omega$ one deduces $(\varphi^t)^* \omega = e^t \omega$. Defining $B_t := \varphi^t(B)$ the map $\varphi^t : (B, \omega) \to (B_t, e^{-t}\omega)$ is, therefore, symplectic and using the invariance and conformality of c_0 we find $e^{-t} c_0(B_t, \omega) = c_0(B_t, e^{-t}\omega) = c_0(B, \omega)$. so that
$$c_0(B_t, \omega) = e^t c_0(B, \omega).$$
The function $C(t) = c_0(B_t, \omega)$ is differentiable at $t = 0$ and thus we have verified that S is of c_0- Lipschitz type.

Theorem 3. Assume $c_0(M, \omega) < \infty$. If the compact hypersurface $S \subset M$ bounds a symplectic manifold and is, moreover, of c_0-Lipschitz type then
$$\mathcal{P}(S) \neq \emptyset.$$

Proof. By assumption we find a parameterized family S_ε with $S = S_0$ such that
$$(4.6) \qquad C(\varepsilon) \leq C(0) + L \cdot \varepsilon$$
for $0 \leq \varepsilon \leq \mu$. For $0 < \tau < \mu$ we introduce the set $\mathcal{F}_\tau$ consisting of smooth functions $f : \mathbb{R} \to (C(0) - \tau L, \infty)$ satisfying

$$f(s) = a \quad \text{if} \quad s \leq 0$$
$$f(s) = b \quad \text{if} \quad s \geq \tfrac{\tau}{2}$$
$$0 < f'(s) \leq c \quad \text{if} \quad 0 < s < \tfrac{\tau}{2}$$

where the constants a, b, c are restricted by the conditions
$$C(0) - L\tau \leq a \leq C(0)$$
$$C(0) + 2L\tau \leq b \leq C(0) + 3L\tau$$
$$c = 10L.$$

4.2 The characteristic line bundle of a hypersurface

Note that $\mathcal{F}_\tau \neq \emptyset$. Fixing τ and recalling the definition of $c_0(B_0)$, we find an admissible function $H \in \mathcal{H}_a(B_0, \omega)$ satisfying $C(0) - L\tau \leq m(H) < C(0)$. Then we choose an $f \in \mathcal{F}_\tau$ with $a = m(H)$ and define the Hamiltonian function F by

$$F(x) = H(x) \quad \text{if} \quad x \in B_0$$
$$F(x) = f(\varepsilon) \quad \text{if} \quad x \in S_\varepsilon,\ 0 \leq \varepsilon \leq \tau$$
$$F(x) = b \quad \text{if} \quad x \notin \overline{B_\tau}.$$

Clearly $F \in \mathcal{H}(B_\tau, \omega)$ and in view of (4.6)

$$m(F) = b \geq C(0) + 2L\tau > C(0) + L\tau \geq C(\tau).$$

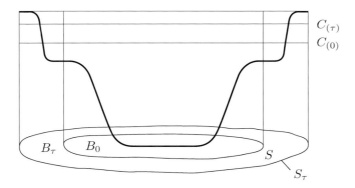

Fig. 4.6

Since $m(F) > c_0(B_\tau, \omega) = C(\tau)$ we deduce from the definition of the capacity function c_0 a nonconstant periodic solution $x(t)$ having period $0 < T \leq 1$ of the Hamiltonian system

$$\dot{x} = X_F(x), \quad x \in B_\tau.$$

By construction this solution cannot be contained in B_0, since the restriction of F onto B_0 is an admissible function, and B_0 being invariant under the flow of X_F, the solution $x(t)$ must be contained in $B_\tau \backslash \overline{B_0}$. As the solution is not constant we find by the properties of f an ε in $0 < \varepsilon < \frac{\tau}{2}$ such that

$$x(t) \subset S_\varepsilon.$$

This argument works for every $\tau > 0$, and choosing a sequence $\tau_j \to 0$ we thus find sequences F_j and ε_j and a corresponding sequence $x_j(t)$ of periodic orbits of X_{F_j} satisfying

$$x_j(t) \subset S_{\varepsilon_j}, \quad \varepsilon_j \to 0$$

and having periods $0 < T_j \leq 1$. These solutions satisfy, in addition, some estimates independent of j. Define on
$$U = \bigcup_{\varepsilon \in I} S_\varepsilon$$
a fixed Hamiltonian function K having the hypersurfaces S_ε as regular energy surfaces by setting
$$K(x) = \varepsilon \quad \text{if} \quad x \in S_\varepsilon.$$
If $x \in S_\varepsilon$ and $0 \leq \varepsilon \leq \mu$ then for every τ_j we have $F_j(x) = f_j(K(x))$ and, therefore, $X_F(x) = f'(K(x)) \cdot X_K(x)$ where we did not indicate the dependence of the functions f and F on j. By construction, the periodic functions solve the equations
$$\dot{x}_j(t) = f'(\varepsilon_j) X_K\Big(x_j(t)\Big)$$
$$x_j(0) = x_j(T_j)$$
and $0 < T_j \leq 1$. Reparameterizing, the functions
$$y_j(t) = x_j\left(\frac{t}{f'(\varepsilon_j)}\right)$$
solve the Hamiltonian equations
$$\dot{y}_j(t) = X_K(y_j) \quad \text{and} \quad K\Big(y_j(t)\Big) = \varepsilon_j.$$
The crucial observation now is that the periods of y_j are given by $f'(\varepsilon_j) \cdot T_j$ and hence, in view of $f'(\varepsilon_j) \leq 10M$, the periods are bounded. By Proposition 1 we, therefore, conclude that there exists a periodic solution $x(t)$ of X_K on the energy surface $K(x) = 0$ which is the hypersurface $S = S_0$. This periodic solution parameterizes the desired closed characteristic on S. ∎

Recall now that the function $C(\varepsilon) = c_0(B_\varepsilon, \omega)$ defined above is monotone. By a theorem due to Lebesgue it is, therefore, differentiable almost everywhere and thus Lipschitz continuous almost everywhere and we deduce from Theorem 3

Theorem 4. Assume the hypersurface $S \subset (M, \omega)$ bounds a compact symplectic manifold. If (S_ε) with $\varepsilon \in I$ is a parameterized family of hypersurfaces modelled on S, then (m denoting Lebesgue measure)
$$m\Big\{\varepsilon \in I \;\Big|\; P(S_\varepsilon) \neq \emptyset\Big\} = m(I),$$
provided $c_0(M, \omega) < \infty$.

A compact and connected hypersurface $S \subset \mathbb{R}^{2n}$ separates the space into a bounded and an unbounded component. (Indeed, every compact smooth hypersurface $M \subset \mathbb{R}^m$ is orientable; the statement follows from the Jordan-Brouwer separation theorem, we refer to E. Lima [142] for a short differential geometric proof in the smooth case.) We, therefore, conclude for the special case of hypersurfaces in the standard symplectic manifold $(\mathbb{R}^{2n}, \omega_0)$ the

4.3 Hypersurfaces of contact type, the Weinstein conjecture

Corollary. For every compact hypersurface $S \subset (\mathbb{R}^{2n}, \omega_0)$ and every parameterized family $(S_\varepsilon), \varepsilon \in I$, of hypersurfaces modelled on S

$$m\left\{\varepsilon \in I \mid \mathcal{P}(S_\varepsilon) \neq \emptyset\right\} = m(I).$$

This result was first proved by M. Struwe in [207] who cleverly modified the proof by H. Hofer and E. Zehnder in [123] of the weaker statement that $\mathcal{P}(S_\varepsilon) \neq \emptyset$ for a dense set of ε in I. As an illustration we take a smooth function $H \in C_c^\infty(\mathbb{R}^{2n})$, i.e., having compact support. Then the set of critical levels is compact in $\mathbb{R}$ and, moreover, by Sard's theorem of measure zero. Consequently, for almost every h in

$$\min H < h < \max H$$

there exists a nonconstant periodic solution $x_h(t)$ of the Hamiltonian vector field X_H having energy $H(x_h(t)) = h$. This is a local analogue to a recent result due to C. Viterbo [218] for mappings instead of vector fields. It states that a symplectic diffeomorphism φ of $(\mathbb{R}^{2n}, \omega_0)$ which is the identity map outside a compact set $K \subset \mathbb{R}^{2n}$ possesses infinitely many periodic orbits contained in K. We should mention that C. Viterbo also deduced his result from the existence of a symplectic invariant for subsets in $(\mathbb{R}^{2n}, \omega_0)$ whose relation to the capacity c_0 is so far not understood.

4.3 Hypersurfaces of contact type, the Weinstein conjecture

We next single out another symplectic property of a hypersurface which guarantees the existence of a closed characteristic. Prompted by the first global results for convex and star-like energy surfaces in $\mathbb{R}^{2n}$, A. Weinstein introduced in [226] the following concept:

Definition. A compact and orientable hypersurface $S \subset M$ in the symplectic manifold (M, ω) is called of contact type if there exists a 1-form α on S satisfying

$$\text{(i)} \quad d\alpha \quad = \quad j^*\omega$$

$$\text{(ii)} \quad \alpha(\xi) \neq 0 \quad \text{for} \quad 0 \neq \xi \in \mathcal{L}_S,$$

where $j : S \to M$ is the inclusion map and $\mathcal{L}_S$ is the canonical line bundle of S.

Since $\xi \in \mathcal{L}_S$ is in the kernel of $\omega|S$ and $\alpha(\xi) \neq 0$, the kernel of α, given by $\ker \alpha(x) = \{\eta \in T_x S \mid \alpha_x(\eta) = 0\}$, is a $(2n-2)$-dimensional subspace of $T_x S$ on which the 2-form ω, and hence also $d\alpha$ is nondegenerate. Consequently, $\alpha \wedge (d\alpha)^{n-1}$ is a volume form on S. Geometrically the 1-form α defines a smooth field of hyperplanes on S which is an example of a contact structure on the odd-dimensional manifold S.

A. Weinstein conjecture (1978): A hypersurface S of contact type and satisfying $H^1(S) = 0$ carries a closed characteristic.

We shall verify the conjecture without the topological assumption $H^1(S) = 0$ but under the symplectic assumption that S has a neighborhood $U \subset M$ of finite capacity: $c_0(U, \omega) < \infty$. We first reformulate the concept of contact type following A. Weinstein [226].

Proposition 4. A compact hypersurface $S \subset M$ is of contact type if and only if there exists a vector field X, defined on a neighborhood U of S, satisfying

$$\text{(i)} \qquad L_X \omega \;=\; \omega \qquad \text{on } U.$$

$$\text{(ii)} \qquad X(x) \;\notin\; T_x S \qquad \text{if } x \in S.$$

i.e., X is transversal to S.

In order to prove the proposition we start with a Poincaré-type lemma.

Lemma 2. Let $\pi : E \to N$ be a vector bundle and let α be a closed k-form on E satisfying $j^* \alpha = 0$, where the inclusion map $j : N \to E$ is the zero section of the bundle. Then there exists a $(k-1)$-form β on E satisfying

$$\alpha \;=\; d\beta \quad \text{and} \quad \beta|N \;=\; 0.$$

Proof. Using the linear structure of the fibers we can define, for $t > 0$, the contraction maps $\varphi^t : E \to E$ along the fibers by multiplication: $\varphi^t(x) = t \cdot x$. With X_t we denote the time-dependent vector field on E generated by the family of diffeomorphisms as usual by $\frac{d}{dt} \varphi^t(x) = X_t(\varphi^t(x))$ if $t > 0$. If we define $X_1(x) = \frac{d}{dt} \varphi^t(x)|_{t=1} \in T_x E$ we have the representation $X_t(x) = \frac{1}{t} X_1(x)$. Moreover, $X_t(x) = 0$ if $x \in N$. Using $d\alpha = 0$ we find by Cartan's formula

$$\frac{d}{dt} (\varphi^t)^* \alpha \;=\; (\varphi^t)^* L_{X_t} \alpha \;=\; d\left\{ (\varphi^t)^* i_{X_t} \alpha \right\},$$

where

$$(\varphi^t)^* i_{X_t} \alpha(x) \;=\; \alpha(t \cdot x) \left(\frac{1}{t} d\varphi^t X_1(x), d\varphi^t \cdot \right).$$

Because $\lim_{t \to 0} (\varphi^t)^* \alpha = \pi^* j^* \alpha = 0$ and $\varphi^1(x) = x$ we conclude that

$$\alpha \;=\; (\varphi^1)^* \alpha - \lim_{t \to 0} (\varphi^t)^* \alpha$$

$$=\; \lim_{t \to 0} \int_t^1 \tfrac{d}{dt} (\varphi^t)^* \alpha \, dt$$

$$=\; d \int_0^1 \left((\varphi^t)^* i_{X_t} \alpha \right) dt \;=\; d\beta$$

for a smooth $(k-1)$-form β on E. If $x \in N$, then $X_t(x) = 0$ and hence $i_{X_t} \alpha = 0$ so that $\beta(x) = 0$. This finishes the proof of the lemma. ■

We shall use the lemma to extend the 1-form α on S to an open neighborhood of S.

4.3 Hypersurfaces of contact type, the Weinstein conjecture

Lemma 3. If $S \subset (M, \omega)$ is of contact type then there exists a 1-form τ on a neighborhood U of S satisfying

$$\text{(i)} \quad d\tau = \omega \quad \text{on } U$$

$$\text{(ii)} \quad j^*\tau = \alpha \quad \text{on } S$$

where $j : S \to U$ is the inclusion mapping.

Proof. Since S is compact and oriented there exists a diffeomorphism $\psi : S \times (-1, 1) \to U$ onto an open neighborhood U of S satisfying $\psi(x, 0) = x$ for all $x \in S$. This has been proved in the previous section. The projection map $(x, t) \mapsto x$ from $S \times (-1, 1)$ onto S induces, therefore, a smooth map $r : U \to S$ which is the identity on S. Define the 1-form μ on U by $\mu = r^*\alpha$. Then $j^*\mu = j^*r^*\alpha = \alpha$. We now consider the 2-form $\omega - d\mu$ on U. Then

$$d(\omega - d\mu) = d\omega = 0,$$

since a symplectic structure is a closed form. Moreover, using that $j^*\omega = d\alpha$

$$j^*(\omega - d\mu) = d\alpha - d(j^*\mu) = d\alpha - d\alpha = 0.$$

Since U is diffeomorphic to $S \times (-1, 1)$, which in turn is isomorphic to the bundle $S \times \mathbb{R}$, we can apply Lemma 2 to the 2-form $\omega - d\mu$ and find a 1-form ϑ on U satisfying $\omega - d\mu = d\vartheta$ and $j^*\vartheta = 0$. Consequently, if we define the 1-form τ on U by $\tau = \mu + \vartheta$ we find $d\tau = d(\mu + \vartheta) = \omega$ and $j^*\tau = j^*\mu = \alpha$ as desired in the lemma. ∎

We are now ready to prove Proposition 4. Assume $S \subset M$ is of contact type. Then taking the 1-form τ of Lemma 3 we can define the vector field X on U by

$$i_X \omega = \tau \quad \text{on} \quad U.$$

This vector field has the desired properties of the proposition. Indeed, by Lemma 3, $\omega = d\tau = d(i_X\omega) = -i_X(d\omega) + L_X\omega = L_X\omega$ and it remains to show that X is transversal to S. If $0 \neq \xi \in \mathcal{L}_S(x)$ then

$$\omega\big(X(x), \xi\big) = \tau(\xi) = \alpha(\xi) \neq 0$$

by the assumption on α. Since ξ, by definition belongs to the kernel of ω restricted to the tangent space T_xS we conclude that $X(x) \notin T_xS$.

Conversely, if the vector field X meets the assumption (i) and (ii) of the proposition we define the 1-form α on U by

$$\alpha = i_X \omega.$$

Then $d\alpha = d(i_X\omega) = L_X\omega = \omega$, and from the transversality condition $X(x) \notin T_xS$ we conclude $0 \neq \omega\big(X(x), \xi\big) = \alpha(\xi)$ for $0 \neq \xi \in \mathcal{L}_S(x)$. We have verified, with this 1-form α, that S is of contact type. This finishes the proof of the proposition. ∎

The significance of the property of S being of contact type lies in the fact that S admits a distinguished parameterized family (S_ε) of hypersurfaces modelled on S. It is defined by the flow φ^t of the special vector field X guaranteed by Proposition 4. Since S is compact and X is transversal to S, the map

$$\psi : S \times (-\varepsilon, \varepsilon) \to U \subset M$$

defined by $\psi(x,t) = \varphi^t(x)$ for $x \in S$ and $|t| < \varepsilon$ is a diffeomorphism onto an open neighborhood U of S provided $\varepsilon > 0$ is sufficiently small. From $L_X \omega = \omega$ we conclude $\frac{d}{dt}(\varphi^t)^*\omega = (\varphi^t)^* L_X \omega = (\varphi^t)^*\omega$. Hence in view of $\varphi^0 = \mathrm{id}$,

$$(\varphi^t)^*\omega = e^t \omega.$$

Assume now that $\xi \in \mathcal{L}_S(x)$, then for all $\eta \in T_x S$

$$0 = \omega(\xi, \eta) = e^t \omega(\xi, \eta) = (\varphi^t)^*\omega(\xi, \eta) = \omega\big(d\varphi^t(x)\xi, d\varphi^t(x)\eta\big),$$

and consequently $d\varphi^t(x)\xi \in \mathcal{L}_{S_t}(\varphi^t(x))$. We see that

$$T\varphi^t : \mathcal{L}_S \to \mathcal{L}_{S_t}$$

is an isomorphism of vector bundles. Therefore, φ^t induces a one-to-one correspondence

$$\mathcal{P}(S) \longleftrightarrow \mathcal{P}(S_t)$$

of the closed characteristics by $P \mapsto \varphi^t(P)$. This leads us to the following definition:

Definition. A compact hypersurface $S \subset (M, \omega)$ is called stable if there exists a parameterized family (S_ε) modelled on S having the following additional property: the associated diffeomorphism $\psi : S \times (-1, 1) \to U$ induces bundle isomorphisms

$$T\psi_\varepsilon : \mathcal{L}_S \to \mathcal{L}_{S_\varepsilon},$$

for every $\varepsilon \in (-1, 1)$.

A hypersurface of contact type is stable as we have just proved. A stable hypersurface need, however, not be of contact type as the following example illustrates. Consider the symplectic manifold $M = N \times \mathbb{R}^2$ with the symplectic structure $\omega = \omega_1 \oplus \omega_0$, where (N, ω_1) is a compact symplectic manifold without boundary. Consider the hypersurface

$$S = N \times \big\{|z|^2 = 1\big\} \subset M$$

and define the parametrization $\psi : S \times (-1, 1) \to M$ by setting $\psi((x, z), \varepsilon) = (x, \varepsilon z)$, so that $S_\varepsilon = N \times \{|z|^2 = \varepsilon^2\}$. Evidently S is stable but S is not of contact type. Otherwise we find a 1-form α on S satisfying $d\alpha = j^*\omega$, with the inclusion $j : S \to M$. Denoting by $i : N \to N \times \{(1, 0)\}$ the inclusion of N in S we infer

4.3 Hypersurfaces of contact type, the Weinstein conjecture

that $i^*(d\alpha) = i^*j^*\omega = \omega_1$. Consequently, the symplectic form $\omega_1 = d(i^*\alpha)$ must be exact. This, however, is not possible. Indeed, by Stokes' theorem

$$\text{vol}(N) = \int_N \omega_1 \wedge \ldots \wedge \omega_1 = 0,$$

hence $\omega_1^{(n)}$ is not a volume form and thus ω_1 is degenerate, contradicting the assumption that ω_1 is a symplectic form.

A stable hypersurface S admits, by definition, a parameterized family (S_ε) having the following property: if $\mathcal{P}(S_\varepsilon) \neq \emptyset$ for a single $\varepsilon \in (-1, 1)$ then $\mathcal{P}(S) \neq \emptyset$. Consequently, we deduce from Theorem 2 the following global existence statement:

Theorem 5. Assume the compact hypersurface $S \subset (M, \omega)$ admits a neighborhood U of finite capacity: $c_0(U, \omega) < \infty$. If S is, in addition, stable, then

$$\mathcal{P}(S) \neq \emptyset.$$

In particular, if S is of contact type, then $\mathcal{P}(S) \neq 0$.

A compact hypersurface in $(\mathbb{R}^{2n}, \omega_0)$ always possesses a bounded neighborhood of finite capacity c_0 and we deduce from Theorem 5 the celebrated solution of A. Weinstein's conjecture due to C. Viterbo [217] in 1987. For another proof we mention [45], 1988.

Theorem 6. (C. Viterbo) Every compact hypersurface $S \subset (\mathbb{R}^{2n}, \omega_0)$ of contact type carries a closed characteristic.

In contrast to the Weinstein conjecture for general symplectic manifolds (M, ω), the condition that $H^1(S) = 0$ is not required here. An example is a compact hypersurface $S \subset \mathbb{R}^{2n}$ which is star-like with respect to an interior point we may assumed to be the origin in $\mathbb{R}^{2n}$ (Fig.4.7).

Denote by $X(x) = \frac{1}{2}x$ the dilatation, then, by definition of star-like

$$X(x) \notin T_x S, \quad x \in S.$$

Moreover, the flow φ^t of X satisfies

$$(\varphi^t)^* \omega_0 = e^t \omega_0,$$

so that in view of the proposition this hypersurface S is of contact type. In particular, a smooth hypersurface bounding a compact and convex domain in $\mathbb{R}^{2n}$ is star-like. Thus convex and star-like hypersurfaces in $(\mathbb{R}^{2n}, \omega_0)$ always carry a periodic Hamiltonian orbit and we have confirmed the pioneering results due to A. Weinstein [225] and P. Rabinowitz [181] in 1978 which marked the beginning of the investigation of global phenomena of Hamiltonian systems in higher dimensions.

We point out that by using the technique of attaching symplectic handlebodies, Y. Eliashberg [72] and A. Weinstein [221] constructed many examples of hypersurfaces of contact type which are topologically intricate. Not every hypersurface

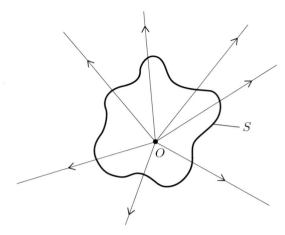

Fig. 4.7

is, of course, of contact type. We illustrate this by an example of a hypersurface S which is of c_0-Lipschitz type but not of contact type. Consider within the closure of the symplectic cylinder $Z(R) \subset (\mathbb{R}^{2n}, \omega_0)$ a hypersurface which is diffeomorphic to the sphere indented, as in the following picture

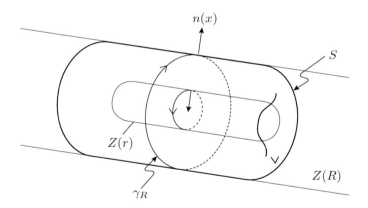

Fig. 4.8

$$Z(R) = \left\{ (x,y) \in \mathbb{R}^{2n} \ \middle| \ x_1^2 + y_1^2 < R^2 \right\}.$$

Denoting by $n(x)$ the outer normal at $x \in S$, we define by $Jn(x) \in T_xS$ a smooth vector field on S which belongs to the characteristic line bundle, $Jn(x) \in \mathcal{L}_S(x)$. Assume now, by contradiction, that S is of contact type. Then there exists a 1-form

4.3 Hypersurfaces of contact type, the Weinstein conjecture

α near S satisfying $d\alpha = \omega_0$ and either

$$\alpha\Big(Jn(x)\Big) > 0, \quad \text{for all} \quad x \in S$$

or $\alpha(Jn(x)) < 0$ for all $x \in S$. Without loss of generality we assume the first alternative to hold. Define the 1-form λ on $\mathbb{R}^{2n}$ by

$$\lambda = \sum_{j=1}^{n} y_j \, dx_j,$$

where $(x, y) \in \mathbb{R}^{2n}$. Then $d\lambda = \omega_0$ and, therefore, $d\lambda = d\alpha$ on S. Hence, in view of $H^1(S) = 0$ we find a smooth function $\rho : S \to \mathbb{R}$ satisfying $\alpha - \lambda = d\rho$. Consequently, for every periodic solution γ of the vector field Jn on S,

$$\int_\gamma \alpha = \int_\gamma \lambda > 0.$$

However, for the two distinguished periodic solutions γ_R and γ_r on the outer respectively inner cylinder, as depicted in the above figure, one finds

$$\int_{\gamma_R} \lambda = \pi R^2 \quad \text{and} \quad \int_{\gamma_r} \lambda = -\pi r^2.$$

This contradiction shows that S is not of contact type. Evidently, there is a parameterized family (S_ε) modelled on S for which $c_0(B_\varepsilon, \omega_0) = \pi(R+\varepsilon)^2$ agrees with the capacity of the cylinders, where $S_\varepsilon = \partial B_\varepsilon$. Therefore, $S = S_0$ is of c_0-Lipschitz type. We remark that the argument above also shows that the actions of all periodic solutions on a compact energy surface S of contact type and satisfying $H^1(S) = 0$ have all the same sign. In particular, the actions never vanish on such hypersurfaces.

We conclude this section with a recent result in dimension 3, which is not obtained by the variational methods described so far. We consider a smooth closed and orientable manifold W of dimension $2n-1$. A contact form on W is a 1-form λ such that $\lambda \wedge (d\lambda)^{n-1}$ is a volume form on W. Such a contact form determines a so-called contact structure which is the $(2n-2)$ dimensional plane field ξ_λ defined by

$$\xi_\lambda = \text{Kern}(\lambda) \subset TW.$$

Moreover, the kernel of $d\lambda$ is 1-dimensional. Thus there exists a unique vector field $X = X_\lambda$ satisfying

$$i_X \, d\lambda \equiv 0 \quad \text{and} \quad i_X \lambda = 1.$$

This distinguished non vanishing vector field on W is called the Reeb vector field associated to λ. Note that we no longer require that $W = \partial M$ and $d\lambda = \omega|M$ for a compact symplectic manifold (M, ω) of dimension $2n$.

We shall say that the Weinstein conjecture holds true for the manifold W, if for every contact form λ on W, the associated Reeb vector field X_λ has a closed orbit.

Theorem 7. (H. Hofer) *The Weinstein conjecture holds true for the three dimensional sphere S^3.*

Actually the Weinstein conjecture holds true also for every closed orientable three manifold W^3 satisfying $\pi_2(W^3) \neq 0$, see H. Hofer [118] (1993). The proof is based on first order elliptic partial differential equations of "Cauchy-Riemann" type and is influenced by Gromov's theory of pseudoholomorphic curves in symplectic manifolds [107] and Eliashberg's outline of filling techniques using holomorphic discs [73].

We point out that there are many contact forms λ on S^3, (namely the so-called overtwisted contact forms) which do not admit an embedding of S^3 into $\mathbb{R}^4$ such that the induced Hamiltonian flow is conjugated to the Reeb flow defined by X_λ.

Here a contact form λ on a closed three dimensional manifold W is called overtwisted by Eliashberg, if there exists an embedded disc $F \simeq D^2$, where $D^2 = \{z \in \mathbb{C} | \, |z| \leq 1\}$, such that

$$T(\partial F) \subset \xi_\lambda | \partial F$$
$$T_x F \not\subset \xi_{\lambda,x} \quad \text{for all} \quad x \in \partial F.$$

In the classical work of Lutz [144] and Martinet, [147] contact structures are established in every compact orientable three-manifold. On S^3, in particular, there exist overtwisted contact structures which are not equivalent by a diffeomorphism (in the sense described below).

A contact form λ which is not overtwisted is called tight. By a deep classification result of Eliashberg [74] there is, up to diffeomorphisms however only one contact structure on S^3 which is tight. More precisely, if λ_1, λ_2 are tight forms on S^3, then according to Eliashberg there exists a smooth function $f : S^3 \to \mathbb{R} \backslash \{0\}$ and a diffeomorphism $\psi : S^3 \to S^3$ satisfying

$$\psi^* \lambda_2 = f \lambda_1.$$

We shall show now that, for a tight contact structure on S^3, the associated Reeb vector field has a periodic solution. Represent S^3 as the subset of $\mathbb{C}^2$ given by $\{(z_1, z_2) | \, |z_1|^2 + |z_2|^2 = 1\}$. Writing $z_j = q_j + i p_j$ with $q_j, p_j \in \mathbb{R}$, we define the one-form λ_0 on $\mathbb{C}^2$ by

$$\lambda_0 = \frac{1}{2} \sum_{j=1}^{2} (p_j dq_j - q_j dp_j)$$

so that $d\lambda_0 = \omega_0$. Due to a fundamental result by Bennequin [20], the special contact form $\lambda_0 | S^3$ is tight. Consequently, given any tight contact form λ on S^3,

we find a nonvanishing function $f : S^3 \to \mathbb{R}\setminus\{0\}$ and a diffeomorphism ψ of S^3 such that
$$\psi^* \lambda = f \cdot (\lambda_0|S^3) \,.$$
Hence the Reeb flows for λ and $f \cdot (\lambda_0|S^3)$ are conjugated. We can use this in order to prove that X_λ has, in this special case, a periodic orbit. Since $X_{f\lambda_0} = -X_{-f\lambda_0}$ we may assume that f is positive. We claim that there is a smooth hypersurface $S \subset \mathbb{C}^2$ bounding a star-shaped domain and a diffeomorphism $\psi : S^3 \to S$ such that
$$T\psi(\mathcal{L}_{f\lambda_0}) = \mathcal{L}_S \,,$$
where $\mathcal{L}_{f\lambda_0}$ is the line bundle of S^3 defined by the Reeb vector field associated to $f\lambda_0$, and where $\mathcal{L}_S$ is the canonical line field induced on the hypersurface S by the standard symplectic structure ω_0 in $\mathbb{C}^2 \cong \mathbb{R}^4$. Indeed, we simply define the hypersurface S by
$$S = \{\sqrt{f(z)}\, z \mid z \in S^3\} \subset \mathbb{C}^2$$
and the diffeomorphism $\psi : S^3 \to S$ by $\psi(z) = \sqrt{f(z)}z$. Observing now that $\lambda_0(z) = \frac{1}{2}\omega_0(z,\cdot), z \in \mathbb{C}^2$ we abbreviate $\sigma = \sqrt{f}$ and compute, for $z \in S^3$ and $h \in T_z S^3$,
$$\psi^*(\lambda_0|S)\,(z;h) = \tfrac{1}{2}\omega_0\Big(\sigma(z)z, (d\sigma(z)h)z + \sigma(z)h\Big)$$
$$= \sigma(z)^2 \tfrac{1}{2}\omega_0(z,h)$$
$$= (f \cdot \lambda_0)(z,h)\,.$$

From this, the claim follows. Using the existence result of P. Rabinowitz proved above, we conclude that the Reeb vector field $X_{f\lambda_0}$ has a periodic solution.

To sum up, we have demonstrated that in the exceptional case of a tight contact structure λ on S^3, the associated Reeb vector field X_λ possesses at least one closed orbit.

4.4 "Classical" Hamiltonian systems

The positions of a conservative system of n degrees of freedom are points in the so-called configuration space, which we assume to be an n-dimensional smooth manifold N. The motion of the system, in the setting of Lagrangian mechanics, is determined by a Lagrangian L defined on the tangent bundle TN of N. This $2n$-dimensional manifold is called the phase space of the system. In our setting of Hamiltonian mechanics, the motion of the system is determined by a Hamiltonian function H defined on the cotangent bundle T^*N of N. This is the basic example of a symplectic manifold. It is equipped with a canonical symplectic structure which we shall recall first.

If N is any manifold of dimension n, and if $T_x N$ is the tangent space at $x \in N$, we denote its dual space, the so called cotangent space by $T_x^* N$. It is the space of linear forms defined on the vector space $T_x N$. The union of all cotangent spaces is called the cotangent bundle of N and denoted by

$$T^*N = \bigcup_{x \in N} T_x^* N.$$

A point α in this set $M = T^*N$ is a linear form α_x in the tangent space $T_x N$ at some point $x \in N$, and we shall sometimes use the notation

$$\xi = (x, \xi_x) \in T^*N$$

for a point in M. One can view T^*N as a differentiable manifold of dimension $2n$ by introducing local coordinates, e.g., as follows. If $(x_1, \ldots, x_n)$ are local coordinates in N, a 1-form $\alpha \in T_x^* N$ is represented by $\alpha = \sum_{j=1}^{n} y_j dx_j$ having the coordinates $(y_1, \ldots, y_n)$ and together $(x_1, \ldots, x_n, y_1, \ldots, y_n)$ form local coordinates in T^*N. In these coordinates one can now define the special 1-form λ on T^*N by

$$(4.7) \qquad \lambda = \sum_{j=1}^{n} y_j \, dx_j,$$

which has, so far only a local meaning. It is remarkable that the above form λ has a global interpretation on T^*N. Since a point of T^*N is represented by a 1-form ξ at a point $x \in N$ we can form $\xi(V)$ for every vector $V \in T_x N$. To define a one-form on T^*N, say β, one has to give its value $\beta(X)$ for every tangent vector X of T^*N. If X is such a tangent vector at $\xi \in T^*N$, one can use the form ξ itself to define

$$(4.8) \qquad \lambda_\xi(X) = \xi(d\pi X), \ X \in T_\xi(T^*N)$$

as a linear functional. Here the projection $\pi : T^*N \to N$ assigns to each $\xi \in T^*N$ its base point $x \in N$ and, therefore, $d\pi$ maps the tangent space $T_\xi(T^*N)$ at ξ onto the tangent space $T_x N$ at x. This 1-form λ on T^*N is sometimes called the tautological form on T^*N since it is defined in terms of itself. It is readily verified that in the above local coordinates the form (4.8) agrees with the form (4.7). In view of the expression (4.7) we find

$$d\lambda = \sum_{j=1}^{n} dy_j \wedge dx_j$$

and conclude that, globally, $\omega = d\lambda$ is a closed and nondegenerate 2-form on T^*N, hence a symplectic form. It is called the canonical symplectic structure on T^*N. The canonical 1-form λ on T^*N is called the Liouville form. As an aside, we observe that λ has the distinguished property that

$$\beta^* \lambda = \beta$$

4.4 "Classical" Hamiltonian systems

for every one-form β on N. Here the one-form β is regarded as a section $\beta : N \to T^*N$. The proof follows from the observation that $\pi \circ \beta = id$ on N. Considering the definition of λ and using the chain rule, we find

$$(\beta^*\lambda)_x = \lambda_{(x,\beta_x)}(d\beta)_x = \beta_x \circ d\pi_{(x,\beta_x)} \circ (d\beta)_x = \beta_x$$

as claimed. Classical mechanical systems are described by Hamiltonian functions on $(T^*N, d\lambda)$ given by a sum of kinetic energy and the potential energy. The kinetic energy is defined by a Riemannian metric k on the configuration space, which associates with every point $x \in N$, a symmetric bilinearform $k_x(v,w)$ for $v, w \in T_xN$ which is positive. This metric defines a bundle isomorphism $\gamma : TN \to T^*N$ by $\gamma(v)(w) = k(v,w)$ which induces the associated Riemannian metric K on the cotangent bundle T^*N by

$$K(\xi, \eta) = k\left(\gamma^{-1}(\xi), \gamma^{-1}(\eta)\right).$$

The potential energy is represented by a smooth function $V : N \to \mathbb{R}$ on the configuration space. On T^*N the Hamiltonian H is now defined by

(4.9) $$H(\xi) = K(\xi) + V\left(\pi(\xi)\right), \quad \xi \in T^*N.$$

In the above local coordinates we have the formulae:

(4.10)
$$H = K(x,y) + V(x)$$
$$\lambda = \sum_{j=1}^{n} y_j \, dx_j$$
$$X_H = \sum_{j=1}^{n} \left(\frac{\partial K}{\partial y_j} \frac{\partial}{\partial x_j} - \frac{\partial H}{\partial x_j} \frac{\partial}{\partial y_j}\right)$$
$$\lambda(X_H) = \sum_{j=1}^{n} y_j \frac{\partial K}{\partial y_j} = 2K,$$

where $K(x,y) = \frac{1}{2}\langle S(x)y, y\rangle$ with a positive definite matrix $S(x)$.

If $\xi(t)$ is a solution of the Hamiltonian vector field X_H contained in the invariant energy surface

(4.11) $$S = \left\{\xi \,\middle|\, H = K + V = E\right\},$$

then the geometric motion of the system in the configuration space is given by $x(t) = \pi(\xi(t)) \in N$ where $\pi : T^*N \to N$ is the projection. Since $K \geq 0$, it is confined in the subset

(4.12) $$N_E = \left\{x \in N \,\middle|\, V(x) \leq E\right\}.$$

If $E < \max V$, this set is an n-dimensional manifold having the smooth boundary $\partial N_E = \{V(x) = E\}$ provided S is a regular energy surface. Indeed, where $K = 0$ we also have $\frac{\partial}{\partial y} K = 0$ so that necessarily $\frac{\partial}{\partial x} V(x) \neq 0$ on $V(x) = E$. However if $E > \max V$, then $N_E = N$ is the whole configuration space and $K > 0$ on S. Therefore $\lambda(X_H) = 2K > 0$ on S and we see that S is a hypersurface of contact type whose contact structure is defined by means of the Liouville form λ. If the capacity $c_0(U)$ is finite for a neighborhood U of S in T^*N we can conclude that S admits a closed characteristic. Unfortunately, nothing is known about the special capacity c_0 on cotangent bundles, except in the special case of the torus which will be treated later.

Periodic solutions on S have been found by using geometrical ideas, as we shall briefly indicate next. Assuming that $E > \max V$, we have

$$T^* N_E = T^* N$$

and we can define the special Riemannian metric G on T^*N by

$$G(\xi) = \frac{K(\xi)}{E - V\big(\pi(\xi)\big)}, \quad \xi \in T^*N.$$

It is the so-called Jacobi metric. We have now two Hamiltonian functions, namely H and G, which describe S as the regular energy surface

$$S = \big\{H(\xi) = E\big\} = \big\{G(\xi) = 1\big\}.$$

The corresponding Hamiltonian vector fields are, therefore, parallel, $X_H = \rho X_G$ on S, with a nonvanishing function ρ defined on S which is actually easily computed in local coordinates. One finds that

$$X_H = K \cdot X_G \quad \text{on} \quad S.$$

Consequently, the vector fields X_H and X_G have, on S, the same orbits up to reparametrization. Geometrically, the vector field X_G generates the geodesic flow defined by the Riemannian metric G on T^*N. The projection $\pi(\xi(t))$ of a flow line is a geodesic line in the configuration space N, i.e., it locally minimizes the length between two points, where the length is measured with respect to the metric g on TN given by

$$g_x(v,w) = \big(E - V(x)\big)^{-1} k_x(v,w), \quad \text{with} \quad v, w \in T_x N.$$

We see that the geodesics for the Riemannian metric G on T^*N agree, up to reparametrization, with the solutions of the Hamiltonian vector field X_H on S. This holds true, of course, only on the distinguished energy surface S. This fact is known as the Euler-Maupertuis-Jacobi principle. It reduces the dynamical problem of

4.4 "Classical" Hamiltonian systems

finding solutions for the Hamiltonian vector field on S to the geometrical problem of finding geodesics for the Jacobi metric on N, a problem with a rich history.

Recall that if N is not simply connected, e.g., a torus, one can consider the family of closed curves in a nontrivial homotopy class and obtain a closed geodesic of shortest length. This was already known to Hadamard 1889. Technically this approach requires that the length functional on the subspace of loops attains its minimum, an analytical problem which was clearly seen and attacked by D. Hilbert (1890), see [112]. To find closed geodesics on a Riemannian manifold N which is simply connected is a difficult and interesting task. In this case the geodesics cannot be found as minima but rather as saddle points of the length functional using the topology of the loop space of the underlying manifold N. G. Birkhoff established a closed geodesic on the two-spheres using a minimax principle which later led to the Ljusternik-Schnirelman theory for critical points. The Morse theory developed at roughly the same time had its origin in the existence problem of geodesics. Extending Birkhoff's idea to higher dimensions, Ljusternik and Fet proved in 1951 [81] that every compact Riemannian manifold (N, g) possesses a closed geodesic. From this geometrical result one deduces immediately by taking the Jacobi metric that the energy surface S above possesses a closed characteristic, provided N is compact and $E > \max V$.

In the case that $E < \max V$ the theory of closed geodesics is not applicable directly, since now the manifold N_E has a boundary ∂N_E where the Jacobi metric blows up. Despite this analytical difficulty, S.V. Bolotin [29] 1978 used the geometrical approach successfully in order to prove:

Theorem 6. Let N be any smooth manifold and assume $H : T^*N \to \mathbb{R}$ describes a classical system of the form (4.9). Then every compact and regular energy surface S_E possess a periodic solution of X_H.

This theorem extends an earlier result by H. Seifert [193] in 1948 which requires that N_E is homeomorphic to an n-cell. In 1917 G. Birkhoff proved the statement for $N_E = S^2$.

Using the geometric approach, Bolotin's result has been rediscovered by H. Gluck and W. Ziller (1983) [106] and later on by V. Benci (1984) [18]. We should mention that in the case $\partial N_E \neq \emptyset$ the periodic solution found by the theorem is a so-called brake orbit. This is one for which the motion $\pi(\xi(t))$ in the configuration space N_E moves back and forth between different points of the boundary ∂N_E but otherwise runs through the interior. Note that the systems considered are reversible: if $\xi(t)$ is a solution of X_H then $-\xi(-t)$ is also a solution so that with a motion $x(t) = \pi(\xi(t))$ in the configuration space the motion $x(-t) = \pi(-\xi(-t))$ traversed backwards also corresponds to a solution of X_H. This situation is, of course, familiar from the 2-dimensional case of a function $H(x, y) = \frac{1}{2}y^2 + V(x)$, as depicted in the following figure 4.9.

There are many results for closed geodesics which, applied to the Jacobi metric lead to closed characteristics of energy surfaces. We refer to Klingenberg's book

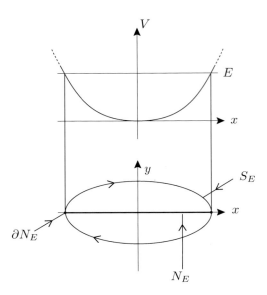

Fig. 4.9

[127] and V. Bangert's survey article [14] for these developments. We should point out that only very recently V. Bangert [15] and J. Franks [100] succeeded in establishing infinitely many closed geodesics on every Riemannian two-sphere!

The geometric approach outlined above is applicable only for a very restricted class of hypersurfaces contained in a cotangent bundle and we would like to mention, that by a quite different analytical approach, C. Viterbo and H. Hofer 1988 in [121] confirmed the Weinstein conjecture for cotangent bundles as follows:

Theorem 7. (Hofer-Viterbo) Let N be a compact manifold with $\dim N \geq 2$, and $M = T^*N$ the canonical symplectic manifold. Assume S is a smooth, compact and connected hypersurface in M having the property that the bounded component of $M \backslash S$ contains the zero section of T^*N. Then $\mathcal{P}(S) \neq \emptyset$, if S is of contact type.

It is, so far, not known whether the proviso about the zero section is really necessary to guarantee a closed characteristic on S.

We next turn to $(\mathbb{R}^{2n}, \omega_0)$ which can be viewed as the trivial cotangent bundle $T^*(\mathbb{R}^n)$. The Liouville form λ is now globally defined by

$$\lambda = \sum_{j=1}^{n} y_j dx_j$$

and $\omega_0 = d\lambda$. We know that the capacity c_0 is finite on bounded sets on $\mathbb{R}^{2n}$. This fact will be used in order to prove some very special existence results based on the following observation:

4.4 "Classical" Hamiltonian systems

Lemma. Assume the smooth Hamiltonian $H: \mathbb{R}^{2n} \to \mathbb{R}$ satisfies
$$\lambda(X_H) > 0 \quad \text{for all} \quad (x,y) \in \mathbb{R}^{2n} \quad \text{with} \quad y \neq 0.$$
Then every compact and regular energy surface $S = \{H(x,y) = E\}$ is of contact type.

Since $\lambda(X_H) = \langle \frac{\partial}{\partial y} H(x,y), y \rangle$, and since S is contained in a neighborhood having finite capacity c_0, we immediately conclude in view of the lemma and Theorem 5:

Theorem 8. If $H \in C^\infty(\mathbb{R}^{2n}, \mathbb{R})$ satisfies
$$\langle \frac{\partial}{\partial y} H(x,y), y \rangle > 0 \quad \text{for all} \quad (x,y) \in \mathbb{R}^{2n} \quad \text{with} \quad y \neq 0,$$
then every compact and regular energy surface S of H possesses a periodic solution for X_H.

The assumption is satisfied e.g. for $H(x,y) = \frac{1}{2}\langle A(x)y, y\rangle + V(x)$, with $A(x) > 0$, i.e., positive definite.

Proof of the Lemma. If S does not contain points $(x,0)$ with $x \in \mathbb{R}^n$, it is of contact type as explained above, the contact structure being given by the Liouville form. Otherwise we simply modify the Liouville form and define a 1-form α on $\mathbb{R}^{2n}$ setting
$$\alpha = \lambda - \varepsilon dF,$$
where the function $F \in C^\infty(\mathbb{R}^{2n}, \mathbb{R})$ is given by
$$F(x,y) = \langle \frac{\partial}{\partial x} H(x,0), y \rangle.$$
Clearly, $d\alpha = d\lambda = \omega_0$, and we shall show for $\varepsilon > 0$ sufficiently small that $\alpha(X_H) > 0$ on S, so that the one-form α meets all the requirements in the definition of contact type. By assumption $\lambda(X_H) = \langle \frac{\partial}{\partial y} H(x,y), y\rangle > 0$ if $y \neq 0$, so that for fixed x the function $y \mapsto \langle \frac{\partial}{\partial y} H(x,y), y \rangle$ has a minimum at $y = 0$ and it follows that $\frac{\partial}{\partial y} H(x,0) = 0$. A computation shows that
$$\alpha(X_H) = \lambda(X_H) + \varepsilon \left\{ \langle \frac{\partial H}{\partial x}(x,y), \frac{\partial H}{\partial x}(x,0) \rangle - \sum_{j,k=1}^n \frac{\partial^2 H}{\partial x_k \partial x_j}(x,0) \frac{\partial H}{\partial y_k}(x,y) y_j \right\}.$$
By assumption, $dH \neq 0$ on S, hence $\frac{\partial}{\partial x} H(x,0) \neq 0$. Consequently, if $(x,y) \in S$ and y is small, the expression in the bracket of the formula above is positive. On the other hand, if $(x,y) \in S$ and y is bounded away from $y = 0$, then the first term is bounded away from zero and, S being compact, we can choose $\varepsilon > 0$ so small that it dominates the second term. Now $\alpha(X_H) > 0$ on S, as we wanted to prove. ■

By the same argument we have also proved the following, more general, statement discovered in 1987 by V. Benci, H. Hofer and P. Rabinowitz [19].

Theorem 9. Assume the energy surface $S = \{(x,y) \,|\, H(x,y) = E\}$ is compact and regular and has the following property: if $(x,y) \in S$ and $y \neq 0$ then $\lambda(X_H) > 0$, and if $(x,0) \in S$ then $\frac{\partial}{\partial x} H(x,0) \neq 0$. Then S is of contact type and hence X_H has a periodic solution of energy E.

The compactness of the energy surfaces is a crucial assumption for the results above. The existence of periodic solutions on non compact energy surfaces requires additional conditions as the example $H(x,y) = \frac{1}{2}|y|^2$ in $\mathbb{R}^{2n}$ which has no periodic solutions shows. We would like to mention a result in the non compact case due to D. Offin, in [171] (1985). It states that if the boundary ∂N_E in the configuration space is disconnected while N_E itself is connected then (under some additional technical assumptions) there exists a periodic orbit of energy E. This orbit is a brake orbit oscillating between two points belonging to different components of ∂N_E. Offin obtains the periodic orbit by a direct variational argument minimizing the energy integral of the corresponding Jacobi metric in the configuration space.

Of special interest in mechanics is the cotangent bundle $T^*(T^n)$ of the torus $T^n = \mathbb{R}^n / \mathbb{Z}^n$ which is isomorphic to $T^n \times \mathbb{R}^n$. This symplectic manifold occurs as the phase space in many mechanical systems; in particular, for systems which are close to integrable ones. We shall briefly explain how it arises, referring to Appendix 1 for details. We start from a particularly simple system, a so-called integrable system. It is characterized by the property of having sufficiently many integrals such that the task of solving (or integrating) the differential equations becomes essentially trivial. More precisely, we consider a Hamiltonian system X_{H_0} on a symplectic manifold (M, ω) of dimension $2n$. It is called integrable (in the sense of Liouville) if there exist $n = \frac{1}{2} \dim M$ functions $F_j : M \to \mathbb{R}, 1 \leq j \leq n$, having the following properties:

(i) $\quad dF_1, dF_2, \ldots, dF_n \quad$ are linearly independent on M,

(ii) $\quad \{F_i, F_j\} = 0 \quad$ for all $\quad i, j$,

(iii) $\quad \{H_0, F_j\} = 0 \quad$ for all $\quad j$,

where the functions $\{F_i, F_j\} = \omega(X_{F_i}, X_{F_j})$ are the Poisson brackets. Assume now that there exists a level set

$$N_{c^*} = \left\{ x \in M \,\bigg|\, F_j(x) = c_j^*, \; 1 \leq j \leq n \right\}$$

which is compact and connected. Then one concludes from (i) and (ii) that $N_{c^*} \cong T^n$ is an embedded n-dimensional torus, which is Lagrangian. Moreover, there exists an open neighborhood U of N_{c^*} foliated into such tori in which one can introduce so-called action and angle variables. This means that there exists a symplectic diffeomorphism

$$\psi : T^n \times D \longrightarrow U$$

$$\psi^* \omega = d\left(\sum_{j=1}^n y_j \, dx_j \right).$$

4.4 "Classical" Hamiltonian systems

Here x_j (mod 1) are the angle variables of T^n and $y \in D$ are the action variables, where $D \subset \mathbb{R}^n$ is an open set. Moreover, the functions $F_j \circ \psi(x,y) =: f_j(y)$ depend only on the action variables. It then follows from assumption (iii) that the Hamiltonian system looks very simple on U, namely

$$H_0 \circ \psi(x,y) = h_0(y),$$

i.e., it depends only on the action variables. What we have described is the Arnold-Jost theorem on the existence of action-angle variables. It extends a local statement which goes back to Liouville, and we refer to the Appendix for a proof. The Hamiltonian equation on $T^n \times D$,

$$\dot{x} = \frac{\partial}{\partial y} h_0(y)$$
$$\dot{y} = 0,$$

can be solved explicitly and for all times. Indeed we read off for the flow on $T^n \times D$

$$\varphi^t(x,y) = (x + t\omega, y)$$

where

$$\omega = \omega(y) = \frac{\partial}{\partial y} h_0(y).$$

Geometrically each torus $T^n \times \{y\}$ is invariant, the motion on it is linear and given by the frequencies $\omega(y) \in \mathbb{R}^n$. For example, if ω is a rationally independent vector, the orbits are dense on the torus and describe quasi-periodic motions of the system. Orbits with $\omega(y) = j \cdot \alpha_0$ for $j \in \mathbb{Z}^n$ are evidently periodic. These periodic solutions have Floquet multipliers all equal to 1 and one expects, therefore, that the integrability is destroyed immediately under the slightest perturbation of the system. Indeed integrable systems are very rare, hence of particular interest; we refer to J. Moser [166] for this topic.

There are, however, many important systems in physics which are close to integrable systems in the sense that

$$|H - h_0|_{C^k(T^n \times D)} \leq \varepsilon,$$

is small, where $H(x,y)$ is a function on $T^n \times D$. In general, the integrability is lost under perturbation and the orbit structure of X_H is extremely intricate. Nevertheless, the celebrated K.A.M.-theory states that many of the quasi-periodic solutions of X_{h_0} can be continued to quasi-periodic solutions of X_H provided only that ε is sufficiently small, the number k of derivatives is sufficiently large and the integrable system is nondegenerate in the sense that

(4.13) $$\det\left(\frac{\partial^2}{\partial y^2} h_0(y)\right) \neq 0.$$

This condition requires the system h_0 to be nonlinear, so that the frequencies depend on the amplitudes. In sharp contrast to these analytically very difficult perturbation results for the existence of special solutions, we shall establish global periodic solutions for every system $H(x,y)$ on $T^n \times \mathbb{R}^n$ assuming, however, that the energy surfaces are compact. This will follow from the following result for the capacities on $T^*(T^n)$ due to Mei-Yue Jiang [125].

Proposition 4. Let $M = T^n \times \mathbb{R}^n$ with $T^n = \mathbb{R}^n/2\pi\mathbb{Z}^n$, equipped with the canonical symplectic structure $\omega = d(\sum_{j=1}^n y_j d\vartheta_j)$. Then every capacity function c is finite on open and bounded subsets of M. Moreover, for every $a > 0$,
$$c(T^n \times (-a,a)^n) \;=\; 4\pi a \;=\; c(T^n \times (-a,a) \times \mathbb{R}^{n-1}).$$

Proof. The proof follows readily from the axioms of a capacity. We start with the symplectic diffeomorphism φ in 2 dimensions:
$$\varphi : S^1 \times (-a,a) \;\longrightarrow\; A = \left\{ (x,y) \in \mathbb{R}^2 \,\middle|\, a < x^2 + y^2 < 5a \right\}$$
given by $y = \sqrt{3a+2r}\cos\vartheta$, $x = \sqrt{3a+2r}\sin\vartheta$, where $0 \le \vartheta \le 2\pi$ and $-a < r < +a$. Then $\varphi^*(dy \wedge dx) = dr \wedge d\vartheta$. Taking a tensor product we find a symplectic diffeomorphism $T^n \times (-a,a)^n \to A \times A \times \cdots \times A \subset \mathbb{R}^{2n}$, where the product of the annuli corresponds to the symplectic splitting $\mathbb{R}^{2n} = \mathbb{R}^2 \oplus \cdots \oplus \mathbb{R}^2$. Therefore it remains to show that $c(A \times A \times \cdots \times A) = 4\pi a$ which is the area of the annulus A. For $\varepsilon > 0$ there exist symplectic embeddings
$$D(2\sqrt{a}-\varepsilon) \;\longrightarrow\; A \;\longrightarrow\; D(2\sqrt{a}+\varepsilon),$$
where $D(R)$ is the two-disk of radius R. They give rise to symplectic embeddings for the product and we conclude that
$$\pi(2\sqrt{a}-\varepsilon)^2 \;\le\; c(A \times A \times \cdots \times A) \;\le\; \pi(2\sqrt{a}+\varepsilon)^2.$$
Since this holds true for every $\varepsilon > 0$ the first equation in the theorem follows; the second one is proved the same way. ∎

In particular, the special capacity c_0 is finite on bounded open sets of $T^*(T^n)$ and we can apply all the qualitative existence results obtained so far. We conclude, for example, that a compact hypersurface $S \subset T^*(T^n)$ which is of contact type always carries a closed characteristic. One can verify, as in the lemma above, that every compact and regular energy surface of a classical system of the form $H = \frac{1}{2}|y|^2 + V(x)$ on $(x,y) \in T^n \times \mathbb{R}^n$, where V is periodic in all its variables, is of contact type and hence possesses a periodic solution of X_H. Nothing is known, however, about the nature of this solution. To which homotopy class of loops on $T^*(T^n)$ does it belong? Is it the brake orbit already guaranteed by Theorem 6? Special periodic solutions for such classical systems can be found by means of the Maupertuis-Jacobi principle, as is shown in the survey article [129] by V.V. Koslov (1985).

4.5 The torus and Herman's Non-Closing Lemma

If $\psi : S \times (-1, +1) \to U \subset M$ is a parameterized family of hypersurfaces modelled on S in a symplectic manifold (M, ω) then U is the union of hypersurfaces $S_\varepsilon = \psi(S \times \{\varepsilon\})$, and we know that

$$c_0(U, \omega) < \infty \quad \Longrightarrow \quad \text{there is an } \varepsilon \text{ with } \mathcal{P}(S_\varepsilon) \neq \emptyset.$$

Conversely, of course, if $\mathcal{P}(S_\varepsilon) = \emptyset$ for every ε then $c_0(U, \omega) = \infty$ and, by the monotonicity property of the capacity $c_0(M, \omega) = +\infty$. We shall next describe an example due to E. Zehnder[231] which illustrates this situation. We consider the manifold

$$M = T^3 \times [0, 1], \quad \text{where} \quad T^3 = \mathbb{R}^3/\mathbb{Z}^3$$

is the 3-dimensional torus and define a distinguished symplectic structure on M by slightly modifying the standard structure ω_0 on $\mathbb{R}^4$. Recall that to every constant matrix A satisfying $\det A \neq 0$ and $A^T = -A$, we can associate the symplectic structure ω by

$$\omega(X, Y) = \langle AX, Y \rangle$$

on $\mathbb{R}^4$, and then, of course, also on M. A Hamiltonian function $H : M \to \mathbb{R}$ is simply a function $H : \mathbb{R}^3 \times [0, 1] \to \mathbb{R}$ such that $H(x_1, x_2, x_3, x_4)$ is periodic in x_1, x_2, x_3 of period 1. The Hamiltonian vector field X_H is, as usual, defined by

$$\omega(X_H, Y) = -dH(Y)$$

for all $Y \in \mathbb{R}^4$, and one obtains

$$X_H = -A^{-1} \nabla H.$$

We now define A by setting $-A^{-1} = J^*$, and

(4.14) $$J^* = \begin{pmatrix} 0 & 1 & 0 & \alpha_1 \\ -1 & 0 & 0 & \alpha_2 \\ 0 & 0 & 0 & 1 \\ -\alpha_1 & -\alpha_2 & -1 & 0 \end{pmatrix},$$

with two real numbers $\alpha_1, \alpha_2 \in \mathbb{R}$. Then $\det J^* = 1$ and the symplectic form $\omega = \omega^*$ is the following 2-form:

(4.15) $\omega^* = dx_2 \wedge dx_1 + dx_4 \wedge dx_3 + \alpha_1 \, dx_3 \wedge dx_2 + \alpha_2 \, dx_1 \wedge dx_3.$

Choosing the Hamiltonian function $H_0 : M \to \mathbb{R}$ defined as

(4.16) $$H_0(x) = x_4,$$

the Hamiltonian vector field on M, given by

$$X_{H_0} = J^* \nabla H_0 = J^* e_4 = (\alpha_1, \alpha_2, 1, 0),$$

is constant. The energy surfaces

(4.17) $$H_0^{-1}(c) = \left\{ x \in M \mid H_0(x) = x_4 = c \right\}$$

are regular and, since $x_j \in S^1$ for $1 \leq j \leq 3$, equal to the 3-dimensional tori $H_0^{-1}(c) = T^3 \times \{c\}$. We introduce the abbreviation

(4.18) $$\alpha = (\alpha_1, \alpha_2, 1) \in \mathbb{R}^3.$$

The flow of the Hamiltonian equations restricted to the energy surfaces $T^3 \times \{c\}$ is linear and determined by

(4.19) $$\dot{x}_k = \alpha_k, \quad 1 \leq k \leq 3.$$

If we choose α rationally independent, requiring

$$\langle \alpha, j \rangle \neq 0 \quad \text{for} \quad 0 \neq j \in \mathbb{Z}^3,$$

then all the orbits of (4.19) are dense on T^3, in view of Kronecker's theorem. Consequently, we have an example of a Hamiltonian system whose energy surfaces are all regular and compact, which, however, does not admit any periodic solution. Incidentally, the energy surfaces are not of contact type.

As an aside, we remark that one can easily construct an embedding $\varphi : T^3 \times [0,1] \to \mathbb{R}^4$. The induced parameterized family of hypersurfaces modelled on T^3 carries no closed characteristic with respect to the pushed forward symplectic structure which we denote again by ω^*. However, ω^* on $\varphi(M) \subset \mathbb{R}^4$ is not equivalent to the standard structure $(\varphi(M), \omega_0)$ induced from $\mathbb{R}^4$. Indeed, ω_0 is an exact symplectic structure, while ω^* is not; moreover, $c_0(\varphi(M), \omega^*) = +\infty$ while $c_0(\varphi(M), \omega_0) < \infty$. In particular, this example fails to represent a counterexample to the conjecture that every hypersurface $S \subset (\mathbb{R}^{2n}, \omega_0)$ has a closed characteristic. We note that our vector field X on T^3 defined by (4.19) satisfies $\beta(X) \neq 0$ for the closed 1-form $\beta = dx_1$ on T^3. Assume now, more generally, that S is any compact manifold without boundary of dim $S = 2n - 1$ and that X is a vector field on S satisfying $\beta(X) \neq 0$ for a closed 1-form β on S. Then S cannot be embedded as a hypersurface $S \subset (M, \omega)$ in any symplectic manifold having an exact symplectic structure, i.e., $\omega = d\lambda$, such that $X \in \mathcal{L}_S$, i.e., belongs to the characteristic line bundle. Indeed, suppose it could be, then we could conclude $\beta \wedge (d\lambda)^{n-1}$ to be a volume form on S, since $\beta(X) \neq 0$ and $X \in \ker d\lambda|S$, so that

$$\text{vol}(S) = \int_S \beta \wedge (d\lambda)^{n-1} \neq 0.$$

4.5 THE TORUS AND HERMAN'S NON-CLOSING LEMMA

This, however, is not possible, since the form $\beta \wedge (d\lambda)^{n-1}$ is equal to $d(\lambda \wedge \beta \wedge (d\lambda)^{n-2})$ and hence is an exact form on S.

We now restrict the symplectic structure ω^* even further by requiring that α be not only rationally independent but satisfies, in addition, the diophantine conditions

(D.C.) $$|\langle \alpha, j \rangle| \geq \gamma |j|^{-\tau}$$

for all $0 \neq j \in \mathbb{Z}^3$ with two constants $\gamma > 0$ and $\tau > 2$. Almost every vector $\alpha \in \mathbb{R}^3$ satisfies such conditions. It then turns out that the above example is actually dynamically stable under perturbations of H_0 small with sufficiently many derivatives. This surprising phenomenon was recently discovered by M. Herman [110, 111], see also J.C. Yoccoz [229]. In our setting he proved

Theorem 10. (M. Herman's non-closing lemma) Consider the symplectic manifold $(M, \omega^*) = (T^3 \times [0,1], \omega^*)$ with the symplectic structure (4.15) and $\alpha \in \mathbb{R}^3$ satisfying the conditions (D.C.) Let $H_0(x) = x_4$. Then there exists a neighborhood W of H_0

$$W = \left\{ H \in C^k(M) \,\middle|\, |H - H_0|_{C^k} \leq \delta \right\},$$

for some small $\delta > 0$ and $k > \tau + 2$, such that for every $H \in W$ and every c in $\frac{1}{2} \leq c \leq \frac{3}{2}$, the regular energy surface

$$H^{-1}(c) \subset M$$

has no periodic solution for X_H.

We note that this dynamical rigidity phenomenon is a property of the special symplectic structure ω^*. Postponing the proof we first point out an important consequence of Herman's theorem.

In view of Poincaré's recurrence theorem (see Chapter 1) almost every point x on a compact energy surface for a Hamiltonian system H, is a recurrent point and one may ask whether there exists a nearby system having a periodic trajectory passing near x. This is the so-called Closing Problem. The answer depends on the notion of what is "nearby". In their work [179] (1983) C.C. Pugh and C. Robinson showed for the Hamiltonian case, that there is a sequence of Hamiltonian systems, H_j and a sequence of corresponding periodic points, x_j for these approximate systems satisfying

$$H_j \to H \quad \text{in} \quad C^2 \quad \text{and} \quad x_j \to x.$$

This is a special case of the celebrated C^1-Closing Lemma in [184]. It is a long standing open question whether H_j can be chosen so that $H_j \to H$ in C^k for $k > 2$. The surprising answer of M. Herman to this question is that on the special manifold (M, ω^*) the C^k-Closing Lemma is false, if k is sufficiently large, namely, roughly larger than the dimension of the manifold in question.

Proof of Theorem 10. Consider the neighborhood of H_0 given by

$$H = x_4 + h(x) \quad \text{and} \quad |h|_{C^k} \quad \text{small}$$

on M. The Hamiltonian vector field X_H is represented as

$$(4.20) \quad X_H(x) = J^* \nabla H(x) = \begin{pmatrix} H_{x_2} + \alpha_1 H_{x_4} \\ -H_{x_1} + \alpha_2 H_{x_4} \\ H_{x_4} \\ -H_{x_3} - \alpha_1 H_{x_1} - \alpha_2 H_{x_2} \end{pmatrix}.$$

The energy surface for c, defined by

$$x_4 + h(x) = H(x) = c,$$

is an embedded torus T^3. Indeed we can solve for x_4, and find $x_4 = K(x_1, x_2, x_3, c)$, where the function K satisfies the identity

$$(4.21) \quad H\big(x_1, x_2, x_3, K(x_1, x_2, x_3, c)\big) = c$$

for all $x \in T^3$. The functions are periodic in $x \in T^3$. To compute the vector field X_H on $H^{-1}(c)$ we differentiate the identity (4.21) and find $H_{x_j} = -H_{x_4} \cdot K_{x_j}$ on $H^{-1}(c)$ for $1 \leq j \leq 3$. Therefore, in view of (4.20) the vector field X_H is, on $H^{-1}(c)$ given, by

$$(4.22) \quad \begin{pmatrix} \dot{x}_1 \\ \dot{x}_2 \\ \dot{x}_3 \end{pmatrix} = \rho \cdot \begin{pmatrix} \alpha_1 - K_{x_2} \\ \alpha_2 + K_{x_1} \\ 1 \end{pmatrix} \quad \text{on } T^3,$$

where $\rho(x) = H_{x_4}(x, K)$ is a positive function. The vector field (4.22) has, up to reparameterization, the same orbits as

$$(4.23) \quad \begin{pmatrix} \dot{x}_1 \\ \dot{x}_2 \\ \dot{x}_3 \end{pmatrix} = V(x) := \begin{pmatrix} \alpha_1 - K_{x_2} \\ \alpha_2 + K_{x_1} \\ 1 \end{pmatrix} \quad \text{on } T^3.$$

Note that this vector field V on T^3 is a perturbation of the constant vector field $\dot{x} = \alpha$ on T^3, and we can apply a special case of the K.A.M.-theory. We make use of the following result due to J. Moser 1966 in [161].

4.5 THE TORUS AND HERMAN'S NON-CLOSING LEMMA

Theorem 11. On the torus $T^n = \mathbb{R}^n/\mathbb{Z}^n$ consider a vector field of the form

$$\dot{x} = \alpha + f(x),$$

and assume the constant vector $\alpha \in \mathbb{R}^n$ to satisfy the diophantine conditions

$$|\langle \alpha, j \rangle| \geq \gamma |j|^{-\tau},$$

for all $0 \neq j \in \mathbb{Z}^n$ with two constants $\gamma > 0$ and $\tau > n - 1$. Then, if $|f|_{C^m}$ is sufficiently small and $m > \tau + 1$, there exist a constant vector $\lambda \in \mathbb{R}^n$ and a C^1-diffeomorphism $u = id + v : T^n \to T^n$ near the identity, such that

$$(du)^{-1}(\alpha + f + \lambda) \circ u = \alpha.$$

In other words, if a vector field is near the constant vector field α together with sufficiently many derivatives, it can be modified by constants $\lambda \in \mathbb{R}^n$, so that this modified vector field is equivalent to α. The proof of this seminal small denominator result is based on an analytically subtle iteration technique in infinitely many spaces and will not be carried out here; we refer to [161].

In general $\lambda \neq 0$, as already the example $f = $ const shows. But in our case the vector field f has an additional structure inherited from the Hamiltonian nature of the problem and this additional structure allows us to conclude that $\lambda = 0$. This will follow from the

Lemma. Let V be a vector field on the torus T^n satisfying div $V = 0$. Assume

$$(du)^{-1} \cdot V \circ u = \alpha,$$

with an irrational vector $\alpha \in \mathbb{R}^n$, and a diffeomorphism $u = id + v$ of the torus. Then u preserves the Lebesgue measure on T^n and

$$\int_{T^n} V = \alpha.$$

Postponing the proof of the lemma we first prove Herman's result. Abbreviating the vector field on the right hand side of (4.23) by $\alpha + f(x)$, we first conclude from Theorem 11 that there exists a $\lambda \in \mathbb{R}^3$ such that the vector field

$$V := \alpha + f(x) + \lambda$$

is transformed into the constant vector field $\dot{x} = \alpha$ on T^3. The vector field f has a special form inherited from the Hamiltonian structure. Namely $\int f = 0$, since the integrations of derivatives of periodic functions vanish. Consequently,

$$\int_{T^3} V = \alpha + \lambda.$$

In addition, V satisfies div $V = 0$. Therefore, we conclude by the lemma that $\alpha + \lambda = \alpha$ and thus $\lambda = 0$. We have proved that the flow of the vector field (4.23) on T^3 is conjugated to the linear flow of $\dot{x} = \alpha$ and hence has no periodic solutions. This holds true for every constant c in $\frac{1}{2} \leq c \leq \frac{2}{3}$. The proof of Theorem 10 is finished, and it remains to prove the lemma.

Integrating $V \circ u = du(\alpha)$ and observing that $u = id + v$ we find

$$\int_{T^n} V \circ u \, d\mu = \alpha$$

since the integral over $dv(\alpha)$ vanishes. In order to prove the lemma we only have to show that $u^*\mu = \mu$ i.e., preserves the Lebesgue measure. The flow ψ^t of the vector field V satisfies

$$(\psi^t)^*\mu = \mu.$$

This is a consequence of the assumption div $V = 0$. Moreover

$$u^{-1} \circ \psi^t \circ u = R_{t\alpha}, \quad t \in \mathbb{R}$$

with the translations $R_{t\alpha} : x \mapsto x + t\alpha$ on the torus. Therefore $\psi^t \circ u = u \circ R_{t\alpha}$ and in view of $(\psi^t)^*\mu = \mu$ we find

$$u^*\mu = (R_{t\alpha})^*(u^*\mu), \quad t \in \mathbb{R}.$$

By assumption, α is irrational so that all the orbits $t \mapsto R_{t\alpha}(x)$ are dense on T^n. Since μ is a constant form we conclude that $u^*\mu = a \cdot \mu$ with a constant a. Integrating over the torus

$$\int_{T^n} \mu = \int_{T^n} u^*\mu = a \int_{T^n} \mu$$

we find $a = 1$, proving that indeed $u^*\mu = \mu$. This finishes the proof of the lemma. ∎

Evidently, also the symplectic manifold (T^4, ω^*) has infinite capacity c_0

$$c_0(T^4, \omega^*) = +\infty$$

if ω^* is the particular symplectic structure defined by (4.15) with $\alpha = (\alpha_1, \alpha_2, 1)$ chosen to be irrational. We know that in the 2-dimensional case $c_0(T^2, \omega) < \infty$ for every volume form ω. It is an open problem whether

$$c_0(T^{2n}, \omega_0) < \infty$$

if $n > 1$. Here ω_0 is the standard symplectic structure induced from $\mathbb{R}^{2n}$.

Chapter 5
Geometry of compactly supported symplectic mappings in $\mathbb{R}^{2n}$

In this chapter we shall study the group $\mathcal{D}$ of those symplectic diffeomorphisms of $\mathbb{R}^{2n}$ which are generated by time-dependent Hamiltonian vector fields having compactly supported Hamiltonians. We shall construct, in particular, an astonishing bi-invariant metric d on $\mathcal{D}$, following H. Hofer [116]. Defining the energy $E(\psi)$ of an element $\psi \in \mathcal{D}$ by means of the oscillations of generating Hamiltonians, the metric d will be defined by $d(\varphi, \psi) = E(\varphi^{-1}\psi)$. It is of C^0-nature. The verification of the property that $d(\varphi, \psi) = 0$ if and only if $\varphi = \psi$, is not easy. It is based on the action principle. The metric d is intimately related, on the one hand, to the capacity function c_0 introduced in Chapter 3 and hence to periodic orbits and, on the other hand, to a special capacity e defined on subsets of $\mathbb{R}^{2n}$ and satisfying $e(U) \geq c_0(U)$. The capacity e is called the displacement energy: $e(U)$ measures the distance between the identity map and the set of $\psi \in \mathcal{D}$ which displaces U from itself, in the sense that $\psi(U) \cap U = \emptyset$. A crucial role in our considerations will be played by the action spectrum, $\sigma(\psi)$, of the fixed points of an element $\psi \in \mathcal{D}$. It turns out to be a compact nowhere dense subset of $\mathbb{R}$. In contrast to the simple variational technique used for a fixed Hamiltonian in Chapter 3, a minimax principle will be designed which is applicable simultaneously to all Hamiltonians generating the elements of $\mathcal{D}$. It singles out a distinguished action $\gamma(\varphi) \in \sigma(\varphi)$. The map $\gamma : (\mathcal{D}, d) \to \mathbb{R}$ is a continuous section of the action spectrum bundle over $\mathcal{D}$; it is the main technical tool in this section. Its properties will also be used in order to establish infintely many periodic orbits for certain elements of $\mathcal{D}$, and to describe the geodesics of $(\mathcal{D}, d)$. We mention that the completion of the group $\mathcal{D}$ with respect to the metric d is not understood.

5.1 A special metric d for a group $\mathcal{D}$ of Hamiltonian diffeomorphisms

We consider the symplectic space $(\mathbb{R}^{2n}, \omega)$ with the standard symplectic structure $\omega = \omega_0$ and denote by $\mathcal{H}$ the vector space of all smooth and compactly supported Hamiltonian functions $H = H(t, x) : [0, 1] \times \mathbb{R}^{2n} \to \mathbb{R}$. Associated to every $H \in \mathcal{H}$ is a time-dependent Hamiltonian vector field X_H on $\mathbb{R}^{2n}$ defined by

$$i_{X_{H_t}}\omega = -dH_t, \text{ where } H_t(x) = H(t, x).$$

Since the vector field X_H has compact support the Hamiltonian equations
$$\dot{x} = X_H(t,x), \quad x(0) = x_0 \in \mathbb{R}^{2n}$$
can be solved over the whole time interval $[0,1]$ for every given initial value $x_0 \in \mathbb{R}^{2n}$. We thus obtain a 1-parameter family of symplectic mappings φ_H^t for $t \in [0,1]$. It is defined by
$$\varphi_H^t(x_0) = x(t),$$
where $x(t)$ solves the equation for the initial value x_0. Clearly $\varphi_H^t(x) = x$ if $|x|$ is sufficiently large. By
$$\varphi_H = \varphi_H^1$$
we shall denote the time-1 map of the flow φ_H^t and define the group $\mathcal{D}$ of compactly supported Hamiltonian diffeomorphisms by
$$\mathcal{D} = \{\varphi_H \mid H \in \mathcal{H}\}.$$
Recall that the support of a map φ is defined as the closed set $\operatorname{supp}(\varphi) = \operatorname{clos}\{x \mid \varphi(x) \neq x\}$. Every compactly supported symplectic diffeomorphism can be interpolated this way by the flow of a time-dependent Hamiltonian vector field; however, it is not known whether the Hamiltonian function can be chosen to be in $\mathcal{H}$ if $n > 2$. It follows from results of M. Gromov [107] that in the special case of $\mathbb{R}^4$ the group $\mathcal{D}$ is contractible. In particular, every compactly supported symplectic diffeomorphism φ in $\mathbb{R}^4$ can be written as $\varphi = \varphi_H$ with $H \in \mathcal{H}$ and, therefore, belongs to $\mathcal{D}$.

It is the aim of this chapter to study some geometric properties of this group $\mathcal{D}$ of Hamiltonian mappings on $\mathbb{R}^{2n}$. We begin with two preliminary formal statements which will be useful later on.

Definition. Given $H, K \in \mathcal{H}$ and $\vartheta \in \mathcal{D}$ we define the functions $\overline{H}, H \# K$ and H_ϑ in $\mathcal{H}$ as follows:
$$\overline{H}(t,x) = -H(t, \varphi_H^t(x))$$
$$(H \# K)(t,x) = H(t,x) + K(t, (\varphi_H^t)^{-1}(x))$$
$$H_\vartheta(t,x) = H(t, \vartheta^{-1}(x)).$$

Proposition 1. For $H, K \in \mathcal{H}$ the following formulae hold true
$$\varphi_{\overline{H}}^t = (\varphi_H^t)^{-1}$$
$$\varphi_{H \# K}^t = \varphi_H^t \circ \varphi_K^t$$
$$\vartheta \circ \varphi_H^t \circ \vartheta^{-1} = \varphi_{H_\vartheta}^t$$
$$(\varphi_H^t)^{-1} \circ \varphi_K^t = \varphi_G^t$$
where $G = \overline{H} \# K = (K - H)(t, \varphi_H^t)$.

5.1 A SPECIAL METRIC d FOR A GROUP $\mathcal{D}$ ...

Proof. The third formula is the transformation law of Hamiltonian vector fields under symplectic transformations, as we know from the introduction. The first formula follows from the second one while the last one is a consequence of the first two. To prove the second formula, we abbreviate the notation and observe that

$$\frac{d}{dt}\varphi^t = X_H \circ \varphi^t \text{ and } \varphi^0 = id$$

$$\frac{d}{dt}\psi^t = X_K \circ \psi^t \text{ and } \psi^0 = id.$$

We have to show that $\varphi^t \circ \psi^t$ is the flow of $X_{H\#K}$. Differentiating, we find

$$\frac{d}{dt}(\varphi^t \circ \psi^t) = \left(\frac{d}{dt}\varphi^t\right) \circ \psi^t + (d\varphi^t \circ \psi^t) \cdot \frac{d}{dt}\psi^t$$

$$= X_H(\varphi^t \circ \psi^t) + \left[d\varphi^t \circ (\varphi^t)^{-1} \circ \varphi^t \circ \psi^t\right] \cdot X_K \circ \left[(\varphi^t)^{-1} \circ \varphi^t \circ \psi^t\right].$$

By the transformation law of Hamiltonian vector fields the second term is equal to

$$X_{K \circ (\varphi^t)^{-1}} \circ (\varphi^t \circ \psi^t),$$

and the claim is proved. ■

By definition, a smooth arc in $\mathcal{D}$ is a map $t \mapsto \psi^t$ from $I = [0, 1]$ into $\mathcal{D}$ such that $(t, x) \mapsto \psi^t(x)$ is a smooth mapping from $I \times \mathbb{R}^{2n}$ into $\mathbb{R}^{2n}$ and such that there exists a compact set in $\mathbb{R}^{2n}$ containing all the supports of ψ^t for $t \in I$. We shall denote by $\mathcal{A}$ the set of smooth arcs $[0, 1] \to \mathcal{D} : t \mapsto \varphi^t$ satisfying $\varphi^0 = id$.

Proposition 2. *The map $H \mapsto \psi_H^t$ from $\mathcal{H}$ into $\mathcal{A}$ is a bijection onto $\mathcal{A}$.*

Proof. If H and K satisfy $\varphi_H^t = \varphi_K^t$ for $t \in I$ we find by differentiation that $X_H = X_K$. Therefore, $dH_t = dK_t$ so that $H(t, x) = K(t, x)$ since the functions have compact support. This proves the injectivity of the map; in order to prove the surjectivity we consider a smooth arc $t \mapsto \psi^t$ in $\mathcal{D}$ satisfying $\psi^0 = id$, and define the time-dependent vector field

$$X_t := \frac{d}{dt}\psi^t \circ (\psi^t)^{-1},$$

so that

$$\frac{d}{dt}\psi^t = X_t(\psi^t) \text{ and } \psi^0 = id.$$

Because of $(\psi^t)^*\omega = \omega$, we have $L_{X_t}\omega = 0$. In view of $d\omega = 0$ we find $d(i_{X_t}\omega) = 0$. The 1-form $i_{X_t}\omega$ being closed is exact on $\mathbb{R}^{2n}$, by the Poincaré lemma. There is a function $H : I \times \mathbb{R}^{2n} \to \mathbb{R}$, determined by

$$H(t, x) = -\int_0^1 \omega(X_t(sx), x) ds,$$

which satisfies $-dH_t = i_{X_t}\omega$. Since X_t has compact support, $dH_t(x) = 0$, for $t \in I$ and $|x| \geq R$, for some large R and there is a $K \in \mathcal{H}$ with $dH_t = dK_t$. It is given by $H(t,x) - H(t,x^*)$ for some x^* with $|x^*| \geq R$. Consequently $X_t = X_{K_t}$ as claimed. ∎

We next define the length of an arc $t \mapsto \psi^t$ belonging to $\mathcal{A}$. In view of Proposition 2 there is a unique $H \in \mathcal{H}$ generating this arc by

$$\psi^t = \varphi_H^t.$$

Define for $H \in C_c^\infty(\mathbb{R}^{2n}, \mathbb{R})$ the norm (called the oscillation)

(5.1) $$||H|| = \sup_x H(x) - \inf_x H(x).$$

Then

$$||H \circ \varphi|| = ||H||$$

for every diffeomorphism φ of $\mathbb{R}^{2n}$. If now $H \in \mathcal{H}$, we can associate with the Hamiltonian vector field X_{H_t} the norm $||X_{H_t}|| := ||H_t||$ and find

$$\int_0^1 ||\dot{\psi}^t||\, dt = \int_0^1 ||X_{H_t}(\psi^t)||\, dt = \int_0^1 ||H_t||\, dt.$$

Introducing the norm in $\mathcal{H}$ by

(5.2) $$||H|| = \int_0^1 ||H_t||\, dt = \int_0^1 [\sup_x H(t,x) - \inf_x H(t,x)]\, dt,$$

which agrees for autonomous H with definition (5.1), we are led to define the length of the arc $\psi^t = \varphi_H^t$ connecting ψ^1 with id, as

(5.3) $$l(\psi^{[0,1]}) = ||H||.$$

Definition. The energy of a symplectic map ψ in $\mathcal{D}$ is the number $E(\psi) \geq 0$ defined by

$$E(\psi) = \inf\{l(\psi^{[0,1]}) \mid \text{the arc } \psi^t \text{ in } \mathcal{A} \text{ satisfies } \psi^1 = \psi\}.$$

Evidently

(5.4) $$E(\varphi_H) \leq ||H||.$$

The Hamiltonian $H = 0$ in $\mathcal{H}$ generates the identity map $\psi = id$ and consequently $E(id) = 0$. Conversely,

(5.5) $$E(\psi) > 0, \text{ if } \psi \neq id.$$

But this crucial fact is not at all obvious. It follows from the following estimates for the energy function in which $c_0(U) = c_0(U, \omega)$ denotes the special capacity introduced in Chapter 3. For notational convenience we set $c_0(\emptyset) = 0$, and recall that $Z(R)$ denotes the symplectic cylinder of radius R.

5.1 A special metric d for a group $\mathcal{D}$ ...

Theorem 1. (energy-capacity inequality) For every $\psi \in \mathcal{D}$

$$\sup\{c_0(U) \mid U \text{ open and } \psi(U) \cap U = \emptyset\} \leq E(\psi)$$

$$E(\psi) \leq 16 \inf\{\pi R^2 \mid \text{there is a } \varphi \in \mathcal{D} \text{ with supp } (\varphi\psi\varphi^{-1}) \subset Z(R)\}.$$

We see, in particular, that the energy $E(\psi)$ is always larger than the capacity $c_0(U)$ of every open set U which can be displaced from itself by the map ψ. It is now easy to deduce the claim (5.5): indeed taking a point x^* satisfying $\psi(x^*) \neq x^*$ there exists an open ball $B(x^*, \varepsilon) = U$ of radius $\varepsilon > 0$, which is displaced from itself, $\psi(U) \cap U = \emptyset$. Using $c_0(U) = \pi\varepsilon^2$ we thus conclude from Theorem 1 that $E(\psi) \geq \pi\varepsilon^2 > 0$ proving the claim (5.5).

The proof of the first estimate in Theorem 1 will be based on the existence of a distinguished critical point of the action functional, it will be carried out in the Sections 5.5 and 5.6 below.

Theorem 2. (C^0-energy estimate) For every $\psi \in \mathcal{D}$

$$E(\psi) \leq C \text{ diameter supp } (\psi) \, |id - \psi|_{C^0}$$

with a constant $C \leq 128$.

This statement expresses the continuity of E at the identity map in the C^0-topology. The proof of Theorem 2 will also be postponed to Section 5.6 below.

Proposition 3. The energy function $E : \mathcal{D} \to \mathbb{R}$ meets the following properties:

(i) $\quad E(\varphi) \geq 0$, and $E(\varphi) = 0 \iff \varphi = id$

(ii) $\quad E(\varphi) = E(\varphi^{-1})$

(iii) $\quad E(\vartheta \varphi \vartheta^{-1}) = E(\varphi)$

(iv) $\quad E(\varphi\psi) \leq E(\varphi) + E(\psi)$,

where $\varphi, \psi \in \mathcal{D}$ and where ϑ is a symplectic diffeomorphism of $\mathbb{R}^{2n}$.

Proof. The proofs of (i) and (ii) follow readily from the definition of E using the nontrivial fact (5.5) proved later on. Clearly

$$l(\varphi_H^{[0,1]}) = l(\varphi_{\overline{H}}^{[0,1]})$$

implying $E(\varphi) = E(\varphi^{-1})$. From

$$\vartheta \circ \varphi_H^t \circ \vartheta^{-1} = \varphi_{H_\vartheta}^t$$

we find

$$l((\vartheta \circ \varphi_H \circ \vartheta^{-1})^{[0,1]}) = l(\varphi_{H_\vartheta}^{[0,1]})$$

implying (iii). Finally, in view of Proposition 1, we have $\varphi_H^t \circ \varphi_K^t = \varphi_{H\#K}^t$ and

$$l(\varphi_{H\#K}^{[0,1]}) = \int_0^1 \{\sup_x (H\#K)_t - \inf_x (H\#K)_t\} \, dt$$

$$\leq \int_0^1 (\sup_x H_t - \inf_x H_t) \, dt + \int_0^1 (\sup_x K_t - \inf_x K_t) \, dt$$

$$= l(\varphi_H^{[0,1]}) + l(\varphi_K^{[0,1]})$$

which implies $E(\varphi_H \circ \varphi_K) \leq E(\varphi_H) + E(\varphi_K)$ and the last claim (iv) follows. ∎

We can derive an estimate for the commutators in $\mathcal{D}$, denoted by

(5.6) $$[\varphi, \psi] := \varphi \psi \varphi^{-1} \psi^{-1},$$

and first claim that

(5.7) $$E([\varphi, \psi]) \leq 2 \min\{E(\varphi), E(\psi)\}$$

for $\varphi, \psi \in \mathcal{D}$. Indeed, in view of the statements (ii)–(iv) in Proposition 3 we can estimate $E(\varphi \psi \varphi^{-1} \psi^{-1}) \leq E(\varphi \psi \varphi^{-1}) + E(\psi^{-1}) \leq 2E(\psi)$ and $E(\varphi \psi \varphi^{-1} \psi^{-1}) \leq E(\varphi) + E(\psi \varphi^{-1} \psi^{-1}) = 2E(\varphi)$, hence proving the claim. We now deduce an "a priori" estimate for the energy of commutators of those maps having their supports in a fixed open set.

Proposition 4. If $U \subset \mathbb{R}^{2n}$ is open and bounded and if $\vartheta \in \mathcal{D}$ satisfies $\vartheta(U) \cap U = \emptyset$, then

$$E([\varphi, \psi]) \leq 4E(\vartheta)$$

for all $\varphi, \psi \in \mathcal{D}$ with $\text{supp}(\varphi)$ and $\text{supp}(\psi)$ contained in U.

Proof. Define

$$\gamma := \varphi \vartheta^{-1} \varphi^{-1} \vartheta \in \mathcal{D}$$

then $\gamma|U = \varphi|U$ since $\vartheta(U) \cap U = \emptyset$ and $\text{supp}(\varphi) \subset U$. Consequently $\varphi \psi \varphi^{-1} \psi^{-1} = \gamma \psi \gamma^{-1} \psi^{-1}$ on U, and since both maps are the identity maps outside of U, they are equal on $\mathbb{R}^{2n}$. Therefore, in view of (5.7) and Proposition 3

$$E([\varphi, \psi]) = E([\gamma, \psi]) \leq 2E(\gamma) \leq 2E(\vartheta^{-1}) + 2E(\vartheta) = 4E(\vartheta)$$

as we wanted to show. ∎

Proposition 3 allows us to define a distinguished metric $d : \mathcal{D} \times \mathcal{D} \to [0, \infty]$, by means of the energy:

(5.8) $$d(\varphi, \psi) := E(\varphi^{-1} \psi).$$

In particular, the energy of $\psi \in \mathcal{D}$, $E(\psi) = d(\text{id}, \psi)$, is the distance from the identity map and we remark that, in view of the last formula in Proposition 1,

(5.9) $$d(\varphi, \psi) = \inf \{ \|H - K\| \mid H \text{ generates } \psi \text{ and } K \text{ generates } \varphi \}.$$

5.1 A special metric d for a group $\mathcal{D}$ ...

Theorem 3. The function d is a bi-invariant metric on $\mathcal{D}$, i.e., it satisfies for all φ, ψ and ϑ belonging to $\mathcal{D}$:

(i) $\quad d(\varphi, \psi) \geq 0$ and $d(\varphi, \psi) = 0 \Leftrightarrow \varphi = \psi$

(ii) $\quad d(\varphi, \psi) \leq d(\varphi, \vartheta) + d(\vartheta, \psi)$ and $d(\varphi, \psi) = d(\psi, \varphi)$

(iii) $\quad d(\vartheta\varphi, \vartheta\psi) = d(\varphi, \psi) = d(\varphi\vartheta, \psi\vartheta)$.

Theorem 3 is an immediate consequence of Proposition 3. In the following we can consider $(\mathcal{D}, d)$ as a topological group. Note that our intrinsic metric on $\mathcal{D}$ is continuous in the C^0-metric provided the supports are uniformly bounded. Indeed, in view of Theorem 1 and Theorem 2,

$$\sup\{c_0(U)|\ U \text{ open and } \psi(U) \cap \varphi(U) = \emptyset\} \leq d(\varphi, \psi)$$

(5.10) $\quad d(\varphi, \psi) \leq 128 \cdot \text{ diameter } (\text{ supp } \varphi^{-1}\psi) \cdot \sup_x |\varphi(x) - \psi(x)|$

for every pair φ and ψ belonging to $\mathcal{D}$. The group $\mathcal{D}$ admits, in addition, an order structure. Define the positive subset $\mathcal{D}^+ \subset \mathcal{D}$ by

$$\mathcal{D}^+ = \{\ \psi_H \mid H \in \mathcal{H} \text{ and } H \geq 0\}$$

and the negative subset $\mathcal{D}^- \subset \mathcal{D}$ by

$$\mathcal{D}^- = \{\ \psi_H \mid H \in \mathcal{H} \text{ and } H \leq 0\}.$$

If $\varphi \in \mathcal{D}^+$ then $\varphi^{-1} \in \mathcal{D}^-$ and $\vartheta\varphi\vartheta^{-1} \in \mathcal{D}^+$ for every symplectic diffeomorphism ϑ of $\mathbb{R}^{2n}$. This follows from Proposition 1. We shall prove later on in Section 5.4 that

(5.11) $\quad\quad\quad\quad\quad\quad\quad\quad \mathcal{D}^+ \cap \mathcal{D}^- = \{id\}.$

A partial order structure $\geq$ on $\mathcal{D}$ can, therefore, as we shall prove, be defined as follows:

(5.12) $\quad\quad\quad\quad\quad\quad \varphi \geq \psi \Leftrightarrow \varphi \circ \psi^{-1} \in \mathcal{D}^+.$

Proposition 5. Assume φ, ψ and $\vartheta \in \mathcal{D}$.

(i) $\quad\quad$ If $\varphi \geq \psi$ and $\psi \geq \varphi$ then $\varphi = \psi$

(ii) $\quad\quad$ If $\varphi \geq \psi$ and $\psi \geq \vartheta$ then $\varphi \geq \vartheta$.

In addition, if $\varphi \geq \psi$, then $\varphi\vartheta \geq \psi\vartheta$.

Proof. If $\varphi \circ \psi^{-1} \in \mathcal{D}^+$ and $\psi \circ \varphi^{-1} \in \mathcal{D}^+$, then $\varphi \circ \psi^{-1} \in \mathcal{D}^+ \cap \mathcal{D}^-$ and hence by (5.11), we conclude that $\varphi = \psi$ proving (i). Next observe that if φ and ψ are in $\mathcal{D}^+$ then also $\varphi \circ \psi \in \mathcal{D}^+$ in view of Proposition 1 implying (ii). ∎

We conclude this section with two observations due to Y. Eliashberg and L. Polterovich [76]. They consider an arbitrary bi-invariant pseudo-metric $\rho : \mathcal{D} \times \mathcal{D} \to [0, \infty)$. By definition this ρ satisfies the properties (i)–(iii) of Theorem 3, except that it is not required that $\rho(\varphi, \psi) \neq 0$ if $\varphi \neq \psi$. Recalling that $E(\psi) = d(id, \psi)$ and arguing as in the proof of Proposition 4, one finds that

$$(5.13) \qquad \rho(id, [\varphi, \psi]) \leq 4\rho(id, \vartheta)$$

for all φ, ψ and $\vartheta \in \mathcal{D}$ satisfying $\vartheta(U) \cap U = \emptyset$ and $\mathrm{supp}(\varphi), \mathrm{supp}(\psi) \subset U$, for some open set $U \subset \mathbb{R}^{2n}$. Following H. Hofer [116] one can associate with ρ the so-called displacement energy $e_\rho(U)$ of an open and bounded subset $U \subset \mathbb{R}^{2n}$ by defining

$$(5.14) \qquad e_\rho(U) := \inf\{ \rho(id, \vartheta) \mid \vartheta \in \mathcal{D} \text{ satisfies } \vartheta(U) \cap U = \emptyset \}.$$

It then follows from (5.13) that

$$4e_\rho(U) \geq \rho(id, [\varphi, \psi])$$

for every pair φ and $\psi \in \mathcal{D}$ having their supports contained in U. If ρ is not only a pseudometric but a metric, then $\rho(id, [\varphi, \psi]) > 0$ provided φ and ψ do not commute. Consequently $e_\rho(U) > 0$ and we have proved

Proposition 6. *If ρ is a bi-invariant metric on $\mathcal{D}$ then the displacement energy e_ρ is positive on open sets.*

This applies in particular to our distinguished bi-invariant metric $\rho = d$. Note, however, that the metric character of this d is concluded from the nontrivial estimate in Theorem 1 from which it follows that $e_d(U) \geq c_0(U) > 0$ for every open set U. The special displacement energy e_d will be studied in more detail later on.

In order to construct the intrinsic bi-invariant metric d on $\mathcal{D}$ we started off from the sup-norm $||H||$ on $C_c^\infty(\mathbb{R}^{2n})$. We can, of course, start off as well from the L^p-norm

$$||H||_p := \left(\int |H(x)|^p \, dx \right)^{\frac{1}{p}}$$

which is evidently invariant under $\mathcal{D}$, i.e., $||H \circ \varphi||_p = ||H||_p$ if $\varphi \in \mathcal{D}$. We thus arrive at $d_p(\varphi, \psi) = E_p(\varphi \psi^{-1})$ which defines a bi-invariant pseudometric on $\mathcal{D}$. It turns out, however, that this d_p for $1 \leq p < \infty$ is not a metric. This has recently been proved by Y. Eliashberg and L. Polterovich [76]. They showed, more precisely, that

$$E_p(\psi) = 0, \text{ if } p = 1,$$

for every $\psi \in \mathcal{D}$ having vanishing Calabi invariant, i.e.,

$$\int_0^1 \int_{\mathbb{R}^{2n}} H(t, x) \, dx \, dt = 0$$

if $\psi = \psi_H$ for $H \in \mathcal{H}$. Moreover, $E_p(\psi) = 0$ for all $\psi \in \mathcal{D}$ provided $1 < p < \infty$.

5.2 The action spectrum of a Hamiltonian map

For $H \in \mathcal{H}$ we shall denote the set of fixed points of the associated Hamiltonian map φ_H by
$$Fix(\varphi_H) = \{x \in \mathbb{R}^{2n} \mid \varphi_H(x) = x\}.$$
If $x_0 \in Fix(\varphi_H)$ then the solution $x(t) = \varphi_H^t(x_0)$ for $0 \leq t \leq 1$ of the Hamiltonian equation $\dot{x} = X_H(x)$ satisfies $x(1) = x(0)$ and hence is a loop in $\mathbb{R}^{2n}$. Its action is the real number $A(x_0, H)$ defined by
$$A(x_0, H) = \int_0^1 \frac{1}{2} \langle -J\dot{x}(t), x(t) \rangle - \int_0^1 H(t, x(t))\, dt.$$
It turns out that $A(x_0, H)$ depends only on the fixed point x_0 and the map φ_H and is independent of the choice of the function H generating the map.

Lemma 1. If H and K in $\mathcal{H}$ generate the same map, i.e., $\varphi_H = \varphi_K$ then
$$A(x_0, H) = A(x_0, K)$$
for every $x_0 \in Fix(\varphi_H) = Fix(\varphi_K)$.

Proof. Define the piecewise smooth arc $t \mapsto \psi^t$ by
$$\psi^t = \varphi_H^t, \quad \text{for } t \in [0, 1]$$
$$\psi^t = \varphi_K^{2-t}, \quad \text{for } t \in [1, 2].$$
Thus for every $x \in \mathbb{R}^{2n}$ the map $t \mapsto \psi^t(x)$ from $[0, 2]$ into $\mathbb{R}^{2n}$ is a loop and we set
$$\Delta(x) = \int_0^2 \frac{1}{2} \langle -J\dot{x}(t), x(t) \rangle\, dt - \int_0^1 H(t, x(t))\, dt + \int_1^2 K(2-t, x(t))\, dt$$
where $x(t) = \psi^t(x)$. This map $x \mapsto \Delta(x)$ is evidently smooth and by differentiation in x we obtain
$$\begin{aligned}
\Delta'(x)h &= \int_0^2 \langle -J\dot{x}(t), d\psi^t(x)h \rangle\, dt \\
&\quad - \int_0^1 \langle \nabla H(t, x(t)), d\psi^t(x)h \rangle\, dt \\
&\quad + \int_1^2 \langle \nabla K(2-t, x(t)), d\psi^t(x)h \rangle\, dt,
\end{aligned}$$
which is equal to 0, since $t \mapsto \psi^t(x)$ is a solution of the Hamiltonian system associated to H between $[0, 1]$ and to $-K(2-t, \cdot)$ between $[1, 2]$. For $|x|$ large we have $\Delta(x) = 0$, since H and K are compactly supported so that $x(t) \equiv x$ for $t \in [0, 2]$. This shows that $\Delta \equiv 0$.
Hence if $x_0 \in Fix(\varphi_H)$ then $A(x_0, H) - A(x_0, K) = \Delta(x_0) = 0$ as claimed. ∎

In view of the lemma we can associate with a fixed point x_0 of the map $\psi \in \mathcal{D}$ the action $A(x_0, \psi)$ defined by
$$A(x_0, \psi) = A(x_0, H), \quad \text{if } \psi = \varphi_H.$$

Definition. The action spectrum of $\varphi \in \mathcal{D}$ is the set $\sigma(\varphi) \subset \mathbb{R}$ defined by

$$\sigma(\varphi) = \{ A(x, \varphi) \mid x \in Fix(\varphi) \}.$$

Clearly $0 \in \sigma(\varphi)$ since the Hamiltonians have compact supports. We shall show that the action spectrum is invariant under symplectic conjugation. More generally, denote by $\mathcal{G}$ the group of conformally symplectic diffeomorphisms. By definition, $\vartheta \in \mathcal{G}$ satisfies

(5.15) $$\vartheta^* \omega = \alpha \omega$$

for some constant $\alpha = \alpha(\vartheta) \in (0, \infty)$. Note that if $\varphi = \varphi_H$, then

(5.16) $$\vartheta \circ \varphi_H^t \circ \vartheta^{-1} = \varphi_{\alpha H_\vartheta}^t, \text{ where } H_\vartheta(t, x) = H\left(t, \vartheta^{-1}(x)\right).$$

Proposition 7. If $\varphi \in \mathcal{D}$ and $\vartheta \in \mathcal{G}$ with $\vartheta^* \omega = \alpha \omega$ then

$$A(\vartheta(x), \vartheta\varphi\vartheta^{-1}) = \alpha A(x, \varphi).$$

In particular,

$$\sigma(\vartheta\varphi\vartheta^{-1}) = \alpha\sigma(\varphi).$$

Proof. Assume $\varphi = \varphi_H$ and choose $x \in Fix(\varphi)$. In view of Lemma 1 and (5.16) we have to show that $A(\vartheta(x), \alpha H_\vartheta) = \alpha A(x, H)$. We set

$$x(t) = \varphi_H^t(x)$$

$$y(t) = \vartheta\left(x(t)\right) = \vartheta \circ \varphi_H^t \circ \vartheta^{-1}\left(\vartheta(x)\right) = \varphi_{\alpha H_\vartheta}^t\left(\vartheta(x)\right).$$

Let λ be a primitve of ω, i.e., $d\lambda = \omega$, then $\vartheta^*\lambda - \alpha\lambda = dF$ for a function F. Since the integral of a closed form over a loop vanishes, we find

$$A\left(\vartheta(x), \alpha H_\vartheta\right) = \int_0^1 \tfrac{1}{2} \langle -J\dot{y}(t), y(t) \rangle \, dt - \int_0^1 \alpha H_\vartheta\left(t, y(t)\right) dt$$

$$= \int_y \vartheta^* \lambda - \int_0^1 \alpha H_\vartheta\left(t, y(t)\right) dt$$

$$= \alpha \int_x \vartheta^* \lambda - \alpha \int_0^1 H\left(t, x(t)\right) dt = \alpha A(x, H). \quad \blacksquare$$

The following property of the action spectrum will be crucial later on.

Proposition 8. The action spectrum $\sigma(\varphi)$ is compact and nowhere dense.

5.2 THE ACTION SPECTRUM OF A HAMILTONIAN MAP

Proof. We first show that $\sigma(\varphi)$ is compact. Assume $\varphi = \varphi_H$ and recall that for $x_0 \in Fix(\varphi)$ the loop $x(t) = \varphi_H^t(x_0)$ is a solution of

$$\dot{x}(t) = X_H(x(t)), \text{ and } x(1) = x(0) = x_0.$$

Since H has compact support we find an $R > 0$ such that $x(t) = x_0$ is constant for $|x_0| \geq R$ and hence $A(x_0, \varphi) = 0$. It is, therefore, sufficient to consider fixed points x_0 for which $|x(t)| \leq R$. It then follows that $A(x_0, \varphi) \leq M$ for all $x_0 \in Fix(\varphi)$ so that $\sigma(\varphi)$ is bounded. To prove the compactness we consider a sequence $x_j(t)$ of solutions corresponding to the fixed points $x_j = x_j(0)$ and satisfying $|x_j(t)| \leq R$. By means of the Arzelà-Ascoli theorem we find a subsequence satisfying $x_j(t) \to x^*(t)$ in $C^\infty([0, 1], \mathbb{R}^{2n})$ and $x^*(0) = x^* \in Fix(\varphi)$. Consequently $A(x_j, \varphi) \to A(x^*, \varphi)$ implying the compactness of $\sigma(\varphi)$.

It remains to prove that $\sigma(\varphi)$ is nowhere dense. We first remark that the loop

$$x(t) = \varphi_H^t(x_0) \in C^\infty([0, 1], \mathbb{R}^{2n})$$

satisfying $x(1) = x(0) = x_0$ is an element of the Sobolev space $E = H^{1/2}(S^1, \mathbb{R}^{2n})$ introduced in Chapter 3. Indeed, its Fourier coefficients $x_k \in \mathbb{R}^{2n}$ are, after two partial integrations, estimated by

$$|x_k| \leq (\frac{1}{2\pi|k|})^2 (|\dot{x}(1) - \dot{x}(0)| + |\ddot{x}|_{C^0}),$$

for $k \in \mathbb{Z}$. Moreover, as in Chapter 3, one verifies that these special loops are precisely the critical points of the variational functional

$$a_H(x) = a(x) - b_H(x), \ x \in E,$$

where now $H \in \mathcal{H}$ and

$$b_H(x) = \int_0^1 H(t, x(t)) \, dt.$$

Consequently $A(x_0, \varphi) = a_H(x)$, with $x(t) = \varphi_H^t(x_0)$, is a critical level of a_H. In order to show that the set $\sigma(\varphi)$ of critical levels is nowhere dense, we follow an idea due to J.C. Sikorav [198] and construct a smooth function on $\mathbb{R}^{2n}$ whose critical levels contain $\sigma(\varphi)$. We extend the function $(t, x_0) \mapsto \varphi_H^t(x_0) : [0, 1] \times Fix(\varphi) \to \mathbb{R}^{2n}$ to a smooth function $\psi : [0, 1] \times \mathbb{R}^{2n} \to \mathbb{R}^{2n}$ satisfying

$$\psi(t, x_0) = \varphi^t(x_0), \quad \text{if } x_0 \in Fix(\varphi)$$

$$\psi(1, x) = \psi(0, x) \quad \text{for all } x \in \mathbb{R}^{2n}.$$

To do so simply take a smooth $f : [0, 1] \to [0, 1]$ which is equal to 1 near 0 and equal to 0 near 1 and set

$$\psi(t, x) = f(t)\varphi^t(x) + [1 - f(t)]\varphi^t((\varphi^1)^{-1}(x)),$$

where $\varphi^t = \varphi_H^t$. For every $x \in \mathbb{R}^{2n}$, the map $t \mapsto \psi(t, x)$ represents a loop belonging to E and we can define the smooth function $\Psi : \mathbb{R}^{2n} \to E$ by setting $\Psi(x)(t) = \psi(t, x)$. Using now the fact (proved in Appendix 3), that $a_H \in C^\infty(E, \mathbb{R})$ we have a smooth function defined by

$$a_H \circ \Psi : \mathbb{R}^{2n} \to \mathbb{R}.$$

It has the property that every $x_0 \in Fix(\varphi)$ is a critical point of $a_H \circ \Psi$. In addition, $A(x_0, \varphi) = a_H \circ \Psi(x_0)$ is a critical level. Since by Sard's theorem the set of critical levels of $a_H \circ \Psi$ is nowhere dense, we conclude that also $\sigma(\varphi)$ is nowhere dense. This finishes the proof of the proposition. ∎

Over the group $\mathcal{D}$ we can define the action spectrum bundle $\mathcal{A} \to \mathcal{D}$ by

$$\mathcal{A} = \bigcup_{\varphi \in \mathcal{D}} \{\varphi\} \times \sigma(\varphi),$$

equipped with the metric induced from $(\mathcal{D}, d) \times \mathbb{R}$. Every fibre is a compact nowhere dense set; a generic fibre consists of isolated points except possibly the point 0. Since $0 \in \sigma(\varphi)$ for every $\varphi \in \mathcal{D}$ there is a trivial continuous section of $\mathcal{A}$. The action spectrum $\sigma(\varphi)$ does not depend continuously on $\varphi \in \mathcal{D}$. In the next two sections we shall construct a nontrivial continuous section

$$\gamma : \mathcal{D} \longrightarrow \mathcal{A},$$

i.e., $\gamma(\varphi) \in \sigma(\varphi)$ and $\gamma \not\equiv 0$. It is, in addition, monotone and invariant under symplectic conjugation. The existence of this distinguished continuous section is established by means of a minimax argument for the action functional.

5.3 A "universal" variational principle

The results in the first section are based on a variational principle to which we will now turn. Our aim is to establish the existence of a special critical point of the action functional

$$a_H(x) = a(x) - b_H(x), \quad x \in E$$

introduced in Chapter 3. In contrast to the principle used there, we shall design a minimax principle which is independent of the choice of $H \in \mathcal{H}$. Recall that E is the Hilbert space $E = H^{1/2}(S^1, \mathbb{R}^{2n})$ which splits into an orthogonal sum $E = E^- \oplus E^0 \oplus E^+$, and, if $x = x^- + x^0 + x^+$, then

$$a(x) = \frac{1}{2}||x^+||^2 - \frac{1}{2}||x^-||^2$$

and

$$b_H(x) = \int_0^1 H(t, x(t))\, dt.$$

5.3 A "UNIVERSAL" VARIATIONAL PRINCIPLE

As in Chapter 3 one verifies that the special solutions $x \in C^\infty([0,1], \mathbb{R}^{2n})$

$$x(t) = \varphi_H^t(x_0) \text{ and } x_0 \in Fix(\varphi_H)$$

of the Hamiltonian equation $\dot{x} = X_H(x)$ agree with the critical points of a_H on E. We shall single out a family $\mathcal{F}$ of subsets $F \subset E$ such that for every $H \in \mathcal{H}$

$$\gamma(H) := \sup_{F \in \mathcal{F}} \inf_{x \in F} a_H(x)$$

is a real number and a critical value of a_H. The crucial point will be that $\mathcal{F}$ is independent of the choice of $H \in \mathcal{H}$ and therefore a "universal" minimax set for $\mathcal{H}$. Later on this will be important for us for the following reason: whenever one has a family of functionals together with a "universal" minimax set $\mathcal{F}$, we conclude from the pointwise estimate $\Phi \leq \Psi$ for two functionals in our family the same inequality for the associated minimax values. Note that

$$a_H(x^+) \to +\infty \text{ as } ||x^+|| \to \infty$$

where $x^+ \in E^+$, while

$$a_H(x^- + x^0) \leq M \text{ for all } x^- + x^0 \in E^- \oplus E^0$$

with a constant M depending on H.

In order to define $\mathcal{F}$ we first introduce a distinguished group of homeomorphisms of E, which is prompted by the structure of the gradient flows of a_H.

Definition. We define a set G of homeomorphisms of E as follows. A homeomorphism h belongs to G if h and h^{-1} map bounded sets onto bounded sets and there exist continuous maps $\gamma^\pm : E \to \mathbb{R}$ and $k : E \to E$ mapping bounded sets into precompact sets and such that there exist $r = r(h)$ satisfying

$$k(x) = 0, \quad \gamma^\pm(x) = 0$$

for all $x \in E^+$ with $||x|| \geq r$. Moreover,

$$h(x) = e^{\gamma^+(x)} x^+ + x^0 + e^{\gamma^-(x)} x^- + k(x)$$

for all $x \in E$.

Lemma 2. G is a group.

Proof. We have to show that with $h, g \in G$ also h^{-1} and $h \circ g \in G$. By assumption on h we have

$$e^{-\gamma^\pm} P^\pm h(x) = x^\pm + e^{-\gamma^\pm} P^\pm k(x)$$

$$P^0 h(x) = x^0 + P^0 k(x).$$

Hence
$$x = x^- + x^0 + x^+$$
$$= e^{-\gamma^-(x)}P^-h(x) + P^0h(x) + e^{-\gamma^+(x)}P^+h(x)$$
$$- (e^{-\gamma^-(x)}P^- + P^0 + e^{-\gamma^+(x)}P^+)k(x).$$

Writing $x = h^{-1}(y)$ and $y = h(x)$ we obtain for $y = y^- + y^0 + y^+$

$$h^{-1}(y) = e^{-\gamma^-(h^{-1}(y))}y^- + y^0 + e^{-\gamma^+(h^{-1}(y))}y^+$$
$$- (e^{-\gamma^-(h^{-1}(y))}P^- + P^0 + e^{-\gamma^+(h^{-1}(y))}P^+)k(h^{-1}(y))$$
$$=: e^{\sigma^-(y)}y^- + y^0 + e^{\sigma^+(y)}y^+ + \hat{k}(y).$$

Using the properties of $\gamma^\pm$ and k together with the fact that h^{-1} maps bounded sets into bounded sets we conclude that $\sigma^\pm$ and $\hat{k}$ have the desired properties and $h^{-1} \in G$. Similarly one verifies that $h \circ g \in G$. ∎

The distinguished family $\mathcal{F}$ of subsets of E is defined by
$$\mathcal{F} = \{\, h(E^+) \mid h \in G\,\}.$$

The crucial topological property of G lies in the following intersection result which is intuitively clear:

Lemma 3. If $h \in G$, then
$$h(E^+) \cap (E^- \oplus E^0) \neq \emptyset.$$

Proof. We have to find $x \in E^+$ satisfying $P^+h(x) = 0$. This is equivalent to $e^{\gamma^+(x)}x + P^+k(x) = 0$ and $x \in E^+$, or
$$x = -e^{-\gamma^+(x)}P^+k(x) =: T(x) , \text{ and } x \in E^+ .$$

The operator $T : E^+ \to E^+$ maps E^+ into a precompact subset of E^+ and hence $T : \overline{B} \to \overline{B}$ for a sufficiently large ball B, so that Schauder's fixed point theorem guarantees a solution proving the lemma. ∎

Theorem 4. Assume $H \in \mathcal{H}$, then $\gamma(H)$, defined by
$$\gamma(H) := \sup_{F \in \mathcal{F}} \inf a_H(F),$$

is a critical value of a_H.

5.3 A "UNIVERSAL" VARIATIONAL PRINCIPLE

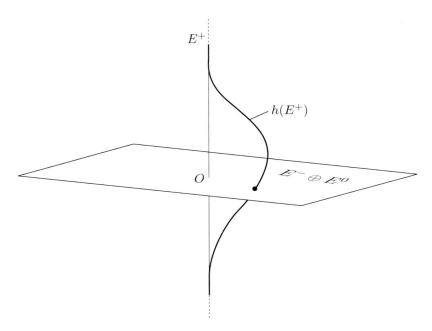

Fig. 5.1

The remainder of this section is devoted to the proof of this existence theorem. We shall proceed in several steps and first derive from the intersection lemma that $\gamma(H)$ is a real number. Define

$$q^+(H) = \int_0^1 \sup_x H(t,x)\, dt$$
$$q^-(H) = \int_0^1 \inf_x H(t,x)\, dt.$$

Then $q^+(H) \geq 0$ and $q^-(H) \leq 0$ since $H \in \mathcal{H}$.

Lemma 4. For every $H \in \mathcal{H}$

$$-q^+(H) \leq \gamma(H) \leq -q^-(H).$$

Proof. Let $F \in \mathcal{F}$, then $F \cap (E^- \oplus E^0) \neq \emptyset$ in view of Lemma 3 and we can estimate

$$\inf a_H(F) \leq \sup a_H(E^- \oplus E^0)$$
$$\leq \sup_{x \in E}[-b_H(x)]$$

$$\leq -\int_0^1 \inf_x H(t,x)\, dt.$$

Since

$$\inf a_H(E^+) \geq -\int_0^1 \sup_x H(t,x)\, dt,$$

the lemma follows. ∎

If $\gamma(H) = 0$, then $\gamma(H)$ is a critical level which contains in particular all the constant solutions outside of the support of H. We, therefore, only have to show that $\gamma(H) \neq 0$ is a critical value of a_H. For this we need the following compactness property.

Lemma 5. Assume $\gamma(H) \neq 0$ and assume the sequence $x_j \in E$ satisfies

$$\nabla a_H(x_j) \to 0, \text{ and } a_H(x_j) \to \gamma(H).$$

Then x_j possesses a convergent subsequence.

Proof. We recall from Chapter 3 that

$$\nabla a_H(x) = x^+ - x^- - \nabla b_H(x),$$

where $\nabla b_H : E \to E$ is globally Lipschitz continuous and maps bounded sets into precompact sets. Moreover $\nabla b_H(x) = j^* \nabla H(x)$ with the L^2-gradient $\nabla H(x)$ of b_H.

Assume that the sequence $x_j \in E$ is bounded in E. Then $\nabla b_H(x_j)$ is precompact and taking a subsequence we may assume that $\nabla b_H(x_j) \to y$ in E. By assumption $\nabla a_H(x_j) \to 0$ so that $x_j^+ - x_j^- \to y$. In addition, since $\dim E^0 < \infty$, we find a subsequence such that also x_j^0 converges and we see that x_j has a convergent subsequence.

Therefore, we only have to prove that the sequence x_j is bounded. Arguing indirectly we may assume by taking a subsequence that $\|x_j\| \to +\infty$. Since $H \in \mathcal{H}$ the sequence $\nabla H(x_j)$ is bounded in L^2 so that $\nabla b_H(x_j) = j^* \nabla H(x_j)$ is precompact in E. Hence, taking a subsequence, $x_j^+ - x_j^- \to y$ in E and also in L^2. This implies that $|x_j^0| \to \infty$. Consequently we find a subsequence such that pointwise

$$|x_j(t)| \to \infty, \text{ for a.e. } t$$

implying that $H(t, x_j(t)) \to 0$ for a.e. t. Hence by Lebesgue's theorem, $b_H(x_j) \to 0$. Also $\nabla H(t, x_j(t)) \to 0$ for a.e. t so that $\nabla H(x_j) \to 0$ in L^2 and consequently $\nabla b_H(x_j) \to 0$ in E. From $\nabla a_H(x_j) \to 0$ we now conclude that $x_j^+ - x_j^- \to 0$. Hence

$$a_H(x_j) = a(x_j) - b_H(x_j) \to 0.$$

This contradicts the assumption that $a_H(x_j) \to \gamma(H) \neq 0$ and shows that the sequence x_j is bounded. The proof of the lemma is finished. ∎

We are ready to finish the proof of Theorem 4 which we reformulate as

5.3 A "UNIVERSAL" VARIATIONAL PRINCIPLE

Lemma 6. The real number $\gamma(H)$ is a critical value of a_H, for every $H \in \mathcal{H}$.

Proof. Recall that $\gamma(H)$ is a real number in view of Lemma 4. If $\gamma(H) = 0$ there is nothing to prove and we assume $\gamma(H) \neq 0$. In this case the Palais-Smale condition as formulated in Lemma 5 holds. Arguing by contradiction we assume that $\gamma(H)$ is not a critical value of a_H. Then, in view of Lemma 5, we find an $\epsilon > 0$ satisfying

$$(5.17) \qquad |\nabla a_H(x)| \geq \epsilon, \text{ if } a_H(x) \in [\gamma(H) - \epsilon, \gamma(H) + \epsilon]$$

In order to first slow down the gradient flow we take a smooth map $\hat{\sigma} : \mathbb{R} \to [0, 1]$ satisfying

$$\hat{\sigma}(t) = 1, \text{ if } t \in [\gamma(H) - \epsilon, \gamma(H) + \epsilon]$$

$$\hat{\sigma}(t) = 0, \text{ if } t \in \mathbb{R} \setminus [\gamma(H) - 2\epsilon, \gamma(H) + 2\epsilon]$$

and define $\sigma(x) = \hat{\sigma}(a_H(x))$, for $x \in E$. The flow of the modified gradient equation

$$(5.18) \qquad \dot{x} = \sigma(x)\nabla a_H(x) =: V(x) \quad \text{on } E$$

exists for all times and we shall denote it by

$$(s, x) \mapsto x \cdot s : \mathbb{R} \times E \to E.$$

Using this flow we pick $T = \frac{2}{\epsilon}$ and define the homeomorphism h of E by setting

$$(5.19) \qquad h(x) = x \cdot T, \ x \in E.$$

By definition of $\gamma(H)$ there exists an $F \in \mathcal{F}$ satisfying

$$F \subset \{x \mid a_H(x) \geq \gamma(H) - \epsilon\},$$

and we shall demonstrate that

$$(5.20) \qquad h(F) \subset \{x \mid a_H(x) \geq \gamma(H) + \epsilon\}.$$

Postponing the proof we first show how the lemma is concluded. By definition of $\mathcal{F}$ we have $F = g(E^+)$. Therefore, if $h \in G$ (a fact which will be proved later on), then also $h \circ g \in G$ since G is a group in view of Lemma 2. Consequently $h(F) = h \circ g(E^+) \in \mathcal{F}$ and we arrive at the contradiction

$$\gamma(H) \geq \inf a_H(h(F)) \geq \gamma(H) + \epsilon,$$

hence concluding the lemma. In order to verify (5.20) we prove that

$$(5.21) \qquad a_H(x) \geq \gamma(H) - \epsilon \Rightarrow a_H(h(x)) \geq \gamma(H) + \epsilon.$$

Arguing by contradiction, we assume that $a_H(h(x)) < \gamma(H) + \epsilon$. Since $s \mapsto a_H(x \cdot s)$ is monotone-increasing we have

$$\gamma(H) + \epsilon > a_H(h(x)) \geq a_H(x \cdot s) \geq a_H(x) \geq \gamma(H) - \epsilon$$

for all $0 \leq s \leq T$. Therefore, $a_H(x \cdot s) \in [\gamma(H) - \epsilon, \gamma(H) + \epsilon]$ so that $\sigma(x \cdot s) = 1$ for $0 \leq s \leq T$. We see that in this time interval $x(s) := x \cdot s$ solves the gradient equation $\dot{x} = \nabla a_H(x)$. Hence using the estimate (5.17) for the gradient:

$$\gamma(H) + \epsilon > a_H(h(x)) = a_H(x \cdot T)$$
$$= a_H(x) + \int_0^T |\nabla a_H(x \cdot s)|^2 ds$$
$$\geq \gamma(H) - \epsilon + T\epsilon^2 = \gamma(H) + \epsilon,$$

which is a contradiction. The statement (5.21) is verified and it remains to show that $h \in G$.

Recall that
(5.22) $$V(x) = -\sigma(x)x^- + \sigma(x)x^+ - \sigma(x)\nabla b_H(x).$$

In order to find a representation formula for the flow of $\dot{x} = V(x)$ on E, we consider for $x \in E$ fixed, the equation

(5.23) $$\frac{d}{ds}u = A_x(s)u + f_x(s) \text{ on } E,$$

where $A_x(s) = -\sigma(x \cdot s)P^- + \sigma(x \cdot s)P^+$ is a time-dependent but linear vector field on E, and where $f_x(s) = -\sigma(x \cdot s)\nabla b_H(x \cdot s)$. If $u(s)$ is a solution of (5.23) with initial value $u(0) = x$, then $u(s) = x \cdot s$ by the uniqueness theorem of ordinary differential equations. The variation of constant formula for the solutions of (5.23) gives now the following representation for the flow of the equation $\dot{x} = V(x)$:

$$x \cdot t = e^{\int_0^t -\sigma(x \cdot s)ds} x^- + x^0 + e^{\int_0^t \sigma(x \cdot s)ds} x^+$$
$$- \int_0^t [e^{\int_s^t -\sigma(x \cdot \tau)d\tau} P^- + P^0 + e^{\int_s^t \sigma(x \cdot \tau)d\tau} P^+] \sigma(x \cdot s)\nabla b_H(x \cdot s)ds$$
$$=: e^{\gamma^-(x)}x^- + x^0 + e^{\gamma^+(x)}x^+ + k(x).$$

We shall show that for every $t \in \mathbb{R}$ this flow map belongs to the group G. Since the vector field V is globally Lipschitz continuous, it maps bounded sets in E into bounded sets. This holds true, of course, also for the inverse map. Note that there exists an $r > 0$ such that if $x \in E^+$ satisfies $||x|| \geq r$ then $a_H(x) \geq \gamma(H) + 2\epsilon$, and, therefore, $\sigma(x) = 0$ and consequently $V(x) = 0$. Hence $\sigma(x \cdot s) = 0$ for all $s \in \mathbb{R}$ and all $x \in E^+$ satisfying $||x|| \geq r$. We conclude that $\gamma^\pm(x) = 0$ and $k(x) = 0$ for $x \in E^+$ satisfying $||x|| \geq r$. Finally, the desired compactness properties for the map k follow from the compactness of the map ∇b_H and we have proved that the flow maps indeed belong to G. This finishes the proof of the lemma and the proof of Theorem 4 is complete. ■

5.4 A continuous section of the action spectrum bundle

Exploiting the fact that the minimax set $\mathcal{F}$ is the same for every $H \in \mathcal{H}$ we first derive some properties of the function

$$\gamma : \mathcal{H} \mapsto \mathbb{R}, \ H \mapsto \gamma(H)$$

where $\gamma(H) = \sup_{F \in \mathcal{F}} \inf a_H(F)$. We first prove that the function γ is nonnegative continuous and monotone-decreasing:

Proposition 9.

(i) $\quad \gamma \geq 0$ and $\gamma(0) = 0$

(ii) $\quad |\gamma(H) - \gamma(K)| \leq ||H - K||$

(iii) $\quad H \leq K \Rightarrow \gamma(H) \geq \gamma(K).$

Proof. For every $x \in E$

$$\begin{aligned}
a_H(x) &= a_K(x) + b_K(x) - b_H(x) \\
&\leq a_K(x) + \int_0^1 \sup_{\mathbb{R}^{2n}} (K - H) \, dt \\
&\leq a_K(x) + \int_0^1 [\sup_{\mathbb{R}^{2n}} (K - H) - \inf_{\mathbb{R}^{2n}} (K - H)] \, dt \\
&= a_K(x) + ||H - K||
\end{aligned}$$

implying the estimate (ii). Since for every $x \in E$ we have $a_H(x) \geq a_K(x)$ if $H \leq K$ the monotonicity (iii) of γ follows. In order to prove $\gamma(0) = 0$ we observe that

$$\gamma(0) = \sup_{F \in \mathcal{F}} \inf a(F) \geq \inf a(E^+) = 0,$$

and, since $F \cap (E^- \oplus E^0) \neq \emptyset$ for $F \in \mathcal{F}$,

$$\inf a(F) \leq \sup a(E^- \oplus E^0) = 0,$$

implying that $\gamma(0) = 0$. Finally, in order to prove that $\gamma(H) \geq 0$ for $H \in \mathcal{H}$, we define for fixed $x_0 \in \mathbb{R}^{2n} \setminus \{0\}$ and $\tau \in \mathbb{R}$ the family $H_\tau \in \mathcal{H}$ of functions by

$$H_\tau(t, x) = H(t, x - \tau x_0).$$

Defining the symplectic map ϑ by $\vartheta(x) = x + \tau x_0$ we have $\varphi_{H_\tau} = \vartheta \varphi_H \vartheta^{-1}$. If $\sigma(\varphi)$ denotes the action spectrum of $\varphi \in \mathcal{D}$ introduced in Section 5.2 we conclude from Proposition 7 that $\sigma(\varphi_{H_\tau}) = \sigma(\varphi_H)$. By definition, $\gamma(H_\tau) \in \sigma(\varphi_{H_\tau})$ and,

therefore, $\gamma(H_\tau) \in \sigma(\varphi_H)$. Since $\tau \mapsto \gamma(H_\tau)$ is continuous with image in a nowhere dense set (Proposition 8) it must be constant, so that
$$\gamma(H) = \gamma(H_\tau) \text{ for all } \tau \in \mathbb{R}.$$
Hence it is sufficient to estimate $\gamma(H_\tau)$ for τ large. Observing that $\mathrm{supp}(H_\tau)$ for $\tau \to \infty$ moves to "infinity" in x we can estimate for τ large
$$H_\tau(t,x) \le \pi |x|^2$$
for all $(t,x) \in [0,1] \times \mathbb{R}^{2n}$. Consequently, recalling that $||x||_{L^2}^2 = \Sigma |x_j|^2$ and $a(x) = \pi \Sigma j |x_j|$, for $x \in E^+$,
$$\gamma(H_\tau) \ge \inf_{E^+}[a(x) - \pi \int_0^1 |x|^2 \, dt] = 0,$$
implying that $\gamma(H) \ge 0$ as claimed. This finishes the proof of Proposition 9. ∎

If, for example, $H \ge 0$, then we conclude by monotonicity that $0 \le \gamma(H) \le \gamma(0) = 0$ implying $\gamma(H) = 0$. This proves the first part of

Proposition 10.

$$\begin{array}{ll} \text{If } H \ge 0 & \text{then } \gamma(H) = 0. \\ \text{If } H \le 0 \text{ and } H \ne 0 & \text{then } \gamma(H) > 0. \end{array}$$

Proof. Assume $H \le 0$ and $H \ne 0$. Then we find a bump function $\alpha \ne 0$ with $0 \le \alpha(t) \le 1$ and a time-independent $G \in C_c^\infty(\mathbb{R}^{2n})$ with $G \le 0$ and $G \ne 0$ such that
$$K(t,x) = \alpha(t) G(x)$$
satisfies, for $\epsilon > 0$ and small,
$$H(t,x) \le \epsilon K(t,x).$$
By monotonicity, $\gamma(H) \ge \gamma(\epsilon K)$ and we shall show that $\gamma(\epsilon K) > 0$. By translation, as in the proof of the previous proposition, we can assume that $G(0) = \inf G < 0$, so that
$$\epsilon \alpha(t) G(x) \le \epsilon \alpha(t) G(0) + \pi |x|^2$$
for all (t,x), provided $\epsilon > 0$ is sufficiently small. Hence
$$\begin{aligned} \gamma(\epsilon K) &\ge \inf_{x \in E^+}[a(x) - \epsilon \int_0^1 \alpha(t) G(x(t)) \, dt] \\ &\ge -\epsilon G(0) \int_0^1 \alpha \, dt + \inf_{E^+}[a(x) - \pi \int_0^1 |x|^2 \, dt] \\ &= -\epsilon G(0) \cdot \int_0^1 \alpha \, dt > 0. \end{aligned}$$
This finishes the proof of the proposition. ∎

5.4 A CONTINUOUS SECTION OF THE ACTION SPECTRUM BUNDLE

The distinguished critical level $\gamma(H)$ of the function a_H belongs, by construction, to the action spectrum of the associated time-1 maps φ_H. We shall show next that it only depends on the time-1 map and not on the particular choice of the Hamiltonian H generating the maps.

Proposition 11. Assume $H, K \in \mathcal{H}$, then
$$\varphi_H = \varphi_K \implies \gamma(H) = \gamma(K).$$

Proof. The proof is based on a homotopy argument and we start with a preliminary observation. Two arcs ψ_0^t and ψ_1^t in $\mathcal{A}$ are called homotopic if there is a smooth map
$$[0,1] \times [0,1] \to \mathcal{D} : (s,t) \mapsto \psi_s^t$$
connecting the arcs in $\mathcal{A}$. We then find a smooth $H : [0,1] \times [0,1] \times \mathbb{R}^{2n} \to \mathbb{R}$ of compact support satisfying
$$\psi_s^t(x) = \varphi_{H_s}^t(x),$$
where $H_s(t,x) = H(s,t,x)$. Assume now $\psi_s^1 = \psi$ for all $0 \le s \le 1$; then
$$\gamma(H_s) \in \sigma(\varphi_{H_s}) = \sigma(\psi).$$

Since $\sigma(\psi)$ is nowhere dense (by Proposition 8) the continuous function $s \mapsto \gamma(H_s)$ is constant so that $\gamma(H_0) = \gamma(H_1)$. The idea of the proof is now as follows, assuming
$$\varphi_H = \varphi_K = \psi$$
we shall homotope φ_H^t and φ_K^t to arcs $\varphi_{H'}^t$ and $\varphi_{K'}^t$ such that
$$|\gamma(H) - \gamma(K)| = |\gamma(H') - \gamma(K')| \le \|H' - K'\|$$
is small. In order to carry out the argument we first, by reparametrizing the time, homotope the arc φ_H^t to an arc φ^t satisfying $\varphi^t = \varphi^1 = \psi$ for $\frac{1}{2} \le t \le 1$ and we do the same for the arc φ_K^t. Hence we can assume that
$$K(t,x) = 0 = H(t,x) \quad \text{if} \quad \frac{1}{2} \le t \le 1.$$

Define now for $0 < s \le 1$
$$H_s(t,x) = s^2 H(t, \frac{1}{s}x)$$
$$K_s(t,x) = s^2 K(t, \frac{1}{s}x).$$

Then
$$\varphi_{H_s}^t(x) = s\varphi_H^t(\frac{1}{s}x)$$
$$\varphi_{K_s}^t(x) = s\varphi_K^t(\frac{1}{s}x).$$

If $\frac{1}{2} \leq t \leq 1$ we have
$$\varphi_{H_s}^t(x) = s\varphi_H^t(\frac{1}{s}x) = s\psi(\frac{1}{s}x) = \varphi_{K_s}^t(x).$$

Take a smooth $\beta : [\frac{1}{2}, 1] \to [0, 1]$ which is equal to 0 near $\frac{1}{2}$ and equal to 1 near 1 and define the arcs
$$\varphi_s^t(x) = \begin{cases} \varphi_{H_s}^t(x) & 0 \leq t \leq \frac{1}{2} \\ (s + (1-s)\beta(t))\psi([s + (1-s)\beta(t)]^{-1}x) & \frac{1}{2} \leq t \leq 1 \end{cases}$$
and similarly $\psi_s^t(x)$ by replacing H_s by K_s. Then
$$\varphi_s^1 = \psi = \psi_s^1 \quad \text{for all } 0 < s \leq 1.$$

If $\hat{H}_s$ generates φ_s^t and $\hat{K}_s$ generates ψ_s^t we conclude that $\gamma(\hat{H}_s) = \gamma(H)$ and $\gamma(\hat{K}_s) = \gamma(K)$. Observe that
$$\hat{H}_s(t, x) = \hat{K}_s(t, x), \quad \tfrac{1}{2} \leq t \leq 1$$
$$\hat{H}_s(t, x) = H_s(t, x), \quad 0 \leq t \leq \tfrac{1}{2}$$
$$\hat{K}_s(t, x) = K_s(t, x), \quad 0 \leq t \leq \tfrac{1}{2},$$
and $|H_s(t, s)| + |K_s(t, x)| \leq s^2(\|H\| + \|K\|)$ if $0 \leq t \leq \frac{1}{2}$ and $x \in \mathbb{R}^{2n}$. Consequently
$$|\gamma(H) - \gamma(K)| = |\gamma(\hat{H}_s) - \gamma(\hat{K}_s)| \leq \|\hat{H}_s - \hat{K}_s\| \leq s^2(\|H\| + \|K\|)$$
for all $0 < s \leq 1$ and hence $\gamma(K) = \gamma(H)$ as claimed. ∎

In view of this result we can associate with every $\varphi \in \mathcal{D}$ the real number
$$\gamma(\varphi) = \gamma(H), \quad \text{if} \quad \varphi = \varphi_H.$$
The qualitative properties of the invariant γ are summarized in the next two propositions.

Proposition 12. *The function $\gamma : \mathcal{D} \to \mathbb{R}$ satisfies*

(i) $\quad \gamma \geq 0$

(ii) $\quad |\gamma(\varphi) - \gamma(\psi)| \leq d(\varphi, \psi)$

(iii) $\quad \psi \geq \varphi \Rightarrow \gamma(\psi) \leq \gamma(\varphi)$

(iv) $\quad \gamma(\vartheta\varphi\vartheta^{-1}) = \gamma(\varphi)$

(v) $\quad \gamma(\varphi) \in \sigma(\varphi)$.

Proof. The statement (i) is proved in Proposition 9 (ii). If $\varphi = \varphi_H$ and $\psi = \varphi_K$ then $|\gamma(\varphi) - \gamma(\psi)| \leq ||H - K||$ in view of Proposition 9. This proves the statement (ii) in view of the definition of the distance $d(\varphi, \psi)$ as reformulated in (5.9). Now we prove (iii). By definition we find $H \leq 0$ satisfying $\varphi_H = \psi^{-1} \circ \varphi$. Take L such that $\psi = \psi_L$, hence $\varphi = \varphi_L \circ \varphi_H = \varphi_{L \# H}$. Since $H \leq 0$ we have $L \# H - L \leq 0$.

Now we make use of the following observation: if $\psi = \varphi_A$ and $\varphi = \varphi_B$ for two Hamiltonians $A, B \in \mathcal{H}$ satisfying $A \geq B$, then $a_A \leq a_B$ pointwise for the associated functionals; consequently we conclude from the minimax characterization of γ that $\gamma(\psi) = \gamma(A) \leq \gamma(B) = \gamma(\varphi)$. Hence the monotonicity statement in (iii) follows. The statement (iv) is proved by the same homotopy argument as in the beginning of the proof of Proposition 11. The last statement is a consequence of the definition of γ, in view of Lemma 6 and Proposition 11. The proof of Proposition 12 is complete. ∎

In view of Proposition 10 we have, in addition,

Proposition 13.

$$\begin{aligned} \gamma(id) &= 0 \\ \gamma(\varphi) &= 0, \text{ for all } \varphi \in \mathcal{D}^+ \\ \gamma(\varphi) &> 0, \text{ for all } \varphi \in \mathcal{D}^- \setminus \{id\}. \end{aligned}$$

Corollary.

$$\mathcal{D}^+ \cap \mathcal{D}^- = \{id\}$$

Recall that the partial order structure on $\mathcal{D}$ introduced in Section 5.1 is based on this fact. Note also that the real number $\gamma(\varphi)$ belongs to the action spectrum $\sigma(\varphi) = \{A(x, \varphi) \mid \varphi(x) = x\}$ associated to the fixed point set of the map φ. If x_0 is a trivial fixed point of $\varphi = \varphi_H$, i.e., satisfies $(t, x_0) \notin \text{supp}(H)$ for $0 \leq t \leq 1$, then $A(x_0, \varphi) = 0$. We see that the statement $\gamma(\varphi) > 0$ refers to nontrivial fixed points of φ.

Summarizing, Proposition 12 and Proposition 13 show that $\gamma : \mathcal{D} \to \mathcal{A}$ is a nontrivial ($\gamma \not\equiv 0$) and continuous section of the action spectrum bundle $\mathcal{A} \to \mathcal{D}$. In addition, γ is monotone with respect to the order structure on the group $\mathcal{D}$ and invariant under symplectic conjugation.

5.5 An inequality between the displacement energy and the capacity

The energy function $E : \mathcal{D} \to \mathbb{R}^+$ is defined by

$$E(\psi) = \inf\{ ||H|| \mid \psi = \varphi_H \},$$

where $H \in \mathcal{H}$ and where, with the notation $H_t(x) = H(t, x)$,

$$||H|| = \int_0^1 (\sup_x H_t - \inf_x H_t) \, dt.$$

The energy is related to the capacity as already mentioned previously and we shall prove

Theorem 5. If $\psi \in \mathcal{D}$, then
$$E(\psi) \geq \sup\{ c_0(U) \mid U \text{ open and } \psi(U) \cap U = \emptyset \} .$$

Here $c_0(U)$ is the special capacity introduced in Chapter 3. In order to prove this theorem we shall proceed in several steps, first defining the positive and the negative energy functions $E^\pm : \mathcal{D} \to \mathbb{R}^+$ by

$$E^+(\psi) = \inf\{ \int_0^1 \sup_x H_t \, dt \mid \psi = \varphi_H \}$$

$$E^-(\psi) = -\sup\{ \int_0^1 \inf_x H_t \, dt \mid \psi = \varphi_H \} = \inf\{ \int_0^1 -\inf_x H_t \, dt \mid \psi = \varphi_H \} .$$

Evidently,
$$E^+(\psi) + E^-(\psi) \leq E(\psi)$$
$$E^+(\psi) = E^-(\psi^{-1}).$$

Proposition 14. If φ, ψ in $\mathcal{D}$, then
$$\gamma(\varphi) - E^+(\psi) \leq \gamma(\varphi \circ \psi) \leq \gamma(\varphi) + E^-(\psi).$$

Setting $\varphi = id$ we find, in particular, by using $E^+(\psi) = E^-(\psi^{-1})$ that
$$E(\psi) \geq \gamma(\psi) + \gamma(\psi^{-1}).$$

Proof. Let $\varphi = \varphi_H$ and $\psi = \varphi_K$, then $\varphi \circ \psi = \varphi_{H\#K}$ and hence

$$\gamma(\varphi \circ \psi) = \sup_{F \in \mathcal{F}} \inf a_{H\#K}(F)$$

$$= \sup_{F \in \mathcal{F}} \inf_{x \in F} [a_H(x) - \int_0^1 K_t((\varphi_H^t)^{-1}(x)) \, dt]$$

$$\geq \gamma(\varphi) - \int_0^1 \sup_x K_t \, dt$$

implying the first inequality; the second one is proved similarly. ∎

Lemma 7. Let $U \subset \mathbb{R}^{2n}$ be open and bounded and assume that $\varphi, \psi \in \mathcal{D}$ satisfy

(i) $\quad \varphi = \varphi_H$, with $\text{supp}(H) \subset [0,1] \times U$.

(ii) $\quad \psi(U) \cap U = \emptyset$.

Then
$$\gamma(\varphi \circ \psi) = \gamma(\psi \circ \varphi) = \gamma(\psi).$$

5.5 An inequality between the displacement energy ...

Proof. Denoting by $\varphi_\tau = \varphi_H^\tau$ the flow of H we have $Fix(\varphi_\tau \circ \psi) = Fix(\psi \circ \varphi_\tau) = Fix(\psi) \subset \mathbb{R}^{2n} \setminus U$. If σ denotes the action spectrum, we claim that

(5.24) $$\sigma(\varphi_\tau \circ \psi) = \sigma(\psi) = \sigma(\psi \circ \varphi_\tau),$$

for $0 \leq \tau \leq 1$. Indeed, fixing τ we can choose, by reparameterizing the time, new Hamiltonians, namely K generating ψ and H_τ generating φ_τ, such that $K(t,x) = 0$ for $\frac{1}{2} \leq t \leq 1$, $H_\tau(t,x) = 0$ for $0 \leq t \leq \frac{1}{2}$ and $\text{supp}(H_\tau) \subset [0,1] \times U$. For the corresponding flows we see that

$$\psi^t \circ \varphi_\tau^t(x) = \psi^t(x) = \varphi_\tau^t \circ \psi^t(x), \quad x \in Fix(\psi),$$

for all $0 \leq t \leq 1$ implying (5.24). The functions $\tau \mapsto \gamma(\psi \circ \varphi_\tau)$ and $\tau \mapsto \gamma(\varphi_\tau \circ \psi)$ being continuous and having values in the nowhere dense set $\sigma(\psi)$ must be constant and hence equal to $\gamma(\psi)$, in view of $\varphi_0 = id$, proving the proposition. ∎

The following immediate consequence of the lemma will be crucial for the proof of Theorem 5.

Proposition 15. Assume that $\varphi, \psi \in \mathcal{D}$ meet the assumptions of Lemma 7, i.e., there is an open bounded subset $U \subset \mathbb{R}^{2n}$ such that $\varphi = \varphi_H$ with $\text{supp}(H) \subset [0,1] \times U$ and $\psi(U) \cap U = \emptyset$. Then

$$\gamma(\varphi) \leq E(\psi).$$

Proof. Since, by Lemma 7, $\gamma(\psi) = \gamma(\varphi \circ \psi)$, we find in view of Proposition 14 that

$$E^-(\psi) \geq \gamma(\psi) = \gamma(\varphi \circ \psi) \geq \gamma(\varphi) - E^+(\psi)$$

and hence $\gamma(\varphi) \leq E^+(\psi) + E^-(\psi) \leq E(\psi)$ as claimed. ∎

It is possible to compute the energy $E(\psi)$ and $\gamma(\psi)$ for those $\psi \in \mathcal{D}$ which are generated by special time-independent Hamiltonians.

Proposition 16. Assume $H \in \mathcal{H}$ is independent of time and assume, moreover, that all T-periodic solutions of $\dot{x} = X_H(x)$ for $0 < T \leq 1$ are constant. Then for $\varphi_H = \varphi$,

$$\gamma(\varphi) = -\inf_x H$$

$$E(\varphi) = \|H\| = \sup_x H - \inf_x H$$

$$E(\varphi) = \gamma(\varphi) + \gamma(\varphi^{-1}).$$

Proof. Abbreviate $\varphi^s = \varphi_H^s$ for $0 \leq s \leq 1$. It follows from the assumption about the periodic solutions that the fixed points of φ^s are the critical points of H and hence
$$\sigma(\varphi^s) = \{-sH(m) \mid dH(m) = 0\}.$$

Abbreviating the set of negative critical levels by $C_H = -\{H(m) \mid dH(m) = 0\}$ we conclude that $\gamma(\varphi^s) \in sC_H$. By Sard's theorem C_H is nowhere dense and since $s \mapsto \frac{1}{s}\gamma(\varphi^s) \in C_H$ is continuous, it is constant. Hence, there exists a critical point m^* of H satisfying

(5.25) $$\gamma(\varphi^s) = -sH(m^*), \quad \text{for all } 0 \leq s \leq 1.$$

We shall show that $H(m^*) = \inf H$. By translating H as in the proof of Proposition 9 we may assume that $H(0) = \inf H$. In view of Proposition 14
$$\gamma(\varphi^s) = \gamma(\varphi_{sH}) \leq E^-(\varphi_{sH}) \leq -sH(0).$$

On the other hand we find $s_0 > 0$ such that $sH(x) \leq sH(0) + \pi|x|^2$ for all $0 \leq s \leq s_0$ and $x \in \mathbb{R}^{2n}$ implying

$$\begin{aligned}\gamma(\varphi_{sH}) &\geq \inf_{x \in E^+} [a(x) - \pi \int_0^1 |x|^2] - sH(0) \\ &= -sH(0)\end{aligned}$$

so that $\gamma(\varphi_{sH}) = -sH(0)$ for all $0 \leq s \leq s_0$ and, hence in view of (5.25) also for $0 \leq s \leq 1$. We see that
$$\gamma(\varphi) = -\inf H$$

as claimed. Now $\varphi_{-H} = (\varphi_H)^{-1}$ and the same argument gives
$$\gamma(\varphi^{-1}) = \gamma(\varphi_{-H}) = -\inf(-H) = \sup H.$$

Consequently, by Proposition 14 again,
$$\begin{aligned}-\inf H &= \gamma(\varphi) \leq E^-(\varphi) \\ \sup H &= \gamma(\varphi^{-1}) \leq E^-(\varphi^{-1}) = E^+(\varphi).\end{aligned}$$

Adding up
$$\|H\| = \sup H - \inf H \leq E^-(\varphi) + E^+(\varphi) \leq E(\varphi_H) \leq \|H\|,$$

so that $E(\varphi) = \|H\|$ as claimed. ∎

5.5 An inequality between the displacement energy ...

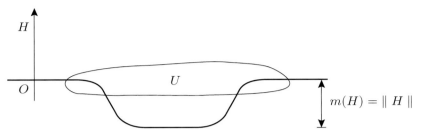

Fig. 5.2

Proof of Theorem 5. Recall first the definition of the capacity $c_0(U)$ where U is an open subset of $\mathbb{R}^{2n}$. Define the set $\mathcal{H}_a(U)$ of admissible Hamiltonian functions as follows: $H \in \mathcal{H}_a(U)$ if $H \in C_c^\infty(\mathbb{R}^{2n})$, $\mathrm{supp}(H) \subset U$ and $H \leq 0$, moreover, $H(x) = -\inf H$ for x in an open set and all the T-periodic solutions of $\dot{x} = X_H(x)$ of periods $0 < T \leq 1$ are constants. Then

$$c_0(U) = \sup\{||H|| \mid H \in \mathcal{H}_a(U)\}.$$

Note that $||H|| = -\inf H = m(H)$ in the notation of Chapter 3.

Assume now that $\psi \in \mathcal{D}$ satisfies $\psi(U) \cap U = \emptyset$. Then U is bounded, since ψ has compact support and hence $c_0(U) < \infty$. Picking $\epsilon > 0$, we find $H \in \mathcal{H}_a(U)$ satisfying

$$||H|| \geq c_0(U) - \epsilon.$$

Setting $\varphi = \varphi_H$ we have, by Proposition 16, that $\gamma(\varphi) = -\inf H = ||H||$ and conclude, in view of Proposition 15,

$$E(\psi) \geq \gamma(\varphi) = ||H|| \geq c_0(U) - \varepsilon.$$

Hence $E(\psi) \geq c_0(U) - \varepsilon$ for every $\varepsilon > 0$ so that $E(\psi) \geq c_0(U)$ implying Theorem 5. ∎

We have already seen that the crucial estimate $E(\psi) > 0$ for $\psi \neq id$ follows from Theorem 5 and we would like to point out another astonishing consequence conjectured by Ya. Eliashberg and proved by H. Hofer in [117] and C. Viterbo in [218].

Theorem 6. Assume ψ_j and $\psi \in \mathcal{D}$ and $\varphi \in C^0(\mathbb{R}^{2n}, \mathbb{R}^{2n})$. If

(i) $\quad d(\psi_j, \psi) \to 0$

(ii) $\quad \lim_{j \to \infty} \psi_j = \varphi$, locally uniformly

then $\varphi = \psi$.

Corollary. Assume $\varphi_{H_j} \in \mathcal{D}$ and $\varphi \in C^0(\mathbb{R}^{2n}, \mathbb{R}^{2n})$ satisfy

$$\text{(i)} \quad H_j \longrightarrow 0 \text{ uniformly}$$
$$\text{(ii)} \quad \psi_{H_j} \longrightarrow \varphi \text{ locally uniformly.}$$

Then $\varphi = id$.

Proof. We may restrict ourselves to the case $\psi = id$. Hence, assuming $d(\psi_j, id) = E(\psi_j) \to 0$ and $\lim_{j \to \infty} \psi_j = \varphi$ we have to show that $\varphi(x) = x$. Arguing indirectly, we find x^* satisfying $\varphi(x^*) \neq x^*$. Hence there is an $\epsilon > 0$ such that $\varphi(B_\epsilon) \cap B_\epsilon = \emptyset$ for $B_\epsilon = B(x^*, \epsilon)$. We conclude that $\psi_j(B_{\epsilon/2}) \cap B_{\epsilon/2} = \emptyset$ for all $j \geq j_0$ so that, by Theorem 5, $E(\psi_j) \geq c_0(B_{\epsilon/2}) > 0$ contradicting the assumption $E(\psi_j) \to 0$ and proving the result. ∎

Given a bounded subset $A \subset \mathbb{R}^{2n}$ one can ask: how much energy is needed in order to deform the set A into a set A' which is disjoint from A?

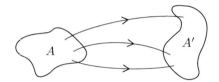

Fig. 5.3

This question prompted the following definition of the displacement energy $e(A)$ introduced by H. Hofer [116]:

$$e(A) = \inf\{ \,||H|| \,\,\,|\, H \in \mathcal{H}, \text{ and } \psi_H(A) \cap A = \emptyset\}$$
$$= \inf\{ \,E(\psi) \,\,\,|\, \psi \in \mathcal{D}, \text{ and } \psi(A) \cap A = \emptyset\}.$$

For unbounded sets B we simply set

$$e(B) = \sup\{\, e(A) \mid A \subset B \text{ and } A \text{ bounded }\}.$$

Clearly e is monotone in the sense that if $A \subset B$ then $e(A) \leq e(B)$. For every symplectic diffeomorphism ϑ of $\mathbb{R}^{2n}$ we have $e(\vartheta(A)) = e(A)$. This follows from $E(\vartheta \circ \psi \circ \vartheta^{-1}) = E(\psi)$. Moreover, $e(U) > 0$ for every open set U. Indeed the following more quantitative statement follows immediately from Theorem 5.

Theorem 7. For every open U

$$c_0(U) \leq e(U).$$

More explicitly, we have the estimate

$$\sup\{\, ||H|| \mid H \in \mathcal{H}_a(U)\} \leq \inf\{\, ||H|| \mid H \in \mathcal{H}, \psi_H(U) \cap U = \emptyset\}.$$

5.5 An inequality between the displacement energy ...

This estimate is much stronger than it looks at first sight. We shall see, in particular, that the existence of the capacity c_0, namely $c_0(B(1)) = \pi = c_0(Z(1))$ is an immediate consequence of the estimate. Recall that this normalization of c_0 was the subtle part in the construction of c_0 which required an existence proof. We point out that the capacity c_0 explained not only all the symplectic rigidity phenomena discussed in Chapter 2 but also the existence phenomena for periodic orbits in Chapter 4.

Corollary. For the open ball $B = B(1)$ and the open cylinder $Z = Z(1)$ we have

$$\pi \leq c_0(B) \leq e(B) \leq e(Z) \leq \pi.$$

Proof. In view of Theorem 7 and the monotonicity of e, only the first and the last inequality have to be proved. The first one is Lemma 3 of Chapter 3 which required the construction of an easy example. It remains to prove that $e(Z) \leq \pi$. For this it is sufficient to show that

$$e(B^2(1) \times B^{2n-2}(R)) \leq \pi$$

for every radius $R > 1$. To displace this set from itself, it is enough to displace $B^2(1) \subset \mathbb{R}^2$ from itself. To do so we first take an area- and orientation-preserving diffeomorphism φ of $\mathbb{R}^2$ mapping $B^2(1)$ into a square of side length $\sqrt{\pi} + \epsilon$.

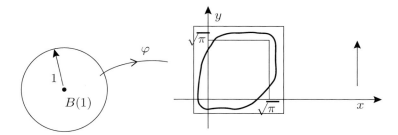

Fig. 5.4

A translation in the y direction which displaces the square from itself requires a Hamiltonian which has a slope close to $\sqrt{\pi}$, for example $H_1(x, y) = (\sqrt{\pi} + 2\epsilon) \cdot x$. By smoothly cutting off this Hamiltonian, we see that we can separate the square from itself by a Hamiltonian $H \in \mathcal{H}$ with norm $\|H\|$ as close as we wish to π. The desired flow $\varphi^{-1} \circ \varphi^t \circ \varphi$ then displaces $B^2(1)$ from itself in time 1, proving the claim and hence the corollary. ∎

As indicated in Fig. 5.5, this construction shows also that, by cutting a given bounded set V into small disjoint cubes and after removing a set $V_\epsilon \subset V$ of arbitrary small measure, the remaining set $V \setminus V_\epsilon$ can be displaced from itself by arbitrary small energy.

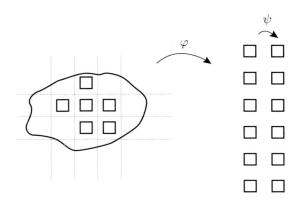

Fig. 5.5

Definition. A special capacity c for open sets in $\mathbb{R}^{2n}$ is a map c, associating to an open set U, a number $c(U)$ or $+\infty$ such that

(A.1) $\qquad c(U) > 0$ if $U \neq \emptyset$ and $c(B^2(1) \times \mathbb{R}^{2n-2}) < \infty$.

(A.2) $\qquad$ If $\psi : \mathbb{R}^{2n} \to \mathbb{R}^{2n}$ is a diffeomorphism satisfying
$\psi^* \omega = \alpha \omega$ for some $\alpha \neq 0$ and $\psi(U) \subset V$
then $|\alpha| c(U) \leq c(V)$.

Evidently every capacity is a special capacity. However, in contrast to the axioms of the capacity in Chapter 2, the maps ψ in (A.2) are required to be defined on all of $\mathbb{R}^{2n}$.

Theorem 8. *The displacement energy e is a special capacity in $\mathbb{R}^{2n}$ satisfying, in addition, $e(B(1)) = \pi = e(Z(1))$.*

Proof. It remains to prove that e satisfies A.2. Assume that ψ satisfies $\psi^*\omega = \alpha\omega$. Then $E(\psi\varphi\psi^{-1}) = |\alpha| E(\varphi)$ for $\varphi \in \mathcal{D}$. If $\varphi(U) \cap U = \emptyset$ then $\psi\varphi\psi^{-1}(\psi(U)) \cap \psi(U) = \emptyset$, implying that $e(\psi(U)) = |\alpha| e(U)$. If now $U \subset V$ then $e(U) \leq e(V)$ so that $\psi(U) \subset V$ implies $e(\psi(U)) = |\alpha| e(U) \leq e(V)$ as desired. ■

It should be mentioned that e is not a capacity as defined in Chapter 2, and $c_0 \neq e$. Indeed, consider an annulus A in $\mathbb{R}^2$.

Then $c_0(A) = \pi(R^2 - r^2)$, i.e., is the area of A. If $\psi(A) \cap A = \emptyset$ for $\psi \in \mathcal{D}$ which is, by definition, defined on all of $\mathbb{R}^2$, then necessarily $\psi(B(R)) \cap B(R) = \emptyset$. Using that ψ is an area-preserving map, this follows from topological reasons, and we see that
$$e(A) = e(B(R)) = \pi R^2 > c_0(A).$$

5.6 Comparison of the metric d on $\mathcal{D}$ with the C^0-metric

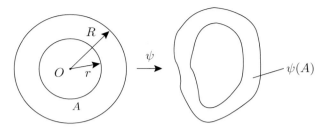

Fig. 5.6

5.6 Comparison of the metric d on $\mathcal{D}$ with the C^0-metric

In order to understand the metric d introduced on $\mathcal{D}$ in Section 5.1, it is useful to compare it with more familiar metrics. One can ask, for example, whether a symplectic map which is C^0-close to the identity mapping can be generated by a C^0-small Hamiltonian. This is not necessarily so as we shall show by means of an example in $\mathbb{R}^2$. We shall construct a sequence $\psi_n \in \mathcal{D}$ satisfying

$$|\psi_n - id|_{C^0} \longrightarrow 0$$
(5.26)
$$d(\psi_n, id) \longrightarrow \infty.$$

We take the Hamiltonians $H_n \in \mathcal{H}$ defined by

$$H_n(x, y) = \tfrac{1}{n} f_n(x) \chi_n(x, y) \quad \text{on} \quad \mathbb{R}^2.$$

Here $f_n \in C^\infty(\mathbb{R})$ satisfies $f_n(x) = x$ for $0 \le x \le n^2$ and looks as in Fig. 5.7:

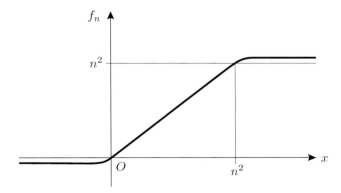

Fig. 5.7

Moreover, $\chi_n \in C_c^\infty(\mathbb{R}^2)$ is a smooth cut-off function which is equal to 1 on the region given by $0 \leq x \leq n^2$ and $0 \leq y \leq \frac{2}{n}$ and decreases slowly such that $|\nabla \chi_n| \leq \frac{1}{n^2}$.

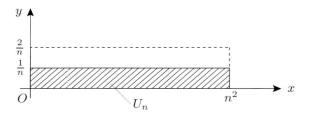

Fig. 5.8

Let $\psi_n = \varphi_{H_n} \in \mathcal{D}$ be the maps generated by H_n. Since the set U_n in Fig. 5.8 is displaced from itself by ψ_n we can estimate, in view of Theorem 8, that $d(\psi_n, id) = E(\psi_n) \geq c_0(U_n) = n$. On the other hand $|\psi_n - id|_{C^0} \leq \frac{C}{n}$ proving the claim (5.26).

For the example it is important that the diameters of the supports of the maps ψ_n grow. Indeed, there is no such example if we keep the supports uniformly bounded. This follows from the following estimate due to H. Hofer in [117]:

Theorem 9. For every $\varphi, \psi \in \mathcal{D}$

$$d(\varphi, \psi) \leq C \ \text{diameter supp} \ (\varphi \psi^{-1}) \ |\varphi - \psi|_{C^0}$$

where C is a constant and $C \leq 128$.

The remainder of this section is devoted to the proof of this statement which compares the intrinsic metric d with the C^0-metric. We need some technical preparations and start with a definition. The notation in the following refers to the symplectic splitting $\mathbb{R}^{2n} = \mathbb{R}^2 \oplus \cdots \oplus \mathbb{R}^2$ and the coordinates $(x_1, y_1, \ldots, x_n, y_n)$. If $a \in \mathbb{R}$ we shall denote by $\{x_1 = a\}$ the hyperplane $\{(x, y) \in \mathbb{R}^{2n} | x_1 = a\}$.

Definition. Denote by $S_1, \ldots, S_k$ any subsets of $\mathbb{R}^{2n}$. We call them properly separated, if for every choice $A_j \subset S_j$ of bounded subsets, there exist a map $\tau \in \mathcal{D}$ and numbers $a_1 < a_2 < \cdots < a_{k-1}$ such that the sets $\tau(A_j)$ lie in different components of $\mathbb{R}^{2n} \setminus \cup_{j=1}^{k-1} \{x_1 = a_j\}$. Relabeling the sets we can assume that $\tau(A_j)$ is to the

5.6 Comparison of the metric d on $\mathcal{D}$ with the C^0-metric

left of $\tau(A_{j+1})$:

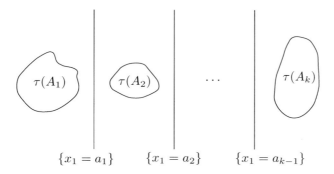

Lemma 8. Assume the sets $S_1, \ldots S_k$ are bounded and already properly separated by parallel hyperplanes. Then, given $R > 0$, there exist a map $\tau = \tau_R \in \mathcal{D}$ and vectors $v_1, v_2, \ldots, v_k \in \mathbb{R}^{2n}$, with $v_1 = 0$, satisfying

$$\tau(x) = x + v_j \quad \text{if} \quad x \in S_j$$

for all $1 \leq j \leq k$ and

$$\text{dist}\left(\tau(S_j), \tau(S_i)\right) \geq R \quad \text{if} \quad i \neq j.$$

Proof. Choosing $0 = \alpha_1 < \alpha_2 < \ldots < \alpha_k$, we introduce the vectors

$$v_j = \alpha_j e_1 \quad \in \mathbb{R}^{2n},$$

where $e_1 = (1, 0, 0, \ldots 0)$ and the translated sets

$$S_j^t := S_j + t v_j.$$

For every $t \geq 0$ these sets are mutually disjoint and separated by hyperplanes. The mutual distances go to infinity as $t \to \infty$. Define the function

$$H : \bigcup_{t \geq 0} \{t\} \times \left(\bigcup_{j=1}^{k} S_j^t\right) \to \mathbb{R}$$

by setting $H(t, x, y) = \alpha_j y_1$ if $(x, y) \in S_j^t$. We can consider H as a restriction of a smooth Hamiltonian function on $\mathbb{R} \times \mathbb{R}^{2n}$ having compact support in the space variables for t in a compact interval. If we take $\alpha_j = R(j-1)$, the time t-map φ_H^t acts on each set S_j as a translation $S_j \cdot t = S_j + t\alpha_j e_1$. Hence the mutual distances between the sets $S_j \cdot t$ are at least tR, thus proving the lemma. ∎

Lemma 9. Assume $\psi_1, \psi_2, \ldots, \psi_k \in \mathcal{D}$ have properly separated supports. Then
$$E(\psi_1 \psi_2 \cdots \psi_k) \leq 2 \max_j E(\psi_j).$$

Note that these maps necessarily commute.

Proof. Since the energy E is invariant under conjugation we may assume that we already have the supports of ψ_j separated by parallel hyperplanes. A priori we cannot assume that the generating Hamiltonians have disjoint supports. We, therefore, use a trick. Let $\psi_j = \varphi_{H_j}$ and denote the support by $S_j = \operatorname{supp}(\psi_j)$. Choose $R > 0$ such that
$$\operatorname{supp}(H_j) \subset [0,1] \times B_R(x_j^*)$$
where $x_j^* \in S_j$. Now choose vectors v_j such that the sets $B_R(S_j + v_j)$ are disjoint. If $\tau \in \mathcal{D}$ is the map associated to the S_j guaranteed by Lemma 8, we set $\hat{\psi}_j = \tau \psi_j \tau^{-1}$ and find
$$E(\psi_1 \psi_2 \cdots \psi_k) = E(\hat{\psi}_1 \hat{\psi}_2 \ldots \hat{\psi}_k).$$
By construction, $\operatorname{supp}(\hat{\psi}_j) = \tau(S_j) = S_j + v_j$. Observe now that miraculously
$$\hat{\psi}_j(x) = \tau \psi_j \tau^{-1}(x) = \psi_j(x - v_j) + v_j = T_j \psi_j T_j^{-1}(x)$$
for all $x \in \mathbb{R}^{2n}$, with the translation $T_j(x) = x + v_j$. Therefore, in view of the transformation law of Hamiltonian vector fields,
$$\hat{\psi}_j = \varphi_{\hat{H}_j} \quad \text{with} \quad \hat{H}_j(t,x) = H_j(t, x - v_j).$$
By construction, $\operatorname{supp}(\hat{H}_j)$ are now mutually disjoint and hence
$$\hat{\psi}_1 \hat{\psi}_2 \cdots \hat{\psi}_k = \varphi_{\hat{H}}$$
with $\hat{H} = \hat{H}_1 + \hat{H}_2 + \cdots + \hat{H}_k$. Given $\varepsilon > 0$, we choose H_j such that $\|H_j\| \leq E(\psi_j) + \varepsilon$, so that
$$E(\psi_1 \psi_2 \cdots \psi_k) = E(\varphi_{\hat{H}}) \leq \|\hat{H}\|.$$
By definition,
$$\|\hat{H}\| = \int_0^1 \left(\sup_x \hat{H}_t - \inf_x \hat{H}_t \right) dt,$$
and since the supports of $\hat{H}_j$ are disjoint we have
$$\|\hat{H}\| \leq 2 \max_j \int_0^1 \left(\sup_x \hat{H}_j - \inf_x \hat{H}_j \right) dt = 2 \max_j \|\hat{H}_j\| \leq 2 \max_j E(\psi_j) + 2\varepsilon.$$
This holds for every $\varepsilon > 0$ and the lemma is proved. ∎

5.6 COMPARISON OF THE METRIC d ON $\mathcal{D}$ WITH THE C^0-METRIC

It is useful for the following to introduce a refinement of the displacement energy. If $S \subset \mathbb{R}^{2n}$ is any subset we define the proper displacement energy $e_p(S)$ as

$$e_p(S) = \inf\Big\{a > 0 \,\Big|\, \text{for every bounded subset } A \subset S \\ \text{there exists } \psi \in \mathcal{D} \text{ satisfying } E(\psi) \leq a \text{ and} \\ A \text{ and } \psi(A) \text{ are properly separated}\Big\}.$$

One verifies, as in the previous section, that e_p is a special capacity for sets in $\mathbb{R}^{2n}$ which satisfies, in addition, $e_p(B(1)) = \pi = e_p(Z(1))$. It is an astonishing observation that the energy of a mapping $\psi \in \mathcal{D}$ can be estimated through the proper displacement energy of the smallest support needed in order to generate the map by a Hamiltonian. This was discovered by J.C. Sikorav in [198]:

Theorem 10. (J.C. Sikorav) If $H \in \mathcal{H}$ satisfies $\mathrm{supp}(H) \subset [0,1] \times U$, then

$$E(\varphi_H) \leq 16 \, e_p(U).$$

Proof. Denote by φ_H^t the time evolution of the map φ_H. Given $\varepsilon > 0$, we take N so large that the maps ψ_k defined by

$$\psi_k = \varphi_H^{t_k}, \quad \text{with } t_k = \frac{k}{N}, \quad 0 \leq k \leq N$$

satisfy

$$\psi_0 = \mathrm{id}, \quad \psi_N = \varphi_H$$

$$\mathrm{supp}\,(\psi_k) \subset S \subset U, \qquad 0 \leq k \leq N$$

$$d(\psi_k, \psi_{k+1}) < \varepsilon \qquad 0 \leq k \leq N-1.$$

Here $S \subset \mathbb{R}^{2n}$ is a bounded subset of U. By definition of e_p there exists a $\varphi \in \mathcal{D}$ such that S and $\varphi(S)$ are properly separated and

$$E(\varphi) \leq e_p(S) + \varepsilon \leq e_p(U) + \varepsilon.$$

We next construct a distinguished sequence of mappings $\varphi_j \in \mathcal{D}$ for $0 \leq j \leq 2N$. If $j = 0, 1$ we set $\varphi_0 = \mathrm{id}$ and $\varphi_1 = \varphi$. In order to define φ_2, we first choose a map $\sigma_2 \in \mathcal{D}$ as in the proof of Lemma 8 which satisfies for $s \in S$

$$\sigma_2(s) = s, \quad \sigma_2\big(\varphi(s)\big) = \varphi(s) + v_2$$

with a vector $v_2 \in \mathbb{R}^{2n}$.

Setting $\varphi_2 = \sigma_2 \varphi \sigma_2^{-1}$, we have for $s \in S$ that $\varphi_2(s) = \sigma_2(\varphi(s)) = \varphi(s) + v_2$. Similarly, we construct all the maps φ_j by

$$\varphi_j = \sigma_j \varphi \sigma_j^{-1}, \quad 2 \leq j \leq 2N$$

choosing the translation vectors v_j so that the sequence S_j of sets defined by
$$S_j = \varphi_j(S), \quad 0 \leq j \leq 2N$$
are disjoint and pairwise properly separated. Defining now the mappings $\alpha_j, \beta_k \in \mathcal{D}$ by
$$\alpha_j = \varphi_{2j-1}\psi_j\varphi_{2j-1}^{-1}, \quad 1 \leq j \leq N$$
$$\beta_k = \varphi_{2k}\psi_k^{-1}\varphi_{2k}^{-1}, \quad 0 \leq k \leq N$$
we see that
$$\operatorname{supp}(\alpha_j) \subset S_{2j-1}, \quad 1 \leq j \leq N$$
$$\operatorname{supp}(\beta_k) \subset S_{2k}, \quad 0 \leq k \leq N.$$
By construction the supports are mutually disjoint so that any two mappings among $\alpha_1, \ldots, \alpha_N, \beta_0, \beta_1 \ldots \beta_N$ commute. Since $\beta_N = \varphi_{2N}\varphi_H\varphi_{2N}^{-1}$ and $\beta_0 = id$ we can use the commutativity of the maps and compute

$$\begin{aligned} E(\varphi_H) &= E(\beta_N) = E\Big(\beta_N \big(\prod_{k=0}^{N-1} \alpha_{k+1}\beta_k\big)\big(\prod_{k=0}^{N-1} \alpha_{k+1}\beta_k\big)^{-1}\Big) \\ &= E\Big(\big(\prod_{k=1}^{N} \alpha_k\beta_k\big)\big(\prod_{k=0}^{N-1} \alpha_{k+1}\beta_k\big)^{-1}\Big) \\ &\leq 2 \max_{1 \leq k \leq N} E(\alpha_k\beta_k) + 2 \max_{0 \leq k \leq N-1} E(\alpha_{k+1}\beta_k), \end{aligned}$$

where we have used Lemma 9.

Observing that $\alpha_k\beta_k = \varphi_{2k-1}(\psi_k\,\varphi_{2k-1}^{-1}\,\varphi_{2k}\,\psi_k^{-1})\varphi_{2k}^{-1}$ and recalling that the maps φ_j are all conjugated to φ, we can estimate
$$E(\alpha_k\beta_k) \leq 4E(\varphi) \leq 4e_p(U) + 4\varepsilon.$$

Similarly, we find by using $\alpha_{k+1}\beta_k = \varphi_{2k+1}(\psi_{k+1}\psi_k^{-1})(\psi_k\varphi_{2k+1}^{-1}\varphi_{2k}\psi_k^{-1})\varphi_{2k}^{-1}$ that
$$E(\alpha_{k+1}\beta_k) \leq 4E(\varphi) + d(\psi_k, \psi_{k+1})$$
$$\leq 4e_p(U) + 4\varepsilon + \varepsilon.$$

Summing up we have $E(\varphi_H) \leq 16 e_p(U) + 18\varepsilon$, and since $\varepsilon > 0$ is arbitrary, the proof of Sikorav's theorem is finished. ∎

The following localization of Sikorav's estimate will be crucial for the proof of Theorem 9.

Lemma 10. Let $\psi \in \mathcal{D}$ with $\psi \neq id$ and let $\delta > |\psi - id|_{C^0}$. For every $Q \subset \mathbb{R}^{2n}$ open and satisfying $Q \cap \operatorname{supp}(\psi) \neq \emptyset$, there exists a $\varphi \in \mathcal{D}$ satisfying

(i) $\quad \varphi|Q = \psi|Q$

(ii) $\quad \operatorname{supp}(\varphi) \subset U$

(iii) $\quad E(\varphi) \leq 16\, e_p(U),$

5.6 Comparison of the metric d on $\mathcal{D}$ with the C^0-metric

where $U = B_\delta(Q) \cap C$. Here C is the convex hull of $\mathrm{supp}(\psi)$ and $B_\delta(Q) = \{x|\ \mathrm{dist}\,(x, Q) < \delta\}$.

Proof. By definition there is a smooth homotopy ψ^t for $0 \le t \le 1$ connecting the identity map with ψ. Choosing $x_0 \in Q \cap \mathrm{int}\ \mathrm{supp}(\psi)$, we find an $R > 0$ such that

(5.27) $$\mathrm{supp}(\psi^t) \subset B_R(x_0), \quad 0 \le t \le 1.$$

Define now the 2-parameter family $\psi_s^t \in \mathcal{D}$ for $0 \le s < 1$ and $0 \le t \le 1$ by setting

$$\psi_s^t(x) = sx_0 + (1-s)\psi^t\left(x_0 + \frac{x-x_0}{1-s}\right).$$

One verifies readily that

(5.28) $$\mathrm{supp}(\psi_s^t) \subset B_{(1-s)R}(x_0)$$

(5.29) $$\mathrm{supp}(\psi_s^1) \subset C.$$

We can fix $s_0 \in (0, 1)$ such that

(5.30) $$B_{(1-s_0)R}(x_0) \subset Q \cap C,$$

and define a new homotopy $\varphi^\tau \in \mathcal{D}$ for $0 \le \tau \le 1$ connecting the identity with ψ by setting

$$\varphi^\tau = \psi_{s(\tau)}^{t(\tau)},$$

where the smooth functions $t(\tau)$ and $s(\tau)$ are chosen such that the curve $\Gamma(\tau) := (t(\tau), s(\tau)) \subset \mathbb{R}^2$ looks as follows:

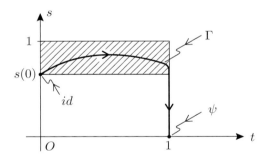

Fig. 5.9

Here the curve Γ runs in the shaded area for $0 \le \tau \le \frac{1}{2}$ and on the vertical part for $\frac{1}{2} \le \tau \le 1$. It now follows from (5.28), (5.29) and (5.30) that $\mathrm{supp}(\varphi^\tau) \subset C$ for all $0 \le \tau \le 1$ and $\mathrm{supp}(\varphi^\tau) \subset Q \cap C$ if $0 \le \tau \le \frac{1}{2}$. Moreover, $|\varphi^\tau - id|_{C^0} \le (1 - s(\tau))|\psi - id|_{C^0}$ if $\frac{1}{2} \le \tau \le 1$, as one verifies readily. Consequently,

(5.31) $$\varphi^\tau(Q) \subset B_d(Q), \quad 0 \le \tau \le 1$$

with $d := |\psi - id|_{C^0}$. Since $\text{supp}(\varphi^\tau) \subset C$ and C is convex, this homotopy is generated by a Hamiltonian $F \in \mathcal{H}$, $\varphi^\tau = \varphi_F^\tau$ having its support contained in $[0,1] \times C$. Cutting off this function smoothly, we find a new Hamiltonian K satisfying

$$K\big|B_d(Q) = F\big|B_d(Q)$$
$$\text{supp}(K) \subset [0,1] \times B_\delta(Q).$$

In view of (5.31) we conclude for the time 1-maps that $\varphi_K|Q = \varphi_F|Q = \psi|Q$. Moreover, $\text{supp}(K) \subset ([0,1] \times B_\delta(Q)) \cap \text{supp}(F) \subset [0,1] \times (B_\delta(Q) \cap C)$ and hence $\text{supp}(\varphi_K) \subset B_\delta(Q) \cap C$. We can apply Sikorav's theorem and find $E(\varphi_K) \leq 16 e_p(B_\delta(Q) \cap C)$. Setting $\varphi = \varphi_K \in \mathcal{D}$ the lemma is proved. ∎

We shall need another consequence of Sikorav's theorem.

Lemma 11. Assume $\psi \in \mathcal{D}$ and $\text{supp}(\psi) \subset U$. If $U = (a_1, a_2) \times (b_1, b_2) \oplus \mathbb{R}^{2n-2}$, then

$$E(\psi) \leq 16(a_2 - a_1)(b_2 - b_1).$$

Proof. Since U is convex we construct, as in the first part of the above proof, a Hamiltonian $H \in \mathcal{H}$ satisfying $\text{supp}(H) \subset [0,1] \times U$ and generating $\psi = \varphi_H$. One verifies easily, by arguing as in the proof of the Corollary to Theorem 7, that $e_p(U) = (a_2 - a_1)(b_2 - b_1)$ so that the lemma is a consequence of Sikorav's theorem above. ∎

Proof of Theorem 9. We consider $\psi \in \mathcal{D}$ with $\psi \neq id$ and choose

$$\delta > d = |\psi - id|_{C^0}.$$

Defining the sequence a_j by $a_0 = 0$ and $a_{k+1} = a_k + 2\delta$, we set

$$Q_k = \Big[(a_k - \varepsilon, a_k + \varepsilon) \times \mathbb{R}\Big] \oplus \mathbb{R}^{2n-2}$$

with $0 < \varepsilon < (\delta - d)$. Since $\text{supp}(\psi)$ is compact, the index set I,

$$I = \Big\{k \,\Big|\, Q_k \cap \text{supp}(\psi) \neq \emptyset\Big\}$$

is a finite set. Choose now $d < \delta_1 < (\delta - \varepsilon)$. In view of the localization Lemma 10 we find, for every $k \in I$, a map $\varphi_k \in \mathcal{D}$ satisfying

(i) $\qquad\qquad \varphi_k|Q_k = \psi|Q_k$

(ii) $\qquad\qquad \text{supp}(\varphi_k) \subset U_k$

(iii) $\qquad\qquad E(\varphi_k) \leq 16\, e_p(U_k)$

5.6 Comparison of the metric d on $\mathcal{D}$ with the C^0-metric

where $U_k = B_{\delta_1}(Q_k) \cap$ convex hull supp(ψ). Since $B_{\delta_1}(Q_k) = (a_k - \varepsilon - \delta_1, a_k + \varepsilon + \delta_1) \times \mathbb{R}^{2n-1}$ and $2(\varepsilon + \delta_1) < 2\delta$, these sets are properly separated and, therefore, the sets supp(φ_k) are also. Abbreviating

$$R := \text{diameter supp } \psi,$$

we find a sequence $b_k \in \mathbb{R}$ such that for every $k \in I$

$$U_k \subset (a_k - \varepsilon - \delta_1, a_k + \varepsilon + \delta_1) \times \left[b_k - \frac{R}{2}, b_k + \frac{R}{2}\right] \oplus \mathbb{R}^{2n-2}.$$

Using Lemma 9 and Lemma 11 we can estimate

$$E\left(\prod_{k \in I} \varphi_k\right) \leq 2 \max_{k \in I} E(\varphi_k)$$

$$\leq 2 \max_{k \in I} 16 \cdot 2(\varepsilon + \delta_1) \cdot R$$

(5.32) $$\leq 64 R \cdot \delta.$$

Define $\vartheta = (\prod_{k \in I} \varphi_k)^{-1}$. In view of (i) above and using the fact that supp(φ_k) are properly separated, we conclude for $x \in Q_k$ that

$$\psi\vartheta(x) = \psi\varphi_k^{-1}(x) = \psi\psi^{-1}(x) = x.$$

We see that the map $\psi\vartheta$ is the identity map on disjoint strips, hence it leaves every gap in between invariant so that it can be written as a composition of maps having their supports in the gaps. More precisely there exists a finite sequence $\hat{\varphi}_j \in \mathcal{D}$, with $1 \leq j \leq l$, satisfying

$$\psi \circ \vartheta = \prod_{j=1}^{l} \hat{\varphi}_j,$$

with supp($\hat{\varphi}_j$) properly separated and

$$\text{supp}(\hat{\varphi}_j) \subset \hat{U}_j,$$

where $\hat{U}_j = [\hat{a}_j - \delta + \varepsilon, \hat{a}_j + (\delta - \varepsilon)] \times [\hat{b}_j - \frac{R}{2}, \hat{b}_j + \frac{R}{2}] \oplus \mathbb{R}^{2n-2}$ for suitable numbers $\hat{a}_j$ and $\hat{b}_j$. We can apply Lemma 9 and Lemma 11 and find, using the estimate (5.32),

$$E(\psi) = E(\psi\vartheta\vartheta^{-1}) \leq E(\psi\vartheta) + E(\vartheta^{-1})$$

$$= E(\prod \hat{\varphi}_j) + E(\prod \varphi_k)$$

$$\leq 2 \cdot 16 \cdot 2(\delta - \varepsilon)R + 64 R\delta$$

$$\leq 128 R \cdot \delta.$$

This holds true for every $\varepsilon > d = |\psi - id|_{C^0}$, and we conclude that
$$E(\psi) \leq 128 \, R \, |\psi - id|_{C^0},$$
where $R = $ diameter $\mathrm{supp}(\psi)$. Observing that $|\varphi\psi^{-1} - id|_{C^0} = |\varphi - \psi|_{C^0}$, we finally arrive at the desired estimate
$$d(\varphi, \psi) = E(\varphi \circ \psi^{-1}) \leq 128 \text{ diameter } \mathrm{supp}(\varphi \circ \psi^{-1}) \, |\varphi - \psi|_{C^0},$$
and the proof of Theorem 9 is finished. ∎

Finally, we shall complete the proof of Theorem 1 by proving the estimate

(5.33) $\quad E(\psi) \leq 16 \inf\{\pi r^2 \mid \text{there is a } \varphi \in \mathcal{D} \text{ with supp} (\varphi\psi\varphi^{-1}) \subset Z(r)\}$,

for every $\psi \in \mathcal{D}$. Given $\varepsilon > 0$, we pick a map $\vartheta \in \mathcal{D}$ and $R > 0$ satisfying $\mathrm{supp}(\vartheta\psi\vartheta^{-1}) \subset Z(R)$ and
$$\pi R^2 - \varepsilon \leq \inf\{\pi r^2 \mid \text{there is a } \varphi \in \mathcal{D} \text{ with } \mathrm{supp}(\varphi\psi\varphi^{-1}) \subset Z(r)\}.$$

Since $E(\psi) = E(\vartheta\psi\vartheta^{-1})$ we may assume without loss of generality that $\mathrm{supp}(\psi) \subset Z(R)$. Choosing now $H \in \mathcal{H}$ with $\psi_H = \psi$, we consider the corresponding arc $t \mapsto \psi_t$ and define for $s > 0$
$$\psi_t^s(x) = s\psi_t(\frac{1}{s}x).$$

For a suitable function $t \mapsto s(t)$ we find an arc $t \mapsto \psi_t^{s(t)} = \bar{\psi}_t$ which is generated by a Hamiltonian function $K \in \mathcal{H}$ satisfying $\mathrm{supp}(K) \subset [0,1] \times Z(R)$ and $\bar{\psi}_1 = \psi$. Therefore, in view of Sikorav's estimate
$$E(\psi) \leq 16\pi R^2,$$
using that the proper displacement energy of $Z(R)$ is equal to πR^2. Since $\varepsilon > 0$ was arbitrarily chosen the announced estimate (5.33) follows and the proof of Theorem 1 is finally completed.

5.7 Fixed points and geodesics on $\mathcal{D}$

A symplectic diffeomorphism $\varphi \in \mathcal{D}$ possesses, as we have demonstrated, a distinguished fixed point $x = \varphi(x)$ guaranteed by the minimax principle. Its action satisfies $A(x, \varphi) = \gamma(\varphi)$. Periodic points are fixed points $x = \varphi^N(x)$ of higher iterates of φ and we call the finite set
$$\mathcal{O}(x) = \{x, \varphi(x), \varphi(\varphi(x)), \ldots, \varphi^N(x) = x\}$$
the orbit of x. The following qualitative existence theorem is due to C. Viterbo [218].

5.7 Fixed points and geodesics on $\mathcal{D}$

Theorem 11. Every $\varphi \in \mathcal{D}^- \setminus \{id\}$ possesses infinitely many periodic points $x_j = \varphi^j(x_j)$ having positive actions

$$A(x_j, \varphi^j) \geq \gamma(\varphi) > 0$$

and disjoint orbits

$$\mathcal{O}(x_j) \cap \mathcal{O}(x_k) = \emptyset, \quad k \neq j.$$

The statement also holds if $\varphi \in \mathcal{D}^+ \setminus \{id\}$. Actually, by using his homological version of a special capacity in $\mathbb{R}^{2n}$, C. Viterbo proved a more general statement assuming merely that $\varphi \in \mathcal{D}$ satisfies $\varphi \neq id$.

Proof. In view of the assumption on φ we find $H \in \mathcal{H}$ satisfying $\varphi = \varphi_H$ and $H \leq 0$ with $H \neq 0$. Hence $\gamma(\varphi) > 0$ in view of Proposition 13. Since $\text{supp}(H) \subset [0,1] \times U$ for some bounded open U, we have

$$0 < \gamma(\varphi) \leq e(U).$$

Indeed, if $\psi \in \mathcal{D}$ satisfies $\psi(U) \cap U = \emptyset$ then, in view of Proposition 15, $\gamma(\varphi) \leq E(\psi)$ implying that $\gamma(\varphi) \leq e(U)$. Since $\varphi \leq id$, we find, using the monotonicity of γ, the estimate

$$0 < \gamma(\varphi) \leq \gamma(\varphi^2) \leq \ldots \leq \gamma(\varphi^j) \leq e(U)$$

for all the iterates of φ. For every j there is a fixed point $x_j = \varphi^j(x_j)$ satisfying $\gamma(\varphi^j) = A(x_j, \varphi^j) \geq \gamma(\varphi) > 0$, and it remains to show that infinitely many of them belong to disjoint periodic orbits. To see this observe that if $x_k = \varphi^l(x_j)$ for some l, then

$$\gamma(\varphi^k) = A(x_k, \varphi^k) = lA(x_j, \varphi^j) = l\gamma(\varphi^j),$$

as the computation shows, so that the theorem follows in view of the above bound for the sequence $\gamma(\varphi^j)$. ∎

If the time-independent Hamiltonian function $H \in C_c^\infty(\mathbb{R}^{2n})$ does not vanish then the Hamiltonian system $\dot{x} = X_H(x)$ possesses many periodic solutions lying on regular energy surfaces of H, as we have seen in Chapter 4. Nothing is known, however, about their periods and we would like to add a qualitative existence statement for fast periodic solutions. Recall that if $H \in C_c^\infty(U)$ satisfies $\|H\| > c_0(U)$, then the system $\dot{x} = X_H(x)$ possesses a nonconstant T-periodic solution where the period is contained in the interval $0 < T \leq 1$, provided $H \leq 0$ and $H = \inf H$ on an open subset. This follows immediately from the definition of the capacity $c_0(U)$. Generalizing this statement we shall prove:

Theorem 12. Assume $H \in C_c^\infty(U)$ does not vanish. Then the Hamiltonian system $\dot{x} = X_H(x)$ has a nonconstant T-periodic solution with $0 < T \leq T^*$, where

$$T^* > \frac{c_0(U)}{|H|_{C^0}}.$$

Proof. By using the arguments of Chapter 5, the proof will follow readily from the definition of the capacity c_0. Let $H^+ = \max H$ and $H^- = -\min H$, then $|H|_{C^0} = \max\{H^+, H^-\}$. Assume $H^+ \geq H^-$ so that $H^+ > 0$ since $H \not\equiv 0$. Choose $\varepsilon > 0$ so small that $T^*(H^+ - \varepsilon) > c_0(U)$. Next choose a smooth function $f : \mathbb{R} \to [0, \infty]$ satisfying $f(s) = 0$ if $s \leq 0$ and $f(s) = H^+ - \varepsilon$ if $s \geq H^+$ and moreover $0 \leq f'(s) \leq 1$. Then define the Hamiltonian K by

$$K = T^* f \circ H.$$

Since $\|K\| > c_0(U)$ there exists a nonconstant τ-periodic solution $x(t)$ of

$$\dot{x} = X_K(x) = T^* f'(H) X_H(x)$$

of period $0 < \tau \leq 1$. The function $T^* f'(H(x(t))) = c$ is a positive constant, hence $y(t) = x(\frac{1}{c}t)$ is a nonconstant periodic solution of $\dot{y} = X_H(y)$ having period $c\tau \leq T^*$. The same argument applies if $H^- \geq H^+$ and the statement is proved. ∎

With a smooth arc $\varphi^t : [a, b] \to \mathcal{D}$, we associate its length by defining

$$\text{length}\left(\varphi^{[a,b]}\right) = \int_a^b [\sup_x H_t - \inf_x H_t] dt.$$

The function H is the unique smooth function on $[a, b] \times \mathbb{R}^{2n}$ of compact support which generates the arc and which is determined by $(\frac{d}{dt}\varphi^t) \circ (\varphi^t)^{-1}(x) = J\nabla H(t, x)$ for $a \leq t \leq b$. In the metric space $(\mathcal{D}, d)$ we then call a smooth path $\varphi^t : [a, b] \to \mathcal{D}$ a geodesic path if, locally, its length is the shortest connection of the endpoints; i.e., if

$$\text{length}\,(\varphi^{[t,s]}) = d(\varphi^t, \varphi^s)$$

for all $a \leq t < s \leq b$ with $|t - s|$ sufficiently small. Examples of geodesics are the flows for time-independent Hamiltonians $H \in C_c^\infty(\mathbb{R}^{2n})$, as we shall show. In this case it follows from the definition that

$$\text{length}\,(\varphi_H^{[s,t]}) = |t - s| \cdot \|H\|.$$

Proposition 17. *If $H \in C_c^\infty(\mathbb{R}^{2n})$ is independent of time, then the flow $\varphi^t = \varphi_H^t$ is a geodesic in $\mathcal{D}$ which moreover satisfies*

$$d(\varphi^t, \varphi^s) = |t - s| \cdot \|H\| = \text{length}\,(\varphi^{[t,s]})$$

for all $t < s$ satisfying $|t - s| < \frac{2\pi}{h}$, where $h := \sup_x |H_{xx}|$.

Proof. We recall first that a $T-$ periodic solution of a time-independent vector field is necessarily a constant solution if the period is too small. Indeed assume $x(t)$ is a $T-$ periodic solution of $\dot{x} = X_H(x)$. Normalizing the period we define $y(t) = x(tT)$, which solves

$$\dot{y} = X_{TH}(y), \text{ and } y(0) = y(1).$$

5.7 Fixed points and geodesics on $\mathcal{D}$

Note that for every 1-periodic loop z in $\mathbb{R}^{2n}$ having mean value zero, we have the L^2- estimate $||z|| \leq \frac{1}{2\pi}||\dot{z}||$. This is seen by looking at the Fourier series. Applying this observation to $\dot{y}$ whose mean value vanishes, we can estimate

$$||\dot{y}|| \leq \frac{1}{2\pi}||\ddot{y}|| = \frac{1}{2\pi}T \cdot ||H_{xx}(y(t))\dot{y}|| \leq \frac{Th}{2\pi}||\dot{y}||,$$

which implies $\dot{y} = 0$ if $hT < 2\pi$, so that $y(t) = y(0)$ is a constant solution. Therefore, if $hT < 2\pi$, we conclude from Proposition 16 that $E(\varphi_{TH}) = T||H||$. Hence assuming $|t - s| < \frac{2\pi}{h}$, we can compute the distance as

$$d(\varphi^s, \varphi^t) = E((\varphi^s)^{-1} \circ \varphi^t)$$
$$= E(\varphi_H^{t-s}) = E(\varphi_{(t-s)H}) = |t - s|\,||H||,$$

as claimed in the proposition. ∎

Evidently, for autonomous Hamiltonians,

$$d(\varphi^t, id) = t||H|| = \text{length}\,(\varphi^{[0,t]})$$

as long as φ^t admits no nontrivial fixed point. By definition, a geodesic in $\mathcal{D}$ minimizes locally the length of the curve. Globally this need not be true as the following example shows.

Proposition 18. *For every $\epsilon > 0$ there exists a geodesic path φ^t in $\mathcal{D}$ for $0 \leq t \leq 1$ starting at $\varphi^0 = id$ and satisfying, for all $0 \leq t \leq 1$,*

$$d(\varphi^t, id) \leq \epsilon \quad \text{and length}\,(\varphi^{[0,t]}) \geq \frac{t}{\epsilon}.$$

Proof. Pick an open neighborhood U of the origin satisfying $16e(U) < \epsilon$ and choose a time-independent $H \in C_c^\infty(U)$ with $||H|| > \frac{1}{\epsilon}$. Then, by Sikorav's estimate, we find if $0 \leq t \leq 1$, that $d(\varphi^t, id) = E(\varphi_{tH}) \leq 16e(U) < \epsilon$, while length $(\varphi^{[0,t]}) = t||H|| \geq \frac{t}{\epsilon}$ as claimed. ∎

We see in particular that the metric space $(\mathcal{D}, d)$ is not a "flat" space. "Locally" however the space $(\mathcal{D}, d)$ is "flat". This has been proved recently by M. Bialy and L. Polterovich [22], who gave a complete description of a geodesic path in $(\mathcal{D}, d)$ which we shall describe next. In order to avoid pathologies, we consider only smooth paths $g : I \to \mathcal{D}$, with $I \subset \mathbb{R}$ an interval, which are regular. This requires that the Hamiltonian $H : I \times \mathbb{R}^{2n} \to \mathbb{R}$ generating the path g satisfies $||H_t|| \neq 0$, for $t \in I$. Recall that $H_t(x) = H(t, x)$ and, as above

$$||H_t|| = \max_x H_t(x) - \min_x H_t(x)\,.$$

Definition. A regular smooth path $g : I \to \mathcal{D}$ is called a minimal geodesic if for all $a, b \in I$, with $a < b$,

$$\text{length } (g|_{[a,b]}) = d\Big(g(a), g(b)\Big) .$$

Recall that the length is defined by

$$\text{length } (g|_{[a,b]}) := \int_a^b ||H_t|| dt ,$$

where $H \in \mathcal{H}$ generates the path g. Consequently $g : I \to \mathcal{D}$ is a geodesic path if for every $c \in I$ there is an interval $I_c \subset I$ such that $g : I_c \to \mathcal{D}$ is a minimal geodesic. In order to characterize geodesics in terms of their generating Hamiltonians, the following concepts introduced in [22] turn out to be useful.

Definition. A smooth Hamiltonian $H : I \times \mathbb{R}^{2n} \to \mathbb{R}$ is called quasi-autonomous if there exists two points $x_-, x_- \in \mathbb{R}^{2n}$ satisfying

$$\max_x H_t(x) = H(t, x_+) \quad \text{for all } t \in I$$

$$\min_x H_t(x) = H(t, x_-) \quad \text{for all } t \in I .$$

Correspondingly a function H on $I \times \mathbb{R}^{2n}$ is called locally quasi-autonomous if, for every $c \in I$, there is an interval $I_c \subset I$ such that H is quasi-autonomous on $I_c \times \mathbb{R}^{2n}$. Special examples are of course autonomous Hamiltonians. Denote by $\mathcal{C} = C_c^\infty(\mathbb{R}^n)$ the linear space of autonomous Hamiltonians equipped with the metric associated to the norm $||H|| = \max H - \min H$. We call a smooth nonvanishing arc $f : I \to \mathcal{C}$ a minimal geodesic of $\mathcal{C}$ if for every $a < b$ in I, we have

$$||f(b) - f(a)|| = \int_a^b ||\frac{d}{ds} f(s)|| ds .$$

In this terminology a Hamiltonian $H \in \mathcal{H}$ is quasi-autonomous precisely if it is the time derivative of a geodesic arc in $\mathcal{C}$, in view of Lemma 13 below.

The following theorem extends Proposition 17 and gives a complete description of geodesics on $(\mathcal{D}, d)$.

Theorem 13. (Bialy-Polterovich) A regular smooth path $g : I \to \mathcal{D}$ is geodesic if and only if it is generated by a locally quasi-autonomous Hamiltonian.

The rest of this section is devoted to the proof of this theorem. The result will be deduced from a criterion for minimal geodesics which we formulate first. Recall that with a map $\varphi \in \mathcal{D}$ we have associated in Section 5.2 its action spectrum $\sigma(\varphi) = \{A(x, \varphi) | x \in Fix(\varphi)\}$. The action spectrum is a compact and nowhere dense set, as we have seen in Proposition 8. In general, it does not depend continuously on

5.7 Fixed points and geodesics on $\mathcal{D}$

φ. To a smooth path $g : [a, b] \to \mathcal{D}$ there belongs a bifurcation diagram $\Sigma(g)$. It is defined as the set

$$\Sigma(g) := \{(t, s) \in \mathbb{R}^2 \mid a < t \leq b \text{ and } s \in \sigma\Big(g(t)g(a)^{-1}\Big)\}.$$

If the path $g : [a, b] \to \mathcal{D}$ is generated by a quasi-autonomous Hamiltonian H, we have the distinguished fixed points $x_+, x_- \in Fix(g(t)g(a)^{-1})$ for every $a \leq t \leq b$. Their actions are the functions $\gamma_+, \gamma_- : [a, b] \to \mathbb{R}$ defined by

$$\gamma_+(t) = -\int_a^t H_t(x_-)dt, \quad \gamma_-(t) = -\int_a^t H_t(x_+)dt.$$

Note that $\gamma_-(t) \leq 0 \leq \gamma_+(t)$. The corresponding graphs $\Gamma_\pm(t) = (t, \gamma_\pm(t))$ are two distinguished curves in the bifurcation diagram which are continuous. Of course, the curve $\Gamma(t) = (t, 0)$ is always contained in $\Sigma(g)$ since the Hamiltonians considered have compact supports.

Definition. Consider a regular path $g : [a, b] \to \mathcal{D}$ generated by a quasi-autonomous Hamiltonian. Then its bifurcation diagram $\Sigma(g)$ is called simple if the following conditions are satisfied.

(i) Whenever $\gamma_+(t)$ does not vanish identically the following holds true for every $\tau > a$. There is no continous curve $(t, u(t)) \subset \Sigma(g), \tau \leq t \leq b$ connecting the point $(\tau, u(\tau)) \in \Gamma_+(\tau)$ with a point $(b, u(b))$ where $\gamma_-(b) < u(b) < \gamma_+(b)$.

(ii) Similarly we require for every $\tau > a$, whenever $\gamma_-(t)$ is not identically 0, that there is no continuous curve $(t, u(t)) \subset \Sigma(g), \tau \leq t \leq b$ connecting $(\tau, u(\tau)) \in \Gamma_-(\tau)$ with $(b, u(b))$ satisfying $\gamma_-(b) < u(b) < \gamma_+(b)$.

The first of the figures in Fig. 5.10 illustrates a diagram which is not simple.

The second figure belongs to a regular path $g : [0, b] \to \mathcal{D}$ starting for simplicity's sake at the identity, whose generating Hamiltonian H does not depend on time t, and which, moreover, does not admit any nonconstant periodic orbits. In this example the fixed points are the critical points x of H and their actions are the lines

$$A\Big(x, g(t)\Big) = -\int_0^t H(x)dt = -tH(x).$$

The corresponding bifurcation diagram is simple because the critical levels of H are nowhere dense.

Theorem 14. (Bialy-Polterovich) Assume the regular smooth path $g : I \to \mathcal{D}$ is generated by a quasi-autonomous Hamiltonian. If the bifurcation diagram $\Sigma(g)$ is simple, then g is a minimal geodesic.

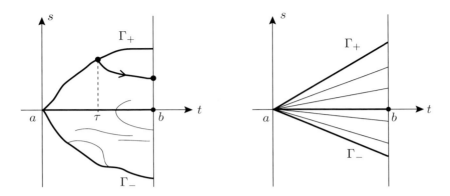

Fig. 5.10

Proof. The proof is based on the properties of the distinguished map $\gamma : \mathcal{D} \to \mathbb{R}$ introduced in Section 5.3 above. This map is, in particular, continuous and hence defines a continuous section of the bifurcation diagram. To simplify matters we assume that $I = [0, 1]$ and $g(0) = id$. The crucial step is the following lemma which extends Proposition 16.

Lemma 12. Assume H is quasi-autonomous and generates a path $g : [0, 1] \to \mathcal{D}$ with a simple bifurcation diagram $\Sigma(g)$. Then

$$\gamma\big(g(1)\big) = \gamma_+(1), \quad \gamma\big(g(1)^{-1}\big) = -\gamma_-(1).$$

Proof. Recall that $\gamma_-(t) \le \gamma(g(t)) \le \gamma_+(t)$ for all t, in view of Lemma 4. First we prove that $\gamma(g(\tau)) = \gamma_+(\tau)$ if $\tau > 0$ sufficiently small. Reparameterizing, we introduce $F(t, x) = \tau H(\tau t, x)$ for $0 \le t \le 1$ and $\tau > 0$. We may assume that $x_- = 0$ so that $F(t, x) \le F(t, 0) + \pi |x|^2$ if τ is sufficiently small. Proceeding as in the proof of Proposition 10, we use the monotonicity of γ and estimate

$$\gamma\big(g(\tau)\big) = \gamma(F) \ge -\int_0^1 F(t, 0) dt = -\int_0^\tau H(t, 0) dt = \gamma_+(\tau).$$

Hence $\gamma(g(\tau)) = \gamma_+(\tau)$ as claimed. Similarly, $\gamma(g(\tau)^{-1}) \ge -\gamma_-(\tau)$ provided τ is sufficiently small. The statements for $t = 1$ now follow from the assumptions on $\Sigma(g)$. Indeed, if $\gamma_+(t) \equiv 0$ then $H(t, x) \ge 0$ and hence by monotonicity of γ (Proposition 10) $\gamma(g(1)) = 0$ as desired. Assume now $\gamma_+(t) \not\equiv 0$. Proceeding indirectly, we assume that $\gamma(g(1)) < \gamma_+(1)$ and claim that $\gamma_-(1) < \gamma(g(1)) < \gamma_+(1)$. Indeed, if $\gamma_-(1) \equiv 0$, then $H(t, x) \le 0$ and hence $\gamma(g(1)) > 0$ by Proposition 10. If $\gamma_-(1) < 0$ we just note that $\gamma \ge 0$ and the claim is proved. Consider now

5.7 Fixed points and geodesics on $\mathcal{D}$

the graph Γ defined by the continuous function $u(t) = \min\{\gamma(g(t)), \gamma_+(t)\}$, for $\tau \leq t \leq 1$. Then $\Gamma \subset \Sigma(g)$ and we arrive at a contradiction to the required simplicity of $\Sigma(g)$. We have proved the first statement in the lemma; similarly one obtains the second statement. ∎

The proof of Theorem 14 follows now in view of Proposition 14; abbreviating $g(1) = \varphi$ we can estimate

$$\gamma_+(1) - \gamma_-(1) \leq \gamma(\varphi) + \gamma(\varphi^{-1}) \leq d(id, \varphi) \leq$$
$$\leq \text{length } (g|_{[0,1]}) = \gamma_+(1) - \gamma_-(1),$$

where we have used in the last equality that the Hamiltonian is quasi-autonomous. We conclude that $d(id, \varphi) = \text{length } (g|_{[0,1]})$. Hence g is a minimal geodesic and the proof of Theorem 14 is completed. ∎

A technically convenient way to describe a small neighborhood of a point $\varphi \in \mathcal{D}$ uses the linear space of generating functions which is explained in Appendix 1. Assume $\varphi \in \mathcal{D}$ is near the identity mapping and, in symplectic coordinates, given by $\varphi : (x, y) \mapsto (\xi, \eta) = \varphi(x, y) \in \mathbb{R}^{2n}$. Then it can implicitly be represented in terms of a single function S by the relations

(5.34)
$$x = \xi + \frac{\partial}{\partial y} S(\xi, y)$$
$$\eta = y + \frac{\partial}{\partial \xi} S(\xi, y),$$

where $S \in C_c^\infty(\mathbb{R}^{2n})$ is also of compact support and small in the C^2-sense. Conversely every such function S defines by means of the formula (5.34), a symplectic map $\varphi \in \mathcal{D}$ near the identity. We, therefore, have a mapping

$$\Phi : S \mapsto \Phi(S) \in \mathcal{D},$$

defined in a small neighborhood of zero in $C_c^\infty(\mathbb{R}^{2n})$. To an arc $S_t(\xi, y) = S(t, \xi, y)$, $t \in I$, there belongs a path $g(t) = \Phi(S_t) \in \mathcal{D}$ for $t \in I$. It is generated by a Hamiltonian $H \in \mathcal{H}$. Its relation to S_t is the Hamilton-Jacobi equation

$$H\left(t, \xi, y + \frac{\partial}{\partial \xi} S(t, \xi, y)\right) = \frac{\partial}{\partial t} S(t, \xi, y),$$

which holds true in our situation for all $(\xi, y) \in \mathbb{R}^{2n}$. This is, of course, well-known, and we refer to the book by C.L. Siegel and J. Moser [196]. In the linear space $C_c^\infty(\mathbb{R}^{2n})$ of generating functions we introduce the norm $||S|| := \max S - \min S$ and the associated metric $d_0(S_0, S_1) := ||S_0 - S_1||$. The arc length of a smooth arc $S_t, a \leq t \leq b$ is, therefore, defined by

$$\text{length } (S_{[a,b]}) = \int_a^b \left|\left|\frac{\partial S}{\partial t}\right|\right| dt.$$

A smooth arc S_t, $t \in I$ is called a minimal geodesic, if for every $a, b \in I$, $a < b$, we have $d_0(S_a, S_b) = \text{length}(S_{[a,b]})$ or, explicitly, if

$$||S_a - S_b|| = \int_a^b ||\frac{\partial S}{\partial t}|| dt \quad \text{for all } a < b,$$

$a, b \in I$. Examples of minimal geodesics are straight lines $(1-t)S_a + tS_b = S_t$ connecting two points S_a and S_b. Lemma 13 below expresses the fact that a smooth arc S_t is a minimal geodesic if and only if $\frac{\partial S}{\partial t}$ is quasi-autonomous. It will turn out that the image of a geodesic under the map Φ is a geodesic in the metric space $(\mathcal{D}, d)$. Indeed we shall deduce from Theorem 14 that Φ is an isometry. This shows, in particular, that the metric space $(\mathcal{D}, d)$ is "locally flat" if one considers the space $(C_c^\infty(\mathbb{R}^{2n}), ||\cdot||)$ as a "flat" space.

Theorem 15.
$$d\Big(\Phi(S_0), \Phi(S_1)\Big) = ||S_0 - S_1||.$$

Proof. Define the arc $S_t = (1-t)S_0 + tS_1$, $0 \le t \le 1$ connecting S_0 with S_1. We shall prove that the path $g(t) := \Phi(S_t)$ is a minimal geodesic in $(\mathcal{D}, d)$. It connects $g(0) = \Phi(S_0)$ with $g(1) = \Phi(S_1)$. In order to verify the assumptions of Theorem 14 we shall prove that $\Sigma(g)$ is simple and the generating Hamiltonian quasi-autonomous. Consider a fixed point $(\xi, \eta) \in \mathbb{R}^{2n}$ of $g(\tau)g(0)^{-1}$, i.e., $g(\tau)g(0)^{-1}(\xi, \eta) = (\xi, \eta)$ for some $0 < \tau \le 1$ fixed. Introducing $(\bar\xi, \bar\eta) = g(0)(x, y)$, this is equivalent to the conditions

$$(\xi, \eta) = g(\tau)(x, y) = g(0)(x, y).$$

Expressed in terms of the generating functions S_τ of $g(\tau)$, respectively, S_0 of $g(0)$ by means of the formula (5.34), this is equivalent to

$$x = \xi + \frac{\partial}{\partial y} S_\tau(\xi, y) = \xi + \frac{\partial}{\partial y} S_0(\xi, y)$$

$$\eta = y + \frac{\partial}{\partial \xi} S_\tau(\xi, y) = y + \frac{\partial}{\partial \xi} S_0(\xi, y).$$

Consequently $\nabla S_\tau(\xi, y) = \nabla S_0(\xi, y)$ and hence $\nabla(S_0 - S_1)(\xi, y) = 0$ and, reversing the steps, $g(t)g(0)^{-1}(\xi, \eta) = (\xi, \eta)$ for all t. We have verified that the periodic solutions of H correspond one-to-one to the critical points of the function $S_0 - S_1$ and are, moreover, all constant. From the Hamilton-Jacobi equation we deduce for the critical points $(\xi, \eta) = g(0)(x, y)$, that

$$H\Big(t, \xi, y + \frac{\partial}{\partial \xi} S_0(\xi, y)\Big) = (S_1 - S_0)(\xi, y).$$

Hence, the Hamiltonian function is quasi-autonomous. The action of a fixed point $P = (\xi, \eta)$ which corresponds to the critical point (ξ, y) of $S_1 - S_0$ is easily com-

5.7 Fixed points and geodesics on $\mathcal{D}$

puted:

$$A\Big(P, g(t)g(0)^{-1}\Big) = -\int_0^t H_s(P)ds = -\int_0^t (S_1 - S_0)(\xi, y)ds$$
$$= -t(S_1 - S_0)(\xi, y).$$

Since the critical values are nowhere dense, the bifurcation diagram $\Sigma(g)$ is simple. We have verified for the path $g(t)$ that the assumptions of Theorem 14 are met. Using that $||H_t|| = ||S_1 - S_0||$ we find

$$d\Big(\Phi(S_0), \Phi(S_1)\Big) = \int_0^1 ||H_t||dt = ||S_1 - S_0||$$

and the proof of Theorem 15 is completed. ∎

In order to deduce Theorem 13 from Theorem 15, we start with two technical observations. In order to check that a function is quasi-autonomous the following lemma is helpful.

Lemma 13. A smooth function $H : [a, b] \times \mathbb{R}^{2n} \to \mathbb{R}$ of compact support is quasi-autonomous if and only if

$$\int_a^b ||H_t||dt = ||\int_a^b H(t, x)dt||.$$

Proof. Clearly, if H is quasi-autonomous, the equation follows from the definition of the norm. Assuming the equality to hold we now show that H is quasi-autonomous. By definition, there exists a compact set $K \subset \mathbb{R}^{2n}$ containing the supports of H_t for every t. Defining the closed sets

$$K_t = \{x \in K \mid H_t(x) = \max H_t\}, \quad a \leq t \leq b,$$

we have to prove that $\bigcap K_t \neq \emptyset$, where $a \leq t \leq b$. Since K has the finite intersection property it suffices to verify that $K_{t_1} \cap \cdots \cap K_{t_N} \neq \emptyset$ for every finite sequence $t_1 < t_2 < \cdots < t_N$. Arguing by contradiction, we assume $K_{t_1} \cap \cdots \cap K_{t_N} = \emptyset$. Thus we find an $\varepsilon > 0$ satisfying

$$\sum_{j=1}^N H_{t_j}(x) + \varepsilon \leq \sum_{j=1}^N \max H_{t_j} \quad \text{for all } x.$$

Consequently, by taking small intervals around the t_j's we find an open set $V \subset [a, b]$ on which

$$\int_V H_t(x)dt + \frac{\varepsilon}{2} \leq \int_V \max H_t dt \quad \text{for all } x,$$

and thus arrive at the contradiction

$$\| \int_a^b H_t \| + \frac{\varepsilon}{2} \leq \int_a^b \|H_t\|.$$

Arguing similarly for the minimum, the proof is completed. ∎

Lemma 14. Consider an arc S_t, $t \in I = [a,b]$ and the corresponding arc $H_t, t \in I$ generating the path $g(t) = \Phi(S_t)$ in $\mathcal{D}$. Then $\frac{\partial S}{\partial t}$ is quasi-autonomous in I if and only if H is quasi-autonomous in I. Moreover, in this case the distinguished points $P_\pm = (\xi_\pm, \eta_\pm)$ respectively $Q_\pm = (\xi_\pm, y_\pm)$ of min and max for H respectively $\frac{\partial S}{\partial t}$, are related by (5.34), the generating function being $S = S_a$. In addition $H(t, P_\pm) = \frac{\partial S}{\partial t}(t, Q_\pm)$ for all $t \in I$.

Proof. If H is quasi-autonomous with respect to the distinguished points $P_\pm$, then $g(t)g(0)^{-1}(P_\pm) = P_\pm$ for all $a \leq t \leq b$. Hence, arguing as in the proof of Theorem 15, $\nabla S_t(Q_\pm)$ is independent of t and so $\nabla S_t(Q_\pm) = \nabla S_a(Q_\pm)$. Consequently $\frac{\partial}{\partial t}S$ is quasi-autonomous in view of the Hamilton-Jacobi equation. Conversely, if $\frac{\partial S}{\partial t}$ takes on its max and min at the fixed points $Q_\pm$, then $S_t - S_a$ also takes on its min and max at these points, for all $t \in I$. Hence $\nabla S_t(Q_\pm) = \nabla S_a(Q_\pm)$ and consequently $g(t)g(0)^{-1}(P_\pm) = P_\pm$. In view of the Hamilton-Jacobi equation, the function H is quasi-autonomous with respect to $P_\pm$. ∎

Proof of Theorem 13. The metric in $(\mathcal{D}, d)$ is bi-invariant and hence we may consider a path $g(t) \in \mathcal{D}$ near the identity $g(0) = id$, where $t \in [-\varepsilon, \varepsilon] = I$ for a sufficiently small $\varepsilon > 0$. It is generated by $H \in \mathcal{H}$ and $g(t) = \Phi(S_t)$, $t \in I$ for an arc S_t of generating functions. We have to show that g is a minimal geodesic in $(\mathcal{D}, d)$ if and only if H, or equivalently $\frac{\partial S}{\partial t}$ (by Lemma 14), is quasi-autonomous. But by Lemma 13, $\frac{\partial S}{\partial t}$ is quasi-autonomous if and only if

$$(5.35) \qquad \| \int_{-\varepsilon}^{\varepsilon} \frac{\partial S}{\partial t} \| = \int_{-\varepsilon}^{\varepsilon} \|\frac{\partial S}{\partial t}\|.$$

The right hand side is, in view of the Hamilton-Jacobi equation, equal to

$$\int_{-\varepsilon}^{\varepsilon} \|\frac{\partial S}{\partial t}\| = \int_{-\varepsilon}^{\varepsilon} \|H_t\| = \text{length } (g|_{[-\varepsilon, \varepsilon]}),$$

while the left hand side is equal to

$$\| \int_{-\varepsilon}^{\varepsilon} \frac{\partial S}{\partial t} \| = \|S_\varepsilon - S_{-\varepsilon}\| = d\Big(g(-\varepsilon), g(\varepsilon)\Big),$$

in view of Theorem 15. Hence (5.35) holds if and only if $g(t)$ is a minimal geodesic for $t \in I$. The proof of Theorem 13 is completed. ∎

Chapter 6
The Arnold conjecture, Floer homology and symplectic homology

In the sixties V.I. Arnold announced several seminal and fruitful conjectures in symplectic topology. This chapter is devoted to his conjecture about fixed points of symplectic mappings which originates in questions of celestial mechanics dating back to the turn of the century. Reformulated dynamically, this conjecture gives a lower bound for the number of global forced oscillations of a time-dependent Hamiltonian vector field on a compact symplectic manifold in terms of the topology of this manifold. Forced oscillations are singled out by the action principle; solutions of the conjecture gave rise to new ideas and techniques in the calculus of variations and in nonlinear elliptic systems. In particular, it prompted A. Floer 1986 to the construction of his homology theory.

We shall describe in the following the history of the Arnold conjecture and start with a proof for the special case of the standard torus T^{2n} in Section 6.2. Here we can make use of the action principle in the $H^{1/2}$ setting as introduced in Chapter 3. But this time the aim is to find all its critical points. Our strategy is inspired by C. Conley's topological approach to dynamical systems: we shall study the structure of the set of all bounded solutions of the regularized gradient equation belonging to the action functional defined on the contractible loops of T^{2n}. This invariant set consists of the critical points together with all their connecting orbits. This way the study of the gradient flow in the infinite dimensional loop space is reduced to the study of a gradient like continuous flow of a compact metric space whose rest points are the desired critical points. It turns out that the compact space contains the cohomology of the torus. The number of rest points is then estimated by the Ljusternik-Schnirelman theory presented in Section 6.3.

A reinterpretation of the geometric ideas underlying the proof of the torus case in terms of partial differential equations will lead us in Section 6.4 to the proof of the Arnold conjecture for the larger class of compact symplectic manifolds (M, ω) satisfying $[\omega]|\pi_2(M) = 0$. In this general case there is no natural regularization and we are forced to investigate the structure of the bounded solutions of the not regularized gradient system. They are smooth solutions of a special system of first order elliptic partial differential equations of Cauchy-Riemann type. These solutions are related to M. Gromov's pseudo holomorphic curves in (M, ω). The crucial compactness of the solution set will be established by means of an analytical technique sometimes called bubbling off analysis. In order to get an insight into the topology of the solution set we shall use an approximation procedure. We approximate the bounded solutions by solutions of a special elliptic boundary value problem, to which we can apply Fredholm theory based on the Cauchy-Riemann operator. Using the continuity property of the Alexander-Spanier cohomology we then deduce the Arnold conjecture by the Ljusternik-Schnirelman theory as in the torus case.

Assuming, in addition, all the forced oscillations to be nondegenerate, the geometric ideas will lead us in Section 6.5 to A. Floer's new approach to Morse theory and Floer homology. It makes use of an additional structure of the set of bounded solutions of the not regularized gradient equation: namely, since M is compact, there are only finitely many nondegenerate forced oscillations and every bounded solution of the gradient equation tends in forward and backward time asymptotically (exponentially) to a specific forced oscillation of the Hamiltonian system. The 1-dimensional connections determine an algebraic complex generated by the forced oscillations, its grading is defined by a Maslov-type index associated with a nondegenerate forced oscillation, which plays the role of the Morse index. We shall merely outline Floer's beautiful ideas. A combination of Floer's construction with our construction of the special capacity c_0 of Chapter 3 results in a symplectic homology theory which, however, is not yet in its final form. It will be sketched, without proofs, in Section 6.6. The technical requirements of these theories, in particular, the intricate transversality arguments are quite advanced and beyond the scope of this book. Chapter 6 is, in contrast to the previous chapters, not self-contained.

6.1 The Arnold conjecture on symplectic fixed points

In his search for periodic solutions in the restricted three body problem of celestial mechanics, H. Poincaré constructed an area-preserving section map of an annulus A on the energy surface. The annulus was bounded by the so-called direct and retrograde periodic orbits. It inspired him in 1912 to formulate the following theorem [177]. It is called the last geometric theorem of Poincaré announced shortly before his death. It was later proved by G. Birkhoff.

Theorem 1. (Poincaré-Birkhoff) Every area preserving and orientation preserving homeomorphism of an annulus $A := S^1 \times [a, b]$ rotating the two boundaries in opposite directions possesses at least 2 fixed points in the interior (Fig.6.1).

The strength of this theorem is that it provides at once infinitely many periodic points if it is applied to the iterates, which leads to infinitely many periodic solutions in applications, see [24]. In 1913 G. Birkhoff [23] succeeded in proving this theorem by an ingenious but strictly two dimensional argument. Note that the theorem is global in nature; there are no interior assumptions required on the mapping. It is not a topological result: it is wrong if the area-preserving assumption is dropped. S. Lefschetz, who undoubtedly would have enjoyed demonstrating to Birkhoff that the famous Poincaré-Birkhoff fixed point theorem is a trivial conse-

6.1 The Arnold conjecture on symplectic fixed points

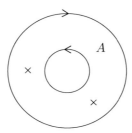

Fig. 6.1

quence of his own topological fixed point theory, had to admit defeat: he could not apply his theory since the Euler characteristic of the annulus vanishes, $\chi(A) = 0$.

The Euler characteristic of the two sphere $(\chi(S^2) = 2)$ does not vanish, hence by Lefschetz's theory every continuous map f of S^2 homotopic to the identity possesses at least one fixed point, as is, of course, well-known. The map f may have only one fixed point, as for example, the translation $z \mapsto z+1$ on the Riemann sphere shows: its only fixed point is the north pole at ∞.

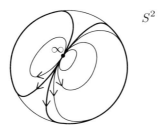

Fig. 6.2

It is, however, a striking fact that f possesses at least 2 fixed points if it is, in addition, area preserving.

Theorem 2. (N. Nikishin, C. Simon, 1974) A homeomorphism f of S^2 homotopic to the identity which preserves a regular measure μ has at least 2 fixed points. In particular, every diffeomorphism of S^2 leaving an area form invariant $f^*\omega = \omega$ possesses at least 2 fixed points.

Proof. The second statement is a consequence of the first one. Indeed, the map has degree 1, hence is homotopic to the identity by H. Hopf's theorem. To prove the first statement, assume $P^* \in S^2$ is the fixed point of f guaranteed by the Lefschetz theory. Then the map $g := f|(S^2 \backslash P^*)$ can be identified with a homeomorphism of the plane $\mathbb{R}^2$. If g had no fixed point then by Brouwer's translation theorem there would be an open set D having the property that $g^j(D) = f^j(D)$ are mutually disjoint for all $j > 0$. But then, since f preserves the measure

$$\sum_{j=0}^{n} \mu\Big(f^j(D)\Big) = (n+1)\,\mu(D) \leq \mu(S^2),$$

for every integer $n \geq 1$ and hence $\mu(D) = 0$. This contradicts the assumption that $\mu(U)$ does not vanish for an open set U. ∎

The translation theorem of Brouwer is a very special two dimensional phenomenon which attracted the attention of many mathematicians, [33, 34, 80, 101, 109, 141, 206]. The same argument as in Theorem 2, used by C. Loewner in the sixties in his Stanford lectures, also shows that a measure-preserving homeomorphism of the open disc in the plane always possesses at least one fixed point. This is clearly false if the homeomorphism does not preserve a regular measure. The proofs by C. Simon [200] and N. Nikishin [169] of the theorem above use a different argument. They show that the fixed point index $j(P)$ for an area preserving diffeomorphism in the plane is less then or equal to 1. Thus the fixed point formula $\chi(S^2) = \sum j(P)$ requires at least 2 fixed points.

V. Arnold observed in [11] that the Poincaré-Birkhoff result could be derived from a fixed point theorem for the two dimensional torus $T^2 = \mathbb{R}^2/\mathbb{Z}^2$ constructed by gluing together two identical annuli along their boundaries; we refer to [46] for details.

$$T^2 = \mathbb{R}^2/\mathbb{Z}^2$$
$$\chi(T^2) = 0.$$

Fig. 6.3

In contrast to S^2, the Lefschetz fixed point theory is not adequate for precisely this surface, since $\chi(T^2) = 0$. There are, indeed, area preserving diffeomorphisms of the torus without fixed points, as the translation maps on the covering space $\mathbb{R}^2$ given by

(6.1) $$X = x + c_1, \quad Y = y + c_2,$$

demonstrate. Consequently, the class of symplectic mappings on T^2 needs to be restricted if it is required that they possess fixed points. With V. Arnold we consider measure-preserving diffeomorphisms ψ of T^2 which are homologous to the identity map, and hence are, on $\mathbb{R}^2$, given by $\psi : (x,y) \mapsto (X,Y)$:

(6.2) $$\begin{aligned} X &= x + p(x,y) \\ Y &= y + q(x,y) \end{aligned}$$

6.1 THE ARNOLD CONJECTURE ON SYMPLECTIC FIXED POINTS

with periodic functions p and q, we require that ψ preserves the center of mass, hence excluding in particular the translations. Summarizing, the diffeomorphisms ψ of T^2 are required to meet the following conditions

(i) ψ is homologous to the identity

(ii) $dX \wedge dY = dx \wedge dy$

(iii) $\int_{T^2} p = 0 = \int_{T^2} q$.

Proposition 1. A diffeomorphism ψ of T^2 satisfying (i)–(iii) possesses ≥ 3 fixed points provided it is sufficiently close (in the C^1 norm) to the identity. If, in addition, all the fixed points are nondegenerate, then ψ possesses ≥ 4 fixed points.

Proof. Since $\psi : (x,y) \mapsto (X,Y)$ is sufficiently close to the identity map it can be represented by a generating function $F : \mathbb{R}^2 \to \mathbb{R}$ in the implicit form

$$X = x + \frac{\partial F}{\partial Y}(x,Y)$$

(6.3)
$$y = Y + \frac{\partial F}{\partial x}(x,Y),$$

see Appendix 1. The gradient of F is periodic and it follows from assumption (iii) that the function F itself is periodic and hence is a function on T^2. We conclude from (6.3) that the fixed points of ψ on T^2 are in one-to-one correspondence with the critical points of the function F on T^2. By the Ljusternik-Schnirelman theory every smooth function on the 2-torus possesses at least 3 critical points (see Section 6.3 below), and by the Morse theory it possesses at least 4 critical points, if a priori we know that all the critical points are nondegenerate. The proposition follows. ∎

There are, of course, symplectic maps on T^2, even ones close to the identity having precisely 3 fixed points. To see this we take a smooth function G on T^2 with precisely 3 critical points. An example of such a function is

$$G(x,y) = \sin \pi x \cdot \sin \pi y \cdot \sin \pi (x+y),$$

whose level lines look as shown in Fig. 6.4.

G has a maximum in the left triangle, a minimum in the right triangle and a monkey saddle in the corners of the unit square. If $\varepsilon > 0$ is small, we define a map $\psi : (x,y) \to (X,Y)$ implicitly by (6.3) with the generating function $F = \varepsilon G$. Then ψ is close to the identity, satisfies (i)–(iii), and has 3 fixed points, namely the critical points of G on T^2. Similarly, the function $G(x,y) = \sin 2\pi x + \sin 2\pi y$ defines a map ψ having 4 fixed points on T^2 which are all nondegenerate. ∎

The idea of relating fixed points of symplectic mappings to critical points of an associated function goes back to H. Poincaré [176] and is used quite frequently

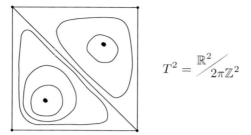

Fig. 6.4

in order to establish local existence results. For example A. Weinstein [222, 224] makes use of it in order to show that a symplectic diffeomorphism of a compact and simply connected symplectic manifold M possesses at least as many fixed points as a function on M possesses critical points provided it is close to the identity map. For more general results and references we refer to J. Moser [165]. In sharp contrast to all these perturbation results, the Arnold conjecture we now turn to asks for fixed points of arbitrary symplectic mappings. For the special case of T^2, V. Arnold conjectured in [10] that Proposition 1 holds without the restrictive assumption that ψ is close to the identity map.

In order to introduce the restricted class of symplectic diffeomorphisms suitable for fixed point theory we first mention that for a diffeomorphism ψ on T^2, the assumptions (i)–(iii) are equivalent to the requirement that ψ belongs to the flow of a time-dependent Hamiltonian vector field X_H on T^2. The proof is not easy and we refer to C. Conley and E. Zehnder, [55]. Hence ψ is a Hamiltonian map of T^2 in view of the following:

Definition. Let (M, ω) be a symplectic manifold. A diffeomorphism ψ of M is called Hamiltonian if it belongs to the flow of a time-dependent Hamiltonian vector field X_H on M. To be more precise there exists a smooth function $H : S^1 \times M \to \mathbb{R}$ such that $\psi = \varphi^1$ where the family $\varphi^t, t \in \mathbb{R}$ of symplectic diffeomorphisms of M satisfies

$$\frac{d}{dt}\varphi^t = X_H(\varphi^t)$$
$$\varphi^0 = id.$$

In his thesis, A. Banyaga [16] shows that a map ψ is a Hamiltonian map if and only if it belongs to the commutator subgroup of the identity component of the group of symplectic diffeomorphisms of M. In addition, he proves that ψ is Hamiltonian if and only if the Calabi invariant of ψ vanishes. A Hamiltonian map being homotopic to the identity has, of course, a fixed point according to Lefschetz's theory provided the Euler characteristic $\chi(M)$ does not vanish.

6.1 The Arnold conjecture on symplectic fixed points

The Arnold conjecture. Every Hamiltonian diffeomorphism ψ of a compact symplectic manifold (M, ω) possesses at least as many fixed points as a function $F : M \to \mathbb{R}$ on M possesses critical points.

In view of the Ljusternik-Schnirelman theory and the Morse theory for functions on M we reformulate this conjecture in the following somewhat weaker form:

$$\#\{\text{ fixed points of } \psi\} \geq \text{cup-length of } M + 1,$$

and, if a priori all the fixed points of ψ are nondegenerate,

$$\geq \text{sum of the Betti numbers of } M.$$

Recall that a fixed point p of ψ is called nondegenerate if 1 is not an eigenvalue of the linearized map $d\psi(p) \in \mathcal{L}(T_p M)$. The cup-length of a manifold M is defined as the longest product in the cohomology ring: $CL(M) = \sup\{k | \exists \alpha_1, \alpha_2, \ldots \alpha_k \in H^*(M) \text{ with } \dim \alpha_j \geq 1 \text{ such that } \alpha_1 \cup \alpha_2 \cup \cdots \cup \alpha_k \neq 0\}$. Let $SB(M)$ denote the sum of the Betti numbers $\beta_k = \dim H^k(M), 0 \leq k \leq \dim M$. Introducing the Poincaré polynomial

$$P(t, M) = \sum_{k=0}^{\dim M} \beta_k t^k, \quad \beta_k = \dim H^k(M)$$

we have $SB(M) = P(1, M)$ and, of course, $P(-1, M) = \chi(M)$ is the Euler characteristic of M. In the special case of a $2n$-dimensional torus T^{2n} these topological invariants are $CL(T^{2n}) = 2n$ and $SB(T^{2n}) = 2^{2n}$ while $\chi(T^{2n}) = 0$. In particular, for T^2 in Proposition 1 we see that $3 = CL(T^2) + 1$ and $4 = SB(T^2)$.

We prefer to reformulate the conjecture dynamically. A Hamiltonian map ψ is, by definition, the time 1-map of the flow of the Hamiltonian equation $\dot{x} = X_H(x)$ on M with a smooth function $H : \mathbb{R} \times M \to \mathbb{R}$ periodic in time, $H(t, x) = H(t + 1, x)$. Consequently, p is a fixed point of ψ if and only if $p = x(0)$ is the initial condition of a solution $x(t)$ of the Hamiltonian equation which is periodic of period 1 and hence satisfies $x(t + 1) = x(t)$ for all $t \in \mathbb{R}$. Such a solution is also called a forced oscillation. Instead of looking for fixed points of a map, we can look as well for periodic solutions of a time-dependent Hamiltonian vector field; the Arnold conjecture formulates a universal lower bound on the number of forced oscillations in terms of topological invariants of the symplectic manifold.

If (M, ω) is a compact symplectic manifold and $H : \mathbb{R} \times M \to \mathbb{R}$ a smooth time periodic function $H(t + 1, x) = H(t, x)$, then for the time-dependent Hamiltonian equation

$$\dot{x} = X_H(x), \quad x \in M$$

the following holds:

$$\#\{\text{ contractible 1-periodic solutions }\} \geq CL(M) + 1.$$

$(\geq SB(M)$ if all the contractible 1-periodic solutions are nondegenerate$)$.

Here a periodic solution x is called nondegenerate if 1 is not a Floquet multiplier of x. Since there are no assumptions on the function H, we only look for periodic solutions which are contractible. Indeed the arc length of a 1-periodic solution $x(t)$ is small if the Hamiltonian vector field is small, so that the orbit is necessarily contractible. This follows from the formula

$$\text{length}(x) = \int_0^1 |\dot{x}(t)| dt = \int_0^1 |X_H(x(t))| dt \,,$$

where we have chosen a Riemannian metric on M.

Take now an almost complex structure J on the symplectic manifold (M, ω) satisfying $\omega_x(v, Ju) = g_x(v, u)$ for all $v, u \in T_x M$, where g is a Riemannian metric on M. Then every Hamiltonian vector field X_H on M is of the form

$$\dot{x} = J_x \nabla H(t, x) \,, \quad x \in M$$

where the gradient of H with respect to x is defined by means of the associated Riemannian metric g. Assume now that $H(t, x) = H(x)$ does not depend on the time and is, moreover, sufficiently small in the C^2 sense. Then every 1-periodic solution $x(t)$ is contractible and hence contained in a Darboux chart so that, by the argument in the proof of Proposition 17 in Chapter 5, it is necessarily constant, i.e., $x(t) \equiv x(0)$. Consequently, the 1-periodic solutions are precisely the critical points of the function $H : M \to \mathbb{R}$ and the Arnold conjecture is reduced, in this special case, to the theories of Ljusternik-Schnirelman and Morse about critical points of functions.

In the special case of a standard torus (T^{2n}, ω_0), the Arnold conjecture was proved in 1983 by C. Conley and E. Zehnder [55]. The proof is based on the action principle for forced oscillations. The same principle allowed A. Floer [82], 1986 and J.C. Sikorav [198] 1985 to confirm the conjecture for large classes of symplectic manifolds admitting compatible Riemannian structures with nonnegative sectional curvatures. These classes include, in particular, all oriented surfaces S_g with positive genus g.

Fig. 6.5

$$CL(S_g) = 3 \,, \quad SB(S_g) = 2 + 2g \,.$$

6.1 THE ARNOLD CONJECTURE ON SYMPLECTIC FIXED POINTS

The earlier proofs by Ya. Eliashberg for surfaces [77] are based on different ideas and were never published. The case $S^2 = \mathbb{C}P^1$ was extended to higher dimensions by B. Fortune [93, 94] in 1985, who verified the conjecture for $\mathbb{C}P^n$ with its standard symplectic structure. Accordingly every Hamiltonian map on $\mathbb{C}P^n$ has $\geq n+1$ fixed points. Note that in this case $CL + 1 = SB = \chi = n+1$, so that in the nondegenerate case the result follows already from the Lefschetz theorem. A. Weinstein pointed out in [227] 1986 that the Arnold conjecture is true for every symplectic manifold (M, ω) if the Hamiltonian map ψ is generated by a Hamiltonian vector field sufficiently small in the C^0 norm. We shall describe the more recent global results and developments in Sections 4 and 5. It will turn out, in particular, that the forced oscillations of a Hamiltonian vector field X_H on a compact symplectic manifold are more intimately related to the topology of the underlying manifold M than it is suggested by the conjecture above. First we study in detail the model case of the standard torus.

Before we turn to the proofs of the Arnold conjecture in higher dimensions, we would like to refer to some recent two dimensional results related to the Poincaré-Birkhoff theorem. As mentioned before, the Arnold conjecture for the two torus T^2 may be viewed as a generalization of this theorem for diffeomorphisms and one may ask whether there is a topological analogue. Quite recently, P. Le Calvez [139] succeeded in proving that every homeomorphism of T^2 homotopic to the identity, preserving a regular measure and preserving the center of gravity (i.e., there is a lift to $\mathbb{R}^2$ satisfying the condition (iii) in (6.2) above) possesses at least three fixed points. Earlier M. Flucher [91] found two fixed points by extending techniques and arguments of J. Franks [96, 97]. Attempts to generalize these results for homeomorphisms of higher dimensional tori must fail: on T^n, $n \geq 3$, there are measure-preserving diffeomorphisms satisfying the assumptions analogous to the T^2 case without any fixed points, see M. Flucher [91]. An interesting modification of the Poincaré-Birkhoff theorem for the annulus has been discovered by J.M. Gambaudo and P. Le Calvez [103] and, independently, by J. Franks [100]:

Theorem 3. A diffeomorphism of the closed annulus $A = \mathbb{R}/\mathbb{Z} \times [0,1]$, isotopic to the identity and preserving a regular measure, possesses infinitely many periodic orbits in the interior of A, provided it has a fixed point.

This result has astonishing consequences. By introducing polar coordinates (r, ϑ) in the neighborhood of a fixed point, one can blow up the fixed point into an invariant circle defined by $r = 0$. Given two fixed points on a sphere S^2, one obtains in this way a map of a closed annulus and hence deduces from Theorem 3 the following:

Corollary. Consider a C^2-diffeomorphism of S^2 preserving the orientation and preserving a regular measure. If it has 3 fixed points then it has infinitely many periodic orbits.

In view of Theorem 2 such a diffeomorphism always has at least two fixed points and it is sufficient to assume one of them to be hyperbolic. Indeed, the blown up invariant circle of a hyperbolic fixed point contains 4 fixed points so that Theorem 3 applies.

J. Franks used Theorem 3 for his breakthrough in an old problem of Riemannian geometry. In [25] G. Birkhoff describes a reduction of the study of the geodesic flow on a Riemannian two-sphere to the study of an area preserving annulus map whose periodic points correspond to the closed geodesics. He showed, for positive Gaussian curvature, that, associated to a simple closed geodesic, such a Birkhoff annulus map is defined. Franks, assuming that S^2 is equipped with a Riemannian metric having a simple closed geodesic for which the associated Birkhoff annulus map is defined, demonstrates, that this map has a fixed point and hence, by Theorem 3, infinitely many periodic points. If, on the other hand, a Riemannian metric on S^2 has no simple closed geodesic for which the Birkhoff annulus map is defined, then V. Bangert [15] established infinitely many closed geodesics by variational methods applied to the energy functional on the loop space of S^2. It is well known that every Riemann sphere possesses a simple closed geodesic. Hence, on a Riemann two-sphere there are always infinitely many closed geodesics which are distinct as point sets.

References related to the Poincaré-Birkhoff theorem and its generalizations are: [4, 35, 38, 49, 48, 61, 91, 95, 96, 97, 98, 99, 103, 109, 138, 140, 141, 139, 200, 206, 143].

6.2 The model case of the torus

We shall prove the Arnold conjecture for the standard torus (T^{2n}, ω_0), where $T^{2n} = \mathbb{R}^{2n}/\mathbb{Z}^{2n}$ and where ω_0 denotes the symplectic structure induced by the standard one from $\mathbb{R}^{2n}$. A Hamiltonian vector field on this torus depending periodically on time is given simply by a smooth function $H : \mathbb{R} \times \mathbb{R}^{2n} \to \mathbb{R}$, $(t, x) \mapsto H(t, x)$ which is periodic in all its variables:

(6.4) $\quad H(t+1, x) = H(t, x) = H(t, x+j), \quad$ all $j \in \mathbb{Z}^{2n}$.

The associated Hamiltonian system is represented by

(6.5) $\quad \dot{x} = J \nabla H(t, x), \quad x \in \mathbb{R}^{2n}$.

A periodic solution $t \mapsto x(t) = x(t+1) \in \mathbb{R}^{2n}$ of (6.5) represents a contractible periodic solution of the vector field on T^{2n}. In contrast, a solution of the form $x(t) = jt + \xi(t)$, with $j \neq 0 \in \mathbb{Z}^{2n}$ and $\xi(t) = \xi(t+1)$ represents a 1-periodic solution on T^{2n} which is not contractible. Yet, such solutions might not exist on T^{2n}, as was explained in the previous section. We shall prove the following qualitative existence statement for Hamiltonian flows:

6.2 THE MODEL CASE OF THE TORUS

Theorem 3. (C. Conley and E. Zehnder) Every smooth time periodic Hamiltonian vector field on the standard torus (T^{2n}, ω_0) possesses at least $2n+1$ contractible 1-periodic solutions (resp. at least 2^{2n} if all of them are nondegenerate).

We shall merely consider the first statement which does not impose any assumptions on the Hamiltonian on T^{2n}. Our proof in the following is based on the action principle and is designed to illustrate, in the analytical framework of Chapter 3, some of the underlying geometric ideas which will be used for a large class of symplectic manifolds described later.

A contractible loop $S^1 \to T^{2n}$ is the projection of a smooth loop $x : S^1 \to \mathbb{R}^{2n}$ represented by its Fourier expansion

$$(6.6) \qquad x(t) = \sum_{k \in \mathbb{Z}} e^{2\pi k Jt} x_k, \qquad x_k \in \mathbb{R}^{2n}.$$

Two such loops x and y on $\mathbb{R}^{2n}$ induce the same contractible loop on T^{2n} iff $x(t) - y(t) = j \in \mathbb{Z}^{2n}$ for all $t \in \mathbb{R}$, so that $x_k = y_k$ for $k \neq 0$ and $x_0 = y_0 + j$. Consequently, we can identify the space of smooth contractible loops on T^{2n} with the space

$$T^{2n} \times E^\infty,$$

where $E^\infty := \{x \in C^\infty(S^1, \mathbb{R}^{2n}) \mid \int_0^1 x(t)dt = 0\}$. This space is the quotient under the $\mathbb{Z}^{2n}$-action of the space $\mathbb{R}^{2n} \times E^\infty \cong C^\infty(S^1, \mathbb{R}^{2n})$. Using the same analytical framework as in Chapter 3, we now replace the space of smooth loops E^∞ by the Sobolev space

$$(6.7) \qquad E = \left\{ x \in H^{1/2}(S^1, \mathbb{R}^{2n}) \mid \int_0^1 x(t)dt = 0 \right\},$$

which has an orthogonal splitting $E = E^- \oplus E^+$ according to the decomposition

$$x = \sum_{k \neq 0} e^{2\pi k Jt} x_k = \sum_{k<0} + \sum_{k>0} = x^- + x^+.$$

We shall denote the loop space of $\mathbb{R}^{2n}$ by $\widehat{\Omega} = \mathbb{R}^{2n} \times E$ and the loop space on T^{2n} by $\Omega = T^{2n} \times E$. The smooth action $\varphi : \widehat{\Omega} \to \mathbb{R}$ is defined by

$$(6.8) \qquad \varphi(x) = a(x) - b(x), \qquad x \in \widehat{\Omega}$$

where

$$(6.9) \qquad a(x) = -\frac{1}{2}\|x^-\|^2 + \frac{1}{2}\|x^+\|^2$$

stems from the symplectic structure, and where

$$(6.10) \qquad b(x) = \int_0^1 H\big(t, x(t)\big) dt$$

contains the given Hamiltonian H. Clearly, φ is invariant under the $\mathbb{Z}^{2n}$ action: $\varphi(x+j) = \varphi(x), x \in \widehat{\Omega}$, and hence defines a smooth function on the loop space $\Omega = T^{2n} \times E$ of contractible loops on the torus. We know from Chapter 3 that a critical point $x \in \widehat{\Omega}$ of φ,

(6.11) $$\nabla \varphi(x) = 0,$$

is actually a smooth loop. The critical points are precisely the desired contractible periodic solutions of (6.5) we are looking for. We recall that the gradient of φ in $H^{1/2}$ is given by

(6.12) $$\nabla \varphi(x) = -x^- + x^+ - b'(x),$$

where we denote by $b'(x)$ the gradient of b in $H^{1/2}$. Also recall that $b'(x) = j^* \nabla b(x)$, with the L^2-gradient $\nabla b(x) = \nabla H(x) \in L^2$ of b, and $j^* : L^2 \to H^{1/2}$ the compact operator introduced in Chapter 3. Since H is a periodic function, its gradient on $\mathbb{R}^{2n}$ is bounded and hence b' maps $\widehat{\Omega}$ into a compact subset of $\widehat{\Omega}$. In order to find the critical points of φ, we shall study the quite artificial gradient flow on the loop space defined by

(6.13) $$\frac{d}{ds} x = -\nabla \varphi(x), \quad x \in \widehat{\Omega}.$$

The solutions $x(s) = x \cdot s$ of this smooth ordinary differential equation exist for all times since the gradient satisfies a uniform Lipschitz estimate (see Chapter 3). They are represented by

(6.14) $$\begin{aligned} x \cdot s &= e^s x^- + x^0 + e^{-s} x^+ - K(s, x) \\ K(s, x) &:= \int_0^s \left(e^{s-\tau} P^- + P^0 + e^{-s+\tau} P^+ \right) b'(x \cdot \tau) d\tau. \end{aligned}$$

The linear operators P^-, P^0 and P^+ denote the orthogonal projections belonging to the splitting $\widehat{\Omega} = E^0 \oplus E^- \oplus E^+$, where $E^0 = \mathbb{R}^{2n}$. In order to motivate our strategy to find critical points, we first consider a bounded orbit of the flow. By definition, such an orbit $x \cdot s$ satisfies

(6.15) $$\sup_{s \in \mathbb{R}} |\varphi(x \cdot s)| < \infty.$$

For every $T > 0$,

(6.16) $$\varphi(x \cdot T) - \varphi\left(x \cdot (-T)\right) = -\int_{-T}^{T} |\nabla \varphi(x \cdot s)|^2 ds.$$

Since $s \mapsto \varphi(x \cdot s)$ is monotone the limit as $T \to +\infty$ exists in view of (6.15), so that

(6.17) $$\int_{-\infty}^{\infty} \|\nabla \varphi(x \cdot s)\|^2 ds < \infty.$$

6.2 THE MODEL CASE OF THE TORUS

Thus we find a sequence $s_n \to \infty$ such that $\nabla\varphi(x \cdot s_n) \to 0$ and $\varphi(x \cdot s_n) \to d \in \mathbb{R}$. Since φ satisfies the P.S. condition, a subsequence of $x \cdot s_n$ converges in $\widehat{\Omega}$ to a critical point y, $\nabla\varphi(y) = 0$, which, as we already pointed out, is a periodic solution of the Hamiltonian equation. Moreover, the set $Cr(\varphi)$ of all critical points of φ is compact and one verifies readily that

$$(6.18) \qquad x \cdot s \longrightarrow Cr(\varphi) \text{ as } s \to \pm\infty$$

for every bounded orbit of (6.13). We see that every bounded orbit comes from and tends to the set of critical points and hence is a so-called connecting orbit.

As an illustration we consider the situation where the Hamiltonian $H \equiv 0$ vanishes. Here the flow on $T^{2n} \times E$ is determined by the symplectic structure alone and is explicitly given by

$$x \cdot s = e^s x^- + x^0 + e^{-s} x^+ ,$$

where $x = x^- + x^0 + x^+ \in \widehat{\Omega}$.

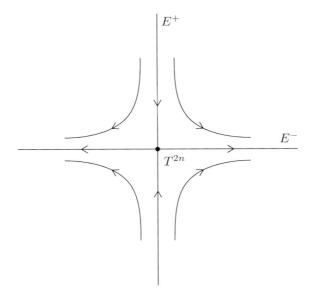

Fig. 6.6

The bounded orbits coincide with the critical points $x^0 \in T^{2n}$ which are the constant orbits. Moreover the torus T^{2n} is a transversally hyperbolic invariant manifold of the flow. The general case is a "compact perturbation" of this picture. The set of bounded orbits is very small, most of the solutions of (6.13) are unbounded in forward and backward time. This is in sharp contrast to the behaviour of the flow of a variational principle bounded from below where every orbit of the negative gradient flow tends to the set of critical points in forward time. The classical example is the energy functional on the loop space of a compact Riemannian manifold in the geometric problem of closed geodesics, see W. Klingenberg [127] and V. Bangert [14].

Our strategy now is to study the structure of the set X_∞ of bounded orbits of the gradient flow (6.13), i.e., the critical points together with the connecting orbits. We shall show that this set is a compact space which inherits the topology of the underlying manifold T^{2n}. On this invariant set X_∞ there is, moreover, a continuous gradient-like flow induced by the gradient flow (6.13), whose rest points are precisely the critical points of the functional. Theorem 3 will then follow from a well-known estimate for the rest points of a continuous gradient-like flow of a compact space which goes back to Ljusternik-Schnirelman.

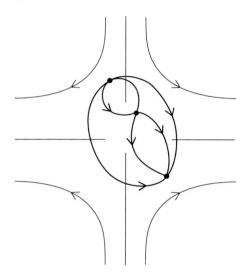

Fig. 6.7

Proof of Theorem 3. We define the set $\widehat{X}_\infty$ of bounded solutions of the negative gradient equation (6.13) by

(6.19)
$$\widehat{X}_\infty = \left\{ u \in C^\infty\left(\mathbb{R}, \mathbb{R}^{2n} \times E\right) \, \Big| \, \tfrac{d}{ds} u(s) + \nabla\varphi\big(u(s)\big) = 0 \text{ for } s \in \mathbb{R}, \ \sup_{s \in \mathbb{R}} \big|\varphi\big(u(s)\big)\big| < \infty \right\}.$$

The set $\widehat{X}_\infty$ is equipped with the topology induced from $C^\infty(\mathbb{R}, \mathbb{R}^{2n} \times E)$: namely that of uniform convergence of all derivatives on compact subsets of $\mathbb{R}$. Note that $\widehat{X}_\infty$ contains, in particular, the constant solutions satisfying $u(s) = u(0)$ for all $s \in \mathbb{R}$; they are critical points of φ. On $\widehat{X}_\infty$ we have a natural action of $\mathbb{Z}^{2n}$. Indeed, since the Hamiltonian is periodic, we conclude that $\varphi(u + j) = \varphi(u)$ and

6.2 THE MODEL CASE OF THE TORUS

$\nabla\varphi(u+j) = \nabla\varphi(u)$ for all j. In the following we shall denote by X_∞ the quotient space. It is defined by (6.20) with $\mathbb{R}^{2n}$ replaced by $\mathbb{R}^{2n}/\mathbb{Z}^{2n} = T^{2n}$. Observe now that the gradient equation (6.13) is independent of time, where the role of the time is played by the parameter s. Consequently, with $s \mapsto u(s)$, also $s \mapsto u(s+\tau)$ is a solution for every fixed $\tau \in \mathbb{R}$. This shift in time induces a continuous $\mathbb{R}$-action on X_∞ defined by

(6.20)
$$\mathbb{R} \times X_\infty \longrightarrow X_\infty, \quad (\tau, u) \mapsto \tau * u$$
$$(\tau * u)(s) = u(s+\tau).$$

This flow has very special property: it is gradient-like with respect to the continuous Ljapunov function

(6.21) $\quad V : X_\infty \longrightarrow \mathbb{R}$ defined by $V(u) = \varphi\bigl(u(0)\bigr).$

In order to prove this we have to show that the function V is strictly monotone decreasing along the nonconstant solutions $\tau * u$ of the flow (6.20). If $\tau > \sigma$, then

(6.22)
$$\begin{aligned} V(\tau * u) - V(\sigma * u) &= \varphi\bigl(u(\tau)\bigr) - \varphi\bigl(u(\sigma)\bigr) \\ &= \int_\sigma^\tau \tfrac{d}{dt}\varphi\bigl(u(t)\bigr) \, dt \\ &= -\int_\sigma^\tau \bigl\|\nabla\varphi\bigl(u(t)\bigr)\bigr\|^2 dt \end{aligned}$$

Consequently $\tau \mapsto V(\tau * u)$ is monotone-decreasing. Assume now that $V(\tau * u) = V(\sigma * u)$ for some $\tau > \sigma$. Then $\varphi(u(\tau)) = \varphi(u(\sigma))$ and hence, by (6.22), $\nabla\varphi(u(s)) = 0$ for all $\sigma \leq s \leq \tau$. Therefore, in view of the uniqueness of the solutions of the gradient equation, $u(s) = u(0)$ for all $s \in \mathbb{R}$ and hence $\tau * u = u$ for all $\tau \in \mathbb{R}$. Consequently, $\tau * u$ is a constant orbit of the flow (6.20) and we have verified that V is a Ljapunov function for the continuous flow (6.20).

Conversely, of course, if we have a constant orbit, i.e., $\tau * u = u$ for $\tau \in \mathbb{R}$, which is sometimes called a rest point of the flow (6.20), then $u(s) \equiv u(0)$ is a constant solution of the gradient flow (6.13) and hence a critical point of our functional φ.

So far we do not even know whether X_∞ is nonempty. In order to get an insight into the topology of the set X_∞, we define the continuous map

(6.23) $\quad \hat{\rho} : C^\infty(\mathbb{R}, \mathbb{R}^{2n} \times E) \to \mathbb{R}^{2n}$

associating with u the meanvalue of the loop $u(0) \in \hat{\Omega}$:

$$\hat{\rho}(u) = \int_0^1 u(0)(t) dt \quad \in \mathbb{R}^{2n}.$$

With $\rho : C^\infty(\mathbb{R}, T^{2n} \times E)$ we shall denote the map on the quotient space induced by $\hat{\rho}$. It turns out, and this is the crucial step in the proof of Theorem 3, that the set X_∞ inherits the topology of the underlying manifold T^{2n}:

Theorem 4. X_∞ is a compact (metric) space and

(6.24) $$(\rho|X_\infty)^* \; : \; \check{H}^*(T^{2n}) \to \check{H}^*(X_\infty)$$

is injective. Here and also in the following $\check{H}^*$ denotes the Alexander-Spanier cohomology with $\mathbb{Z}_2$ coefficients.

Postponing for a moment the proof of this theorem, we first observe that Theorem 3 can now be deduced from the following qualitative result of dynamical systems:

Theorem 5. (Ljusternik-Schnirelman) Consider a continuous gradient-like flow on a compact metric space X; then

$$\# \; \{ \text{ rest points } \} \geq CL(X) + 1 \, .$$

We shall prove Theorem 5 in the next section devoted to Ljusternik-Schnirelman theory. Recalling that the critical points of φ are the rest points of the gradient-like flow (6.20), we conclude from Theorem 5 applied to the compact space $X = X_\infty$, that #{critical points of φ} $\geq CL(X_\infty) + 1$. Now in view of Theorem 4, $CL(X_\infty) \geq CL(T^{2n})$. Since $CL(T^{2n}) = 2n$ we have established at least $2n + 1$ critical points of φ. They correspond to the desired $2n + 1$ periodic solutions of the Hamiltonian vector field and Theorem 3 is proved. ∎

Proof of Theorem 4. The remainder of this section is devoted to the proof of Theorem 4. In order to get an insight into the topology of the set X_∞ of bounded orbits of the gradient equation (6.13) we look for smooth compact submanifolds which "approximate" the set X_∞ and whose topology can be estimated. The desired cohomology of the set X_∞ will then be concluded by means of the continuity property of the Alexander-Spanier cohomology (see Appendix 7).

If $T > 0$ we define a set $\widehat{X}_T$ consisting of solutions of a very special boundary value problem for the gradient equation:

(6.25)
$$\widehat{X}_T := \left\{ u \in C^\infty \left([-T, T], \mathbb{R}^{2n} \times E \right) \, \middle| \right.$$
$$\tfrac{d}{ds} u(s) = -\nabla\varphi\bigl(u(s)\bigr), \quad -T \leq s \leq T$$
$$u(-T) \; \in \; \mathbb{R}^{2n} \times E^-,$$
$$\left. u(T) \; \in \; \mathbb{R}^{2n} \times E^+ \right\} .$$

We have the natural $\mathbb{Z}^{2n}$ action on $\widehat{X}_T$ and denote by X_T the quotient. It is defined by (6.25) with $\mathbb{R}^{2n}$ replaced by T^{2n}. We shall investigate the set $\widehat{X}_T$ using standard

6.2 THE MODEL CASE OF THE TORUS

methods of Fredholm theory. For this purpose we introduce the Banach spaces

(6.26)
$$\widehat{\mathcal{B}} := \left\{ u \in C^1\left([-T,T], \mathbb{R}^{2n} \times E\right) \; \Big| \; \begin{aligned} u(-T) &\in \mathbb{R}^{2n} \times E^- \\ u(T) &\in \mathbb{R}^{2n} \times E^+ \end{aligned} \right\}$$

$$\widehat{\mathcal{E}} := \left\{ u \in C^0\left([-T,T], \mathbb{R}^{2n} \times E\right) \right\},$$

equipped with the usual C^1 respectively C^0 norms. We denote by $\mathcal{B}$ and $\mathcal{E}$ the quotients under the $\mathbb{Z}^{2n}$ action. These may be identified with the Banach manifolds defined by (6.26) with $\mathbb{R}^{2n}$ replaced by T^{2n}. Now define the nonlinear smooth map

(6.27)
$$\widehat{L} : \widehat{\mathcal{B}} \longrightarrow \widehat{\mathcal{E}}, \quad u \mapsto \widehat{L}(u)$$
$$\widehat{L}(u)(s) = \tfrac{d}{ds}u(s) + \nabla\varphi\big(u(s)\big), \quad -T \leq s \leq T.$$

Then $\widehat{L}(j*u) = \widehat{L}(u)$ and we shall denote by $L : \mathcal{B} \to \mathcal{E}$ the induced map. In the following we shall use the abbreviation $\widehat{L}(u) = \dot{u} + \nabla\varphi(u)$. Since the solutions of the smooth ordinary differential equation are smooth in time, we can represent the sets $\widehat{X}_T$ (respectively X_T) as the solution set of $\widehat{L}(u) = 0$ (respectively $L(u) = 0$).

(6.28)
$$\widehat{X}_T = \widehat{L}^{-1}(0) \text{ and } X_T = L^{-1}(0).$$

Proposition 2. $\widehat{L}$ is a nonlinear Fredholm map and the Fredholm index of the linearized map is
$$\text{ind } \widehat{L}'(u) = 2n, \quad u \in \widehat{\mathcal{B}}.$$

Proof. From $\widehat{L}(u) = \dot{u} + u^+ - u^- - b'(u)$, we obtain for the linearized map at $u \in \widehat{\mathcal{B}}$

(6.29)
$$\widehat{L}'(u)h = \dot{h} + h^+ - h^- - b''(u)h,$$

and we shall show first that the linear operator $b''(u) \in \mathcal{L}(\widehat{\mathcal{B}}, \widehat{\mathcal{E}})$ is compact for every $u \in \widehat{\mathcal{B}}$. Since the derivative of a smooth compact map is compact, it is sufficient to verify that the map $u \mapsto b'(u) : \widehat{\mathcal{B}} \to \widehat{\mathcal{E}}$ is compact, i.e., maps bounded sets into precompact sets. Fix $u \in \widehat{\mathcal{B}}$; then recalling that $b'(u(s)) = j^*\nabla b(u(s))$ we can estimate

$$\|b'\big(u(s)\big) - b'\big(u(t)\big)\|_{1/2} =$$
$$= \|j^*\big(\nabla b(u(s)) - \nabla b(u(t)))\big)\|_{1/2}$$
$$\leq M \|\nabla b(u(s)) - \nabla b(u(t))\|_{L^2}$$
$$\leq M \|H\|_{C^2} \cdot \|u(s) - u(t)\|_{L^2}$$
$$\leq C \|u(s) - u(t)\|_{1/2} \leq C\|u\|_{\widehat{\mathcal{B}}} \cdot |t - s|,$$

for a constant C depending only on the second derivative of the Hamiltonian function H. We conclude that b' maps a bounded subset of $\widehat{\mathcal{B}}$ into a set in $\widehat{\mathcal{E}}$ which is equi-continuous, and since, for every fixed s, the set $\{b'(u(s))|u \in \widehat{\mathcal{B}}\} = \{j^*\nabla b(u(s))|u \in \widehat{\mathcal{B}}\}$ is precompact, we conclude by the Arzelà-Ascoli theorem that b' maps indeed bounded sets into precompact sets.

Using now that neither the Fredholm character nor the Fredholm index of a linear operator changes if we add a compact linear operator, it is sufficient to prove that the linear operator

$$(6.30) \qquad L_0 : \widehat{\mathcal{B}} \to \widehat{\mathcal{E}}, u \mapsto \dot{u} + u^+ - u^-$$

is Fredholm of index $2n$. For its kernel one verifies readily that $\ker(L_0) = \{u \in \widehat{\mathcal{B}} | u(s) = x^0 \in \mathbb{R}^{2n}\}$ consists of the constant functions with values in $\mathbb{R}^{2n} \times \{0\}$, so that $\dim \ker(L_0) = 2n$. Hence it suffices to show that L_0 is surjective. Define for $g \in \widehat{\mathcal{E}}$ the function u by

$$u(s) = \int_{-T}^{s} \left\{ e^{-s+\tau} g^+(\tau) + g^0(\tau) + e^{s-\tau} g^-(\tau) \right\} d\tau - \int_{-T}^{T} e^{s-\tau} g^-(\tau) d\tau \ .$$

Then u satisfies the boundary conditions, i.e., $u \in \widehat{\mathcal{B}}$ and moreover $L_0(u) = g$ as is readily verified. This finishes the proof of the proposition. ∎

Proposition 3. $\widehat{L} : \widehat{\mathcal{B}} \to \widehat{\mathcal{E}}$ is a proper map modulo the $\mathbb{Z}^{2n}$ action, i.e., if u_k is a sequence satisfying $\widehat{L}(u_k) \to y$, then there exists a sequence $j_k \in \mathbb{Z}^{2n}$ such that $u_k + j_k$ is precompact in $\widehat{\mathcal{B}}$.

Proof. Recall that $\widehat{L} = L_0 + b'$ and set $\widehat{L}(u_n) = y_n$. Then $L_0(u_n) = y_n - b'(u_n)$. Since $\dim \ker (L_0) < \infty$ there is a continuous splitting $\widehat{\mathcal{B}} = \widehat{\mathcal{B}}^0 \oplus \widehat{\mathcal{B}}^1$ with $\widehat{\mathcal{B}}_0 = \ker(L_0)$, and $L_0|\widehat{\mathcal{B}}_1$ is an isomorphism of Banach spaces. Now set $u_n = u_n^0 + u_n^1 \in \widehat{\mathcal{B}}^0 \oplus \widehat{\mathcal{B}}^1$. Since $b'(u_n)$ and hence $y_n - b'(u_n)$ is bounded in $\widehat{\mathcal{E}}$ we conclude that u_n^1 is bounded in $\widehat{\mathcal{B}}$. Pick a sequence $j_n \in \mathbb{Z}^{2n}$ such that $u_n^0 + j_n$ is bounded in $\mathbb{R}^{2n}$. Then, by the compactness of b', the sequence

$$y_n - b'(u_n) = y_n - b'(u_n + j_n)$$

is precompact in $\widehat{\mathcal{B}}$ so that u_n^1 is precompact in $\widehat{\mathcal{B}}$. Hence after taking a subsequence we conclude $u_n + j_n \to u$ in $\widehat{\mathcal{B}}$ as claimed. ∎

Passing to $\mathcal{B}$ and $\mathcal{E}$, we conclude from Proposition 2 and Proposition 3 that the map $L : \mathcal{B} \to \mathcal{E}$ is a proper Fredholm map of index $2n$. Similarly, the map

$$(6.31) \qquad (L, \rho) : \mathcal{B} \to \mathcal{E} \times T^{2n}$$

defined by $u \mapsto (L(u), \rho(u))$ is a proper Fredholm map having Fredholm index equal to zero. This class of maps possesses a special degree, the Smale degree

6.2 THE MODEL CASE OF THE TORUS

mod 2, denoted by

(6.32) $$\deg_2 \Big((L, \rho), (y, m) \Big)$$

for every point $(y, m) \in \mathcal{E} \times T^{2n}$. The degree "counts" mod 2 the number of preimages under a proper and smooth Fredholm map of index zero and is defined as follows. Assume first that $x \equiv (y, m)$ is a regular value of $F \equiv (L, \rho)$. If u solves $F(u) = x$, the derivative $F'(u)$ is a surjective map, and since its index is zero, it is bijective. Thus the map F is locally one-to-one near the solutions u. Since it is proper, it follows that there are only finitely many solutions u and the degree is defined as the number of these solutions (mod 2)

(6.33) $$\# \{u \in \mathcal{B} \mid (L, \rho)(u) = (y, m)\} \mod 2 .$$

If (y, m) is not a regular value, its degree is defined by taking any regular value nearby using the Sard-Smale theorem. That this is well-defined can be proved the same way as for the Brouwer degree of mappings between manifolds of the same dimensions, see J. Milnor [157]. The Smale degree mod 2 enjoys the same useful properties as the Brouwer degree. In particular, it is invariant under proper homotopies in the class of maps considered. For complete proofs we refer to the forthcoming book by H. Brezis and L. Nirenberg [32]. For a survey on Fredholm maps and corresponding degrees we refer to [30].

Proposition 4.
$$\deg_2 \Big((L, \rho), (y, m) \Big) = 1 \text{ for all } (y, m) .$$

Proof. We consider the smooth homotopy of Fredholm operators $\Phi : [0, 1] \times \mathcal{B} \to \mathcal{E} \times T^{2n}$ defined by

(6.34) $$\Phi_s(u) = \Big(\dot{u} + u^+ - u^- - s\, b'(u), \rho(u) \Big) .$$

Since the compactness properties proved above depend only on the C^2 norm of the Hamiltonian function, we conclude that Φ is a proper homotopy, so that the degree with respect to a point (y, m) is independent of s and hence

$$\deg_2 \Big((L, \rho), (y, m) \Big) = \deg_2 \Big(\Phi_0, (y, m) \Big) ,$$

where $\Phi_0(u) = (\dot{u} + u^+ - u^-, \rho(u))$. Similarly we can homotope the point (y, m) to $(0, m)$ and find that the degree is equal to

(6.35) $$\deg_2 \Big(\Phi_0, (0, m) \Big) .$$

We claim that the number of solutions u of $\Phi_0(u) = (0, m)$ is equal to one. Indeed, if $u \in \mathcal{B}$ solves

$$\dot{u} + u^+ - u^- = 0 , \quad \rho(u) = m$$
$$u(-T) \in T^{2n} \times E^- , \quad u(+T) \in T^{2n} \times E^+,$$

then necessarily $u(s) = m \in T^{2n}$ is the unique solution.

Consequently, $\deg_2(\Phi_0,(0,m)) = 1$ provided $(0,m)$ is a regular value of Φ_0. To see this we observe that the linearized map at $u(s) = m$ is the linear operator

(6.36) $$h \mapsto \left(\dot{h} + h^+ - h^-, \int_0^1 h(0)(t)dt\right)$$

between the corresponding tangent spaces. It clearly is an isomorphism so that $(0,m)$ is indeed a regular value and the proof is finished. ■

So far we know that the set $X_T = L^{-1}(0) \subset \mathcal{B}$ is compact and a smooth manifold of dimension $2n$. perhaps empty, if 0 is a regular value. Consider now the continuous map $\rho : X_T \to T^{2n}$, $u \mapsto \rho(u)$ induced by the map $\hat{\rho}$ defined in (6.23).

Proposition 5. The map $\rho : X_T \to T^{2n}$ induces an injective map in cohomology:
$$\rho^* : \check{H}^*(T^{2n}) \to \check{H}^*(X_T).$$

In particular, the set X_T is not empty.

Proof. Fix an open neighborhood U of X_T in $\mathcal{B}$. If $y \in \mathcal{E}$ is sufficiently close to 0, then $L^{-1}(y) \subset U$ because L is a proper map. Since, by the Sard-Smale theorem for Fredholm maps, the set of regular points is dense we can assume, in addition, that y is a regular value of L. The aim is to prove that

(6.37) $$\left(\rho|L^{-1}(y)\right)^* : \check{H}^*(T^{2n}) \to \check{H}^*\left(L^{-1}(y)\right)$$

is injective. Postponing the proof we conclude from the commutativity of the diagram

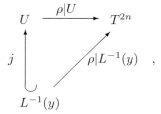

where j denotes the inclusion map, that $(\rho|U)^*$ is injective as well. This holds for every open neighborhood U of the compact set X_T. Hence, using the continuity property of the Alexander-Spanier cohomology

(6.38) $$\check{H}^*(X_T) = \operatorname{dir.lim} \check{H}^*(U),$$

where this direct limit is taken over all neighborhoods U of X_T (see Appendix 7). We obtain that $(\rho|X_T)^*$ is indeed injective as claimed in the proposition. It remains to prove the injectivity of (6.37).

6.2 THE MODEL CASE OF THE TORUS

Since y is a regular value of L, the map $\rho := \rho|L^{-1}(y) : L^{-1}(y) \to T^{2n}$ is a map between two compact manifolds of dimension $2n$. We shall prove that

$$\deg_2(\rho) = 1 \,. \tag{6.39}$$

Pick a regular value $m \in T^{2n}$ of ρ. We claim that (y, m) is a regular value of (L, ρ). Indeed, if $(L, \rho)(u) = (y, m)$ then $L'(u) : \widehat{\mathcal{B}} \to \widehat{\mathcal{E}}$ is surjective since y is, by assumption a regular value of L. Observe that $\ker L'(u) = T(L^{-1}(y))$. Since by assumption m is regular for ρ the linear map $\rho'(u) : T(L^{-1}(y)) \to T(T^{2n})$ is a bijection. Consequently the linear map $(L'(u), \rho'(u)) : \widehat{\mathcal{B}} \to \widehat{\mathcal{E}} \times \mathbb{R}^{2n}$ between the corresponding tangent spaces is a bijection as well, so that (y, m) is regular as claimed. By Proposition 4 we therefore have

$$\begin{aligned} 1 &= \#\{(L, \rho)^{-1}(y, m)\} \mod 2 \\ &= \#\{(\rho|L^{-1}(y))^{-1}(m)\} \mod 2 \,. \end{aligned}$$

Since m is regular, it follows that $\deg_2(\rho) = 1$ as claimed in (6.39).

Abbreviating the compact $2n$-dimensional manifold $L^{-1}(y)$ by M, we denote by o_M and o_T the $\mathbb{Z}_2$ fundamental classes of M and T^{2n} respectively. In view of (6.39) we conclude by the homological definition of the $\mathbb{Z}_2$ degree (see Appendix 7)

$$\rho_*(o_M) = o_T \,. \tag{6.40}$$

We point out that on smooth compact manifolds the usual cohomology theories (like the singular, Alexander-Spanier, Čech) are naturally isomorphic, see F. Warner [219]. Hence, we may consider the singular cohomology instead of the Alexander-Spanier cohomology. Then by the Poincaré duality we have the isomorphisms

$$H^i(M) \xrightarrow{\cap o_M} H_{2n-i}(M) \quad : \quad \alpha \mapsto \alpha \cap o_M \,.$$

$$H^i(T^{2n}) \xrightarrow{\cap o_T} H_{2n-i}(T^{2n}) \quad : \quad \beta \mapsto \beta \cap o_T \,.$$

Pick $\alpha \in H^i(T^{2n})$, we have to show that $\rho^*(\alpha) = 0$ implies $\alpha = 0$. For this purpose we consider the composition of maps

$$H^i(T^{2n}) \xrightarrow{\rho^*} H^i(M) \xrightarrow{\cap o_M} H_{2n-i}(M) \xrightarrow{\rho_*} H_{2n-i}(T^{2n}) \xrightarrow{(\cap o_T)^{-1}} H^i(T^{2n}) \,.$$

Using the naturality property of the cap product (see A. Dold [62]) and using (6.40) we compute

$$\begin{aligned} (\cap o_T)^{-1} \, \rho_*\{(\rho^*\alpha) \cap o_M\} &= \\ (\cap o_T)^{-1} \, \{\alpha \cap \rho_*(o_M)\} &= \\ (\cap o_T)^{-1} \, \{\alpha \cap o_T\} &= \alpha \,. \end{aligned}$$

Consequently if $\rho^*\alpha = 0$ then $\alpha = 0$ proving that ρ^* is injective. The proof of Proposition 5 is completed. ∎

We now return to the set X_∞ of bounded solutions and show that it is a compact set, which can be "approximated" by X_T for T large.

Proposition 6. The space X_∞ is a compact (metric) space.

Proof. If $u \in X_\infty$ then $u(s) \to Cr(\varphi)$ as $s \to \pm\infty$. Since $\sup\{|b(u)|, u \in \widehat{\Omega}\} < \infty$ it follows from the monotonicity of $\varphi(u(s))$ that there exists a constant $C > 0$ satisfying

$$-C \le \varphi\big(u(s)\big) \le C \tag{6.41}$$

for all $u \in X_\infty$ and all $s \in \mathbb{R}$. Consequently the set X_∞ is closed. From the variation of constants formula (6.14) we conclude, for every $T > 0$

$$u(0)^+ = e^{-T} u(-T)^+ + P^+ \int_{-T}^{0} e^s b'\big(u(s)\big) \, ds. \tag{6.42}$$

Since $u(s) \to Cr(\varphi)$ as $|s| \to \infty$, the set $u(\mathbb{R})$ is bounded, and hence we conclude, as $T \to \infty$,

$$u(0)^+ = P^+ j^* \int_{-\infty}^{0} e^s \nabla b\big(u(s)\big) \, ds. \tag{6.43}$$

A similar formula holds for $u(0)^-$ and hence we find a compact set $K \subset T^{2n} \times E$ such that $u(0) \in K$ for all $u \in X_\infty$. Consequently the set $\{u(0) | u \in X_\infty\}$ is compact. Since in the C^∞-topology the solutions u depend continuously on their initial conditions, we conclude that X_∞ is compact as well and the proof is finished. ∎

If $u \in X_T$, then u is by definition a solution of the gradient equation for $|s| \le T$, hence it has a unique continuation to a solution $u(s)$ for all times $s \in \mathbb{R}$ in view of the global Lipschitz continuity of $\nabla \varphi$. This defines an inclusion map $j : X_T \to C^\infty(\mathbb{R}, T^{2n} \times \mathbb{R})$ by which in the following, we can consider X_T as a subset of $C^\infty(\mathbb{R}, T^{2n} \times E)$.

Proposition 7. If U is an open neighborhood of X_∞, then

$$X_T \subset U$$

for T sufficiently large.

6.2 THE MODEL CASE OF THE TORUS

Proof. Since b maps $\widehat{\Omega}$ into a bounded set, there exists a constant $C > 0$ such that

$$\varphi|E^- \leq C \text{ and } \varphi|E^+ \geq -C. \tag{6.44}$$

Consequently, if $u \in X_T$, then

$$-C \leq \varphi\big(u(s)\big) \leq C, \tag{6.45}$$

for all $u \in X_T$, all $|s| \leq T$ and all $T > 0$. This follows because $u(s)$ is a solution of the negative gradient equation (6.13) and hence $\varphi(u(s))$ is a monotone-decreasing function. We conclude also, that there exists a constant $M > 0$ such that

$$\|u(-T)\| + \|u(T)\| \leq M \tag{6.46}$$

for $u \in X_T$ and all $T > 0$. In order to finish the proof of the proposition we argue indirectly and assume there is a sequence $T_n \to \infty$ and a sequence $u_n \in X_{T_n}$ satisfying $u_n \notin U$. We shall show that a subsequence of $u_n(0)$ converges to a point $x \in T^{2n} \times E$ through which a bounded orbit $u \in X_\infty$ passes. Indeed, it follows from (6.46) together with (6.42), as in the proof of Proposition 6, that $\text{dist}(u_n(0)^+, K^+) \to 0$ as $n \to \infty$, for a compact set K^+. Similarly for $u_n(0)^-$; hence taking a subsequence $u_n(0) \to x \in T^{2n} \times E$. In view of (6.45) we have $-C \leq \varphi(u_n(s)) \leq C$, for $|s| \leq T_n$ for every u where the constant C is independent of n. We conclude for the solution $u(s)$ through $x = u(0)$ that $-C \leq \varphi(u(s)) \leq C$ for all $s \in \mathbb{R}$, hence $u \in X_\infty$. Consequently, $u_n \in U$ for n large in view of the topology of $C^\infty(\mathbb{R}, T^{2n} \times E)$ and the continuous dependence of the solutions on the initial conditions. This contradiction proves the proposition. ∎

End of the proof of Theorem 4. In order to prove the injectivity of $(\rho|X_\infty)^*$ we take any open neighborhood U of X_∞. By Proposition 6 we have $X_T \subset U$. By Proposition 5 the map $(\rho|X_T)^* : \check{H}^*(T^{2n}) \to \check{H}^*(X_T)$ is injective and therefore $(\rho|U)^* : \check{H}^*(T^{2n}) \to \check{H}^*(U)$ is injective. This holds true for every open neighborhood of the compact set X_∞ and, again, we conclude by the continuity of the Alexander-Spanier cohomology that $(\rho|X_\infty)^* : \check{H}^*(T^{2n}) \to \check{H}^*(X_\infty)$ is injective. The proof of Theorem 4 is completed. ∎

Remarks: It should be pointed out that the torus case permits more elementary proofs. The original proof by Conley and Zehnder in [55] is also based on the Hamiltonian variational function on the loop space. However, using the linear structure of the covering space, the very special symplectic structure ω_0 and the compactness of the torus the problem of finding critical points on the infinite dimensional loop space was reduced to the equivalent problem of finding the critical points of a related functional on a finite dimensional bundle $T^{2n} \times \mathbb{R}^N$ of the torus. Here the dimension N is large and depends, of course, on the Hamiltonian function H considered. The reduction used is the so-called Ljapunov-Schmidt reduction, a well-known technique in functional analysis. The question arises whether there is an intrinsic finite dimensional approach. Indeed M. Chaperon found in [43] (1984) another variational functional which is directly defined on a finite-dimensional vector bundle of the torus by means of a finite sequence of generating functions. He also applied his "method of broken geodesics" to the geometric and global intersection problem of Lagrangian submanifolds in $T^*(T^n)$. He showed that the zero section $0^*_{T^n}$, which is diffeomorphic to T^n, can never be separated from itself by a Hamiltonian map φ of $T^*(T^n)$. More precisely he proved that

$$\#\{T^n \cap \varphi(T^n)\} \geq \text{cup-length } (T^n) + 1,$$

($\geq SB(T^n)$ if the intersection is transversal). This was a long-standing open question raised by V.I. Arnold in [9]. For modifications and further applications of this variational approach we refer to [41]–[44]. In this connection we also mention A.B. Givental [105], M. Brunella [36] and Yu. V. Chekanov [47]. The related finite dimensional approach due to J.C. Sikorav [197] (1987) is based on the concept of a generating phase function for a Lagrangian manifold. The concept appeared earlier in the theory of Fourier integral operators and we refer to L. Hörmander [124]. This elegant approach turned out to be particularly useful for the special symplectic manifolds $(T^*M, d\lambda)$ as demonstrated by the new proof of Hofer's Lagrangian intersection theorem in T^*M for general compact manifolds M generalizing the torus case above, see [197, 136]. The result was previously discovered by H. Hofer [114] (1985). It states that

$$\#\{M \cap \varphi(M)\} \geq \text{cup-length } (M) + 1,$$

for every Hamiltonian map φ of $(T^*M, d\lambda)$, where we have identified M with the zero section 0^*_M.

In this context of finite dimensional variational arguments we should also mention, that in his work "Symplectic topology as the geometry of generating functions" [218] (1992), C. Viterbo defined symplectic capacities for Lagrangian submanifolds as critical values of the associated phase functions. In this way he defined the capacities for open subsets of $(\mathbb{R}^{2n}, \omega_0)$ by using Lagrangians which are graphs of compactly supported Hamiltonian mappings. In his 1992 lectures Ya. Eliashberg outlined how Viterbo's approach could be extended to an alternate construction of the Floer-Hofer symplectic homology theory described in Section 6.6, for the special case of open subsets of $(\mathbb{R}^{2n}, \omega_0)$. The details of his construction were carried out quite recently by L. Traynor in the preprint "Symplectic homology via generating functions" [212].

It seems that the finite dimensional variational approaches introduced so far are not adequate for general symplectic manifolds. For this reason we based our proof above for the torus case on the action principle in the loop space. Our proof demonstrates that the crucial information of the variational principle is contained in the set of bounded solutions of the artificial gradient flow. The set consists of the rest points together with the connecting orbits. This observation will be relevant in Floer's approach to the Morse theory of functionals whose gradients do not even generate a flow.

6.3 Gradient-like flows on compact spaces

In the following X is a compact metric space. A flow on X is a continuous map $\varphi : \mathbb{R} \times X \to X : (t, x) \mapsto \varphi(t, x) = \varphi^t(x)$ satisfying

(6.47)
$$\varphi^t\left(\varphi^s(x)\right) = \varphi^{t+s}(x)$$
$$\varphi^0(x) = x$$

for all $t, s \in \mathbb{R}$ and $x \in X$. An example is the flow of a differentiable vector field on a compact manifold. We shall use the notation

$$\varphi^t(x) = x \cdot t$$

and abbreviate $M \cdot B := \{x \cdot t \mid x \in M \text{ and } t \in B\}$ for subsets $M \subset X$ and $B \subset \mathbb{R}$. If $x \in X$, then the map

$$t \mapsto \varphi^t(x) = x \cdot t$$

from $\mathbb{R}$ into X is called the solution through the point $x \in X$. The image of the positive (respectively negative) solution through x will be denoted by $\gamma^{\pm}(x) = x \cdot \mathbb{R}^{\pm}$ and we set $\gamma(x) = x \cdot \mathbb{R}$. Every point $x \in X$ belongs, in view of (6.47), to precisely one orbit $\gamma(x)$. The distinguished constant solutions, characterized by

$$x \cdot t = x, \quad t \in \mathbb{R}$$

are called rest points or fixed points of the flow, clearly $\gamma(x) = \{x\}$ in this case. In order to describe the global orbit structure of a flow, it is useful to recall the concept of the positive limit set $\omega(x)$ and the negative limit set $\omega^*(x)$ of the solution through $x \in X$.

Definition.

$$\omega(x) = \{y \in X \mid \exists\, t_j \to +\infty \text{ with } x \cdot t_j \to y\}$$
$$\omega^*(x) = \{y \in X \mid \exists\, t_j \to -\infty \text{ with } x \cdot t_j \to y\}.$$

We summarize the well-known properties of these limit sets. Recall that X is compact.

Proposition 8. *If $x \in X$ then the limit sets $\omega(x)$ and $\omega^*(x)$ are not empty, compact, connected and invariant under the flow, i.e., $\omega(x) \cdot t = \omega(x)$ for all $t \in \mathbb{R}$. Moreover,*

(6.48)
$$x \cdot t \to \omega(x) \quad \text{as} \quad t \to +\infty$$
$$x \cdot t \to \omega^*(x) \quad \text{as} \quad t \to -\infty.$$

Proof. If $t_j \to \infty$ then the sequence $x \cdot t_j$ has a convergent subsequence so that $\omega(x) \neq \emptyset$. Assume that (6.48) does not hold; then there exists an open neighborhood $U \supset \omega(x)$ and a sequence $t_j \to \infty$ such that $x \cdot t_j \in X \setminus U$. Hence for a subsequence $x \cdot t_j \to y \in X \setminus U$. By definition $y \in \omega(x)$, in contradiction to $\omega(x) \subset U$. If $\omega(x)$ is not connected, then $\omega(x) = \omega_1 \cup \omega_2$ for two nonempty compact and disjoint subsets $\omega_j \subset X$. Take disjoint open neighborhoods U_j of $\omega_j, j = 1, 2$; then, by (6.48), $x \cdot t \to \omega(x)$ and hence for $t \geq t^*$, we have $x \cdot t \in U_1 \cup U_2$ and $\gamma^+(x \cdot t) \cap U_j \neq \emptyset$ for $j = 1, 2$. Consequently the set $\gamma^+(x \cdot t)$ is not connected contradicting the fact that a continuous image of an interval in $\mathbb{R}$ is connected. The invariance of the limit sets is an immediate consequence of the continuity of the flow. ∎

We assume now, in addition, that the flow is gradient-like, i.e., we require that there exists a continuous function $V : X \to \mathbb{R}$, which is strictly monotone-decreasing along every nonconstant solution:

$$t > s \implies V(x \cdot t) < V(x \cdot s).$$

Such a function is usually called a Ljapunov function for the flow. A well-known special example is the flow generated by a smooth gradient vector field $\dot{x} = -\nabla V(x)$ on a compact Riemannian manifold.

Proposition 9. If the flow is gradient-like the limit sets for every solution consists of rest points:

$$\omega(x) \text{ and } \omega^*(x) \subset \{ \text{ rest points } \},$$

for every $x \in X$.

Proof. Since $V(x \cdot t)$ is monotone-decreasing, the limit

$$\lim_{t \to +\infty} V(x \cdot t) = \inf\{V(x \cdot t) \mid t \geq 0\} =: d$$

exists in $\mathbb{R}$. If $y \in \omega(x)$ we conclude by the continuity of V that $V(y) = d$ and hence in view of the invariance of $\omega(x)$ under the flow, $V(y \cdot t) = d$ for every $t \in \mathbb{R}$. By assumption V is strictly decreasing along nonconstant solutions, hence y must be a rest point as claimed. ∎

If the gradient-like flow possesses only isolated rest points, then there are only finitely many of them, say $\{x_1, \ldots, x_N\}$ and we conclude from the proposition that every nonconstant solution connects two distinct rest points. Every point $x \in X$ tends under the flow to a distinguished rest point in forward and backward time.

It is not surprising that the rest points of the flow are intimately related to topological properties of X. Indeed, we shall conclude from Proposition 9 the following existence statement formulated in Section 6.2 as Theorem 5.

Theorem 6. (Ljusternik-Schnirelman) For every gradient-like flow on a compact metric space we have

$$\#\{\text{rest points}\} \geq \text{cup-length}(X) + 1.$$

6.3 Gradient-like flows on compact spaces

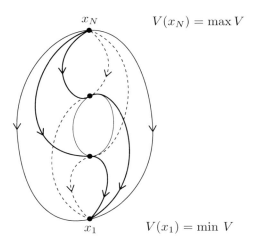

Fig. 6.7a

The theorem will follow immediately from a general qualitative statement for Morse decompositions of flows which are not necessarily gradient-like. The statement is taken from [55].

Definition. A Morse decomposition of a flow is a finite collection $\{M_j\}_{j \in I}$ of disjoint, compact and invariant subsets $M_j \subset X$ which can be ordered, say $(M_1, M_2, \ldots, M_k)$, such that the following property holds: for every

$$x \in X \setminus \bigcup_{j \in I} M_j$$

there exists a pair of indices $i < j$ such that $\omega(x) \subset M_i$ and $\omega^*(x) \subset M_j$.

Theorem 7. Assume $\{M_j\}_{j \in I}$ is a Morse decomposition of a flow on a compact space X. Then

(6.49) $$CL(X) + 1 \leq \sum_{j \in I} \{CL(M_j) + 1\}.$$

In particular, if $|I| < CL(X)+1$ then some compact invariant set M_j has nontrivial Alexander-Spanier cohomology and so contains a continuum of points.

We have abbreviated the cup-length of a space S by $CL(S)$. The flow is not required to be gradient-like. Postponing the proof of this theorem, we first prove Theorem 6.

Proof of Theorem 6. Denote by V the Ljapunov function of the gradient-like flow and assume it has only finitely many rest points. Then the rest points can be ordered such that $V(x_1) \leq V(x_2) \leq \cdots \leq V(x_k)$. Since V is strictly decreasing along the nonconstant solutions, the rest points $\{x_1, \ldots, x_k\}$ constitute, in view of Proposition 9, a Morse decomposition of the flow. Since $CL(\{x_j\}) = 0$ we conclude from Theorem 7

$$CL(X) + 1 \leq \sum_{1 \leq j \leq k} 1$$

so that the number of rest points must indeed be larger than or equal to $CL(X)+1$, thus proving Theorem 6. ∎

Proof of Theorem 7. First observe that any Morse decomposition of X can be obtained by first decomposing X into two sets, then decomposing one of these and continuing until the decomposition is reached. Indeed assume $\{M_1, M_2, \ldots, M_k\}$ is an ordered Morse decomposition of X. Then define the subset $X_1 = \{x \in X | \omega(x) \text{ and } \omega^*(x) \subset M_1 \cup M_2 \cup \cdots \cup M_{k-1}\}$. This set is compact and invariant under the flow so that $\{X_1, M_k\}$ constitutes an ordered Morse decomposition of X into two sets. Therefore one only needs to prove the theorem for decompositions into two sets. So let $\{M_1, M_2\}$ be an ordered Morse decomposition of X. From the definition we conclude that there is a compact neighborhood S_1 of M_1 and a compact neighborhood S_2 of M_2 in X satisfying $S_1 \cup S_2 = X$ and, moreover,

$$M_1 = \bigcap_{t>0} S_1 \cdot t \quad \text{and} \quad M_2 = \bigcap_{t>0} S_2 \cdot (-t) .$$

Consequently, observing that $\check{H}^*(S_1 \cdot t) = \check{H}^*(S_1)$ for every $t \in \mathbb{R}$ (similarly for S_2), we conclude by the continuity property of the Alexander-Spanier cohomology that $\check{H}^*(S_1) = \check{H}^*(M_1)$ and $\check{H}^*(S_2) = \check{H}^*(M_2)$. It is, therefore, sufficient to prove that $c(S_1) + c(S_2) \geq c(X)$ for $S_1 \cup S_2 = X$, abbreviating $c(S) = CL(S) + 1$. This will follow from the following topological observation:

Lemma 1. Let $S_1 \cup S_2 \subset S$ be three compact sets. Denote by $i_1 : S_1 \to S, i_2 : S_2 \to S$ and $i : S_1 \cup S_2 \to S$ the inclusion maps. Assume $\alpha, \beta \in \check{H}^*(S)$. Then $i_1^*(\alpha) = 0$ and $i_2^*(\beta) = 0$ imply $i^*(\alpha \cup \beta) = 0$.

Proof of Lemma 1. Consider the following diagram:

$$\begin{array}{ccccc}
\check{H}^*(S, S_1) & \otimes & \check{H}^*(S, S_2) & \xrightarrow{\cup} & \check{H}^*(S, S_1 \cup S_2) \\
\downarrow j_1^* & & \downarrow j_2^* & & \downarrow j^* \\
\check{H}^*(S) & \otimes & \check{H}^*(S) & \xrightarrow{\cup} & \check{H}^*(S) \\
\downarrow i_1^* & & \downarrow i_2^* & & \downarrow i^* \\
\check{H}^*(S_1) & & \check{H}^*(S_2) & & \check{H}^*(S_1 \cup S_2) .
\end{array}$$

The vertical sequences are exact. Hence if $\alpha \in \check{H}^*(S)$ satisfies $i_1^*(\alpha) = 0$ then there is an $\hat{\alpha} \in \check{H}^*(S, S_1)$ with $j_1^*(\hat{\alpha}) = \alpha$. Similarly, if $i_2^*(\beta) = 0$ then $j_2^*(\hat{\beta}) = \hat{\alpha}$ for some

6.3 Gradient-like flows on compact spaces

$\hat{\beta} \in \check{H}^*(S, S_2)$. By the commutativity of the diagram, $j^*(\hat{\alpha} \cup \hat{\beta}) = j_1^*(\hat{\alpha}) \cup j_2^*(\hat{\beta}) = \alpha \cup \beta$. and by the exactness, $i^*(\alpha \cup \beta) = i^* \circ j^*(\hat{\alpha} \cup \hat{\beta}) = 0$, as claimed. ∎

Returning to the proof of Theorem 7, we assume $\alpha_1, \alpha_2, \ldots \alpha_l \in \check{H}^*(X)$ satisfy $\alpha_1 \cup \alpha_2 \cup \cdots \cup \alpha_l \neq 0$. Let the α's be ordered so that $\alpha_1 \cup \cdots \cup \alpha_r$ is the longest product not in the kernel of i_1^*. Then, by definition of the cup-length, $c(S_1) \geq r+1$ and $i_1^*(\alpha_1 \cup \cdots \cup \alpha_r \cup \alpha_{r+1}) = 0$. Since $X = S_1 \cup S_2$ we conclude from Lemma 1 that $i_2^*(\alpha_{r+2} \cup \cdots \alpha_l) \neq 0$ and, therefore, $c(S_2) \geq l - (r+1) + 1 = l - r$. Consequently $c(S_1) + c(S_2) \geq l + 1$. We have shown that if X admits a nontrivial product with l factors, i.e., if $c(X) \geq l+1$, then $c(S_1) + c(S_2) \geq l+1$ so that Theorem 7 is proved. ∎

We defined a gradient-like flow by a Ljapunov function which is strictly monotone decreasing along the nonconstant orbits of the flow. This tradition comes from the study of semiflows, defined only for positive time, on noncompact manifolds. In our situation of a flow on a compact space, we could have taken a Ljapunov function instead, which is strictly increasing along nonconstant orbits and we would have reached the same conclusions by the same arguments. Actually, we shall make use of this remark in the next section, where for different reasons we consider the positive instead of the negative gradient equation, which leads to a increasing Ljapunov function. This is not relevant since our functional is bounded neither from below nor from above.

Remarks: In this connection we would like to point out another useful inequality associated with a Morse decomposition $\{M_1, \ldots, M_k\}$ of a flow on X which is not assumed to be gradient-like. It relates dynamical invariants of the invariant sets M_j to topological invariants of the space X and is due to C. Conley. Every set M_j in a Morse decomposition is an isolated invariant set in the sense that it is the maximal invariant set contained in a compact neighborhood. Hence it possesses a so-called Conley-index $h(M_j)$ which is the homotopy type of a pointed compact space and which is determined by the flow near M_j, see C. Conley [54]. Denoting by $P(t, h(M_j))$ the Poincaré polynomial of the pointed compact space one has

$$\text{(6.50)} \qquad \sum_j P\bigl(t, h(M_j)\bigr) = P(t, X) + (1+t)Q(t) ,$$

P being the Poincaré polynomial of the space X and Q being a power series containing only nonnegative integer coefficients, for a proof we refer to [57]. This Conley-Morse equation relates the dynamical invariants of the local invariant sets $M_j \subset X$ to the global topological invariant $P(t, X)$. Consider, for example, the special case of a gradient flow defined by $\dot{x} = -\nabla V(x)$ on a compact Riemannian manifold X. If the critical points x_j of the smooth function V on X are not only isolated but, in addition, nondegenerate, one can show that

$$h(\{x_j\}) = \left[\dot{S}^{m(x_j)}\right]$$

is the homotopy type of a pointed sphere whose dimension $m(x_j)$ is the Morse index of the critical point x_j. Consequently $P(t, h(\{x_j\})) = t^{m(x_j)}$ is a monomial, see e.g. [56]. Therefore, if all the critical points of V are nondegenerate they constitute a Morse-decomposition of X and the equation (6.50) for this special flow is the familiar Morse equation

$$(6.51) \qquad \sum_{\{x|\nabla V(x)=0\}} t^{m(x)} = P(t,X) + (1+t)Q(t) .$$

It relates the nature of the isolated critical points x of the differentiable function V to those topological invariants of X which are described by the Betti numbers.

C. Conley designed his homotopy index theory in view of its invariance properties under deformations. These are crucial for applications. He developed his theory for continuous flows on locally compact metric spaces. For a concise presentation of this theory, we refer to D. Salamon [187]. Later on K. Rybakowski in a series of papers [184, 185, 186] extended Conley's index theory to a restricted class of continuous semiflows in an infinite dimensional setting. It includes flows generated by partial differential equations of parabolic type. In this more general setting there is also a Morse equation associated with a Morse decomposition of an isolated compact invariant subset of the space, as has been demonstrated in [186].

6.4 Elliptic methods and symplectic fixed points

We are going to prove the Arnold conjecture for general compact symplectic manifolds (M,ω) satisfying $[\omega]|\pi_2(M) = 0$. The proof, based on the action functional on the loop space, is inspired by the geometric ideas underlying the proof for the standard torus in Section 6.2. But instead of looking at O.D.E. problems, we shall now be confronted with first order elliptic systems.

For the following it is useful to first recall our approach to the torus case. The covering map $\mathbb{R}^{2n} \to T^{2n}$ and the special symplectic structure permit by extending the action functional, originally defined on smooth contractible loops, to the Hilbert space $H^{1/2}$. The gradient of the extended functional in $H^{1/2}$ is smooth and can be viewed as a regularization of the L^2 gradient of the action. The gradient equation on $H^{1/2}$ determines a unique global flow, which has, in addition, the necessary compactness properties. For these reasons we can find all the critical point by using O.D.E.-methods as follows. The study of the gradient flow on the infinite dimensional space $H^{1/2}$ of contractible loops can be reduced to the study of an (induced) gradient-like flow on a compact metric space X_∞, an almost classical situation. The set X_∞ here is the set of all bounded solutions of the gradient flow, consisting of all the critical points of the action functional together with their connecting orbits. It is, in general, not a smooth manifold. We saw that X_∞ contained the topology of the underlying manifold T^{2n}. The Ljusternik-Schnirelman theorem, applied to the gradient-like flow on X_∞, then

6.4 ELLIPTIC METHODS AND SYMPLECTIC FIXED POINTS

guaranteed the required number of rest points. In order to get an insight into the topology of the set X_∞, we approximated X_∞ by a set X_T of solutions of a very special O.D.E. boundary value problem using the special symplectic structure. Applying Fredholm theory to this boundary value problem we approximated X_T by smooth compact manifolds of dimension $2n$ whose topology could be estimated. By the continuity property of the Alexander-Spanier cohomology we finally found the desired cohomology of the set X_∞.

Since the space $H^{1/2}$ is not contained in C^0, the loops in $H^{1/2}$ have no local meaning, and it is not easy to define $H^{1/2}$ loops on a general manifold. It can be done but the space obtained does not carry a natural manifold structure. The main idea in the case of a general symplectic manifold (M, ω) is to replace the regularized gradient (of the torus case) by the unregularized gradient of the action functional on the space of smooth contractible loops of M. The associated gradient equation, however, does not determine a flow: its Cauchy initial value problem is ill posed, as we shall see. But analogous to the torus case, we can still define the set X_∞ of all bounded solutions $u : \mathbb{R} \times S^1 \to M$ of the unregularized gradient equation; they are now solutions of a system of first order elliptic partial differential equations of Cauchy-Riemann type, related to the pseudo-holomorphic curves of M. Gromov [107]. On this solution set X_∞ there is, moreover, a natural $\mathbb{R}$-action defining a gradient-like flow. Hence, following our old strategy, we shall prove that this set of solutions is compact and contains the cohomology of the underlying manifold M. The Arnold conjecture then follows by applying the Ljusternik-Schnirelman theorem as in the torus case. In order to carry out this programme the O.D.E. methods of the torus case have to be replaced by P.D.E. methods. In order to determine the topology of X_∞ we shall approximate the set X_∞ by the set of solutions X_T of a distinguished elliptic boundary value problem, to which we apply Fredholm theory based on the Cauchy-Riemann operator on compact surfaces. In this way we can approximate X_T by smooth compact manifolds of dimension $2n$ whose topology can be determined. The easy compactness arguments for the torus case, based on O.D.E. methods, will be replaced by an intricate bubbling off analysis based on elliptic estimates.

In order to find the crucial substitute for X_T used in Section 6.2 we begin with the analysis of the torus case $T^{2n} = \mathbb{R}^{2n}/\mathbb{Z}^{2n}$, this time from the point of view of the unregularized gradient. The aim is to reinterpret the approach of Section 6.2 from a P.D.E. point of view. We recall that to a periodic function $H \in C^\infty(S^1 \times T^{2n}, \mathbb{R})$, we associated the action functional Φ_H, defined on the space of smooth contractible loops x of T^{2n} by

$$(6.52) \qquad \Phi_H(x) = \int_0^1 \left\{ \frac{1}{2} \langle -J\dot{x}, x \rangle - H\big(t, x(t)\big) \right\} dt \ .$$

On the covering space $\mathbb{R}^{2n}$ the contractible loops x are simply the smooth functions $x \in C^\infty(S^1, \mathbb{R}^{2n})$. By extending Φ_H to a smooth functional φ_H on the

Hilbert space $H^{1/2}(S^1, \mathbb{R}^{2n})$ we obtained for the $H^{1/2}$ gradient, which we now denote by φ'_H, the representation

(6.53) $$\varphi'_H(x) = j^* \Phi'_H(x) \text{ if } x \in C^\infty(S^1, \mathbb{R}^{2n}).$$

Here $\Phi'_H(x)$ is the L^2-gradient of Φ_H given by

(6.54) $$\nabla \Phi_H(x) = -J\dot{x} - \nabla H(t, x),$$

and j^* is the adjoint of the inclusion mapping $j : H^{1/2} \to L^2$ defined explicitly in Chapter 3. We may view φ'_H as the regularization of the L^2-gradient Φ'_H by means of the positive and compact operator j^* on L^2. Instead of the gradient equation $\frac{d}{ds}x = -\varphi'_H(x)$ on $H^{1/2}$ studied previously, we now study the unregularized equation $\frac{d}{ds}x = \Phi'_H(x)$. Observe that we now consider the positive gradient instead of the negative one considered in Section 6.2. For indefinite functionals this is irrelevant. However, this way we are led, as we shall see, in a natural way to the familiar Cauchy-Riemann operator instead of the anti-Cauchy-Riemann operator. Setting $u(t, s) = x(s)(t)$ for the loops at time s, we obtain the partial differential equation

(6.55) $$\frac{\partial u}{\partial s} = -J \frac{\partial u}{\partial t} - \nabla H(t, u),$$

for a smooth function $u \in C^\infty(\mathbb{R} \times S^1, \mathbb{R}^{2n})$. Abbreviating the notation in the following, we write $u' = -J\dot{u} - H'(t, u)$ instead of (6.55), where $H' = \nabla H$ stands for the gradient of the function H in the x variable. The operator

(6.56) $$u' + J\dot{u} = \frac{\partial u}{\partial s} + J \frac{\partial u}{\partial t}$$

is nothing but the well-known Cauchy-Riemann operator for maps $u : \mathbb{R}^2 \to \mathbb{R}^{2n}$. To see this it is convenient to identify $\mathbb{R}^{2n}$ with $\mathbb{C}^n$ as follows. We split $\mathbb{R}^{2n} = \mathbb{R}^2 \oplus \cdots \oplus \mathbb{R}^2$ into symplectic subspaces $\mathbb{R}^2$, having the symplectic structure $\omega_0(a, b) = \langle Ja, b \rangle$ with the matrix J given by

(6.57) $$J = \begin{pmatrix} 0 & 1 \\ -1 & 0 \end{pmatrix}.$$

We identify $\mathbb{R}^2$ with $\mathbb{C}$ by means of the linear isomorphism $U : \mathbb{C} \to \mathbb{R}^2$ given by $z = (y + ix) \mapsto (x, y) = (\text{Im } z, \text{Re } z)$. On $\mathbb{C}$, the complex structure is now the multiplication by $i = \sqrt{-1}$, since

(6.58) $$U^{-1}JU = i,$$

and ω_0 becomes $\omega_0(z, \xi) = -Im\langle z, \bar{\xi} \rangle$. Setting $z = s + it \in \mathbb{C}$ we shall view the map $u : \mathbb{R}^2 \to \mathbb{R}^{2n}$ as a map $u : z \mapsto u(z) \in \mathbb{C}^n$ so that the operator (6.56) becomes the familiar Cauchy-Riemann operator on every copy of $\mathbb{C}$,

(6.59) $$\frac{\partial u}{\partial s} + i \frac{\partial u}{\partial t} = \bar{\partial} u.$$

6.4 Elliptic methods and symplectic fixed points

Similarly the operator $\frac{\partial u}{\partial s} - J\frac{\partial u}{\partial t}$ is, in our complex notation, the operator

(6.60) $$\frac{\partial u}{\partial s} - i\frac{\partial u}{\partial t} = \bar{\partial} u.$$

Note that for convenience we drop the factor $\frac{1}{2}$ in the usual definition of the Cauchy-Riemann operator.

The Cauchy-Riemann operator is a first order elliptic partial differential operator and ∇H is a lower order perturbation. There is no well-defined initial value problem associated with the elliptic equation (6.55). Indeed, assume $H = 0$. Then every solution u of (6.55) solves $\bar{\partial} u(z) = 0$ and hence is holomorphic so that the initial data $u(0,t) = x(t)$ at $s = 0$ has to be at least a real analytic loop. But even working in the space of real analytic loops does not help since there are analytic loops which are not restrictions of globally defined holomorphic maps.

Assume now that $u \in C^\infty(\mathbb{R} \times S^1, T^{2n})$ solves the partial differential equation (6.55), and define the map $s \mapsto u(s)$ from $\mathbb{R} \to C^\infty(S^1, T^{2n})$ by setting $u(s)(t) := u(s,t)$. From (6.52) we compute readily

(6.61) $$\frac{d}{ds}\Phi_H\big(u(s)\big) = \|-J\dot{u}(s) - H'\big(t, u(s)\big)\|_{L^2}^2.$$

Hence the map $s \to \Phi_H(u(s))$ is nondecreasing. The following nontrivial fact will become crucial.

Lemma 2. Assume $u \in C^\infty(\mathbb{R} \times S^1, T^{2n})$ is a solution of (6.55). If for some $s_0 \in \mathbb{R}$

$$\frac{d}{ds}\Phi_H\big(u(s)\big)\big|_{s=s_0} = 0,$$

then $u(s) = u(s_0)$ for all $s \in \mathbb{R}$ hence is independent of s. Moreover $x \in C^\infty(S^1, M)$ defined by $x(t) := u(s_0)(t)$ is a 1-periodic solution of the Hamiltonian system associated to H.

Proof. Considering our geometric approach, this statement seems obvious, yet the proof is not easy. It is based on Carleman's similarity principle proved in Appendix 5. Without loss of generality we may assume that $s_0 = 0$ so that

$$-J\dot{u}(0) - H'\big(t, u(0)\big) = 0, \quad \text{for all } t \in S^1.$$

This means that $x(t) := u(0)(t)$ defines a 1-periodic solution of the Hamiltonian vector field $\dot{x} = X_H(x)$. We denote by ψ_t the 1-parameter family of symplectic diffeomorphisms solving $\frac{d}{dt}\psi_t = X_H(\psi_t)$ and $\psi_0 = \text{Id}$. Extending the map u periodically in t we shall view it as a map $u : \mathbb{R} \times \mathbb{R} \to M$. Now we define a new map $v : \mathbb{R} \times \mathbb{R} \to M$ by

$$u(s,t) = \psi_t\big(v(s,t)\big), \quad \text{for all } (s,t) \in \mathbb{R}^2,$$

and compute,

$$\begin{aligned}
0 &= \tfrac{\partial u}{\partial s} + J\tfrac{\partial u}{\partial t} + H'(t,u) \\
&= T\psi_t(v)\tfrac{\partial v}{\partial s} + JT\psi_t(v)\tfrac{\partial v}{\partial t} + J\left(\tfrac{d}{dt}\psi_t\right)(v) + H'\left(t,\psi_t(v)\right) \\
&= T\psi_t(v)\left[\tfrac{\partial v}{\partial s} + \left(T\psi_t(v)^{-1}JT\psi_t(v)\right)\tfrac{\partial v}{\partial t}\right].
\end{aligned}$$

Hence, introducing the almost complex structure

$$J(s,t) := T\psi_t(v(s,t))^{-1}JT\psi_t(v(s,t))$$

the map $v : \mathbb{R}^2 \to M$ solves

$$\frac{\partial v}{\partial s} + J(s,t)\frac{\partial v}{\partial t} = 0,$$

and, since $u(0,t) = \psi_t(v(0,0))$ by the assumption on u,

$$v(0,t) = v(0,0), \quad \text{for } t \in \mathbb{R}.$$

Let $v(0,0) = m \in M$. We take a chart around m, say $\varphi : U \to \varphi(U) \subset \mathbb{R}^{2n}$ satisfying $\varphi(m) = 0$. Then the map $w : \mathbb{R}^2 \to \mathbb{R}^{2n}$, defined by $w = \varphi \circ v$, satisfies, near $0 \in \mathbb{R}^2$, the partial differential equation

$$\frac{\partial w}{\partial s} + T\varphi\left(v(s,t)\right)J(s,t)T\varphi\left(v(s,t)\right)^{-1}\frac{\partial w}{\partial t} = T\varphi(v)\left[\frac{\partial v}{\partial s} + J(s,t)\frac{\partial v}{\partial t}\right].$$

Moreover, $w(0,t) = 0$ for t close to $t = 0$. To sum up, introducing the almost complex structure $\hat{J}(s,t) := T\varphi(v(s,t))J(s,t)T\varphi(v(s,t))^{-1}$, the map $w : \mathbb{R}^2 \to \mathbb{R}^{2n}$ solves, near $0 \in \mathbb{R}^2$, the equations

$$\frac{\partial w}{\partial s} + \hat{J}(s,t)\frac{\partial w}{\partial t} = 0, \quad w(0,t) = 0 \text{ for } t \text{ close to } 0.$$

In particular, the point $(0,0) \in \mathbb{R}^2$ is a cluster point of zeroes $w(s,t) = 0$ of the solution w. In this situation we can apply the generalized Carleman similarity principle proved in Appendix 5. It states that w is of the form $w(z) = \Phi(z)\sigma(z)$ for all $z = s + it$ in a neighborhood U of $z = 0 \in \mathbb{C}$, where $\Phi : U \to \mathcal{G}l_\mathbb{R}(\mathbb{C}^n)$ is continuous and $\sigma : U \to \mathbb{C}^n$ is holomorphic. We conclude that $w = 0$ in a full neighborhood of $0 \in \mathbb{R}^2$. This implies that v is constant in a neighborhood of zero and, therefore,

$$u(s,t) = x(t) \text{ for } s \text{ and } t \text{ close to } 0.$$

Consider now the following set of points:

$$\begin{aligned}
\Sigma = \{(s,t) \in \mathbb{R} \times \mathbb{R} \mid u(s,t) &= x(t) \\
\text{and there exists a sequence } (s_k, t_k) &\to (s,t), \\
(s_k, t_k) \neq (s,t) \text{ satisfying } u(s_k, t_k) &= x(t_k)\}.
\end{aligned}$$

6.4 ELLIPTIC METHODS AND SYMPLECTIC FIXED POINTS

This set is closed, and by the previous discussion nonempty. If $(s_0, t_0) \in \Sigma$ we can apply the previous argument and deduce that (s_0, t_0) is an interior point of Σ. Consequently $\Sigma = \mathbb{R} \times \mathbb{R}$, and hence $u(s,t) = x(t)$ for all $(s,t) \in \mathbb{R} \times S^1$. This implies that $u(s) = u(0)$ for all $s \in \mathbb{R}$, hence $u(s)$ is a rest point representing the periodic orbit $x(t) = u(0)(t)$. The proof of Lemma 2 is complete. ∎

The space X_∞ of "bounded solutions" of the (unregularized) gradient equation is defined by

(6.62) $\quad X_\infty = \left\{ u \in C^\infty(\mathbb{R} \times S^1, T^{2n}) \,\middle|\, u \text{ satisfies (6.63) below} \right\}.$

Here

(6.63)
$$u_s + Ju_t + H'(t,u) = 0$$
$$-\infty < \inf_s \Phi_H\!\left(u(s)\right) \leq \sup_s \Phi_H\!\left(u(s)\right) < +\infty.$$

We equip the space X_∞ with the topology induced from $C^\infty(\mathbb{R} \times S^1, T^{2n})$. There is a continuous $\mathbb{R}$-action on X_∞ defined by

(6.64)
$$\mathbb{R} \times X_\infty \longrightarrow X_\infty$$
$$(\tau, u) \longmapsto u \cdot \tau$$

where $(u \cdot \tau)(s,t) = u(s+\tau, t)$. The gradient is independent of time; hence with $u(s,t)$ also $u(s+\tau, t)$ is a solutions of (6.63). The flow (6.64) on X_∞ is gradient-like. Indeed, define the continuous function V by

$$V : X_\infty \longrightarrow \mathbb{R}, \quad u \mapsto \Phi_H\!\left(u(0)\right).$$

Then it follows from Lemma 2 that V is a Ljapunov function, i.e., $\tau \mapsto V(u \cdot \tau)$ strictly increases with τ, along nonconstant orbits. In addition, the constant orbits of the flow are the 1-periodic solutions of the Hamiltonian equation $\dot{x} = X_H(x)$. Hence the rest points are the same as the ones for the set X_∞ previously introduced in Section 6.3. But the new set X_∞ differs from the old one. If the set X_∞ is compact and if its cohomology is sufficiently rich, we can conclude from Theorem 1 a certain number of rest points of the flow. Before we turn to the compactness proof of X_∞ which is closely related to elliptic regularity theory, we first address the question how we can replace T^{2n} by a general compact symplectic manifold (M, ω).

We choose an almost complex structure J compatible with ω, in the sense that g_J defined by

(6.65) $\quad g_J(m)(h,k) = \omega(m)\!\left(h, J(m)k\right)$

for $m \in M$ and $h, k \in T_m M$, is a Riemannian metric on M. We consider the partial differential equation for $u \in C^\infty(\mathbb{R} \times S^1, M)$.

(6.66) $\quad \dfrac{\partial u}{\partial s} + J(u)\dfrac{\partial u}{\partial t} + H'(t,u) = 0\,.$

Here H' is the gradient associated to g_J with respect to the x-variable. Again we assume that for fixed s, the loop $u(s)(\cdot)$ in $C^\infty(S^1, M)$ is contractible. Hence $s \to u(s)$ defines a map from $\mathbb{R}$ into $C_c^\infty(S^1, M)$, where C_c^∞ denotes the set of contractible loops on M. Here the first difficulty arises. What should the functional Φ_H be? There is only one way to produce such a Φ_H: if x is a contractible loop in M. We pick an extension $\bar{x} : D \to M$, where D is the closed unit disc bounded by $S^1 = \partial D$ such that $\bar{x}|\partial D = x$. We now define the action by

$$(6.67) \qquad \Phi_H(x) = \int_D \bar{x}^* \omega - \int_0^1 H\big(t, x(t)\big) dt .$$

Unfortunately, this is in general not well-defined. Namely, if we take another extension $\tilde{x}$ of x the numbers $\int_D \bar{x}^* \omega$ and $\int_D \tilde{x}^* \omega$ need not be the same. In fact they may differ by a number $r \in \Gamma \subset \mathbb{R}$, where Γ, the so-called period group, is the image of the group homomorphism

$$\sigma : \pi_2(M) \to \mathbb{R} : [u] \to \int_{S^2} u^* \omega , \text{ where } u : S^2 \to M.$$

We note that the map σ is well-defined by Stokes's theorem; moreover, $\Gamma = \text{im}(\sigma)$ is a subgroup of $(\mathbb{R}, +)$. Summing up we only have a functional

$$\Phi_H : C_c^\infty (S^1, M) \to \mathbb{R}/\Gamma .$$

Since for $M = T^{2n}$ we have $\pi_2 = 0$, we conclude that $\Gamma = \{0\}$ implying $\mathbb{R}/\Gamma = \mathbb{R}$. In contrast to this special example, the subgroup Γ can be dense as the following example shows. Take $M = S^2 \times S^2$ with the symplectic form $\Omega \oplus r\Omega$ where Ω is a volume form of total volume 1 on S^2 and where $r > 0$ is an irrational number. Then $\Gamma = \{a + br \mid a, b \in \mathbb{Z}\}$, which is dense in $\mathbb{R}$. We point out that the condition $\Gamma = \{0\}$, which for example is satisfied if $\pi_2(M) = 0$, is a very restrictive condition. Indeed, since M is compact we know that the cohomology class of the symplectic structure ω, say $[\omega]$, is nontrivial in $H^2(M, \mathbb{R})$. If $\pi_1(M) = 0$ the Hurewicz-homomorphism

$$\pi_2(M) \to H_2(M, \mathbb{Z})$$

is an isomorphism. Therefore the homomorphism σ is nontrivial, implying that $\Gamma \neq \{0\}$. We see that $\Gamma = \{0\}$ implies $\pi_1(M) \neq \{0\}$. We want to point out that the case $\Gamma \neq \{0\}$ can also be treated at least partially. In fact, in order to do Morse theory, one only needs a hypothesis on the first Chern class of TM, which is vacuous in dimensions 2,4 and 6. Under this condition no hypothesis on the symplectic form is needed. So, in particular, the Arnold conjectures for the Morse theory hold in dimensions 2,4 and 6. We refer the reader to H. Hofer and D. Salamon [119] and K. Ono [172].

6.4 Elliptic methods and symplectic fixed points

In the following we shall assume $\Gamma = \{0\}$ hence requiring that

$$(6.68) \qquad \int_{S^2} u^*\omega = 0 \text{ for } u : S^2 \to M .$$

Now we have a well-defined functional $\Phi_H : C_c^\infty(S^1, M) \to \mathbb{R}$ defined on the contractible loops by (6.67). In order to prove the Arnold conjecture we need to show the compactness of X_∞ and to compute its cohomology. The compactness will be studied by using elliptic regularity theory. The cohomology will be computed by using an approximation argument, replacing the sets X_T previously used in Section 6.2 by suitable substitutes.

The first aim is to replace the O.D.E. boundary value problem X_T in Section 6.2 by a boundary value problem for an elliptic P.D.E. To find the appropriate modification for the general symplectic manifold, we go back to the torus case and consider $x \in T^{2n} \times E^+$. Identifying $\mathbb{R}^{2n}$ with $\mathbb{C}^n$, a loop $x \in T^{2n} \times E^+$ is represented as

$$(6.69) \qquad x(t) = a_0 + \sum_{k=1}^{\infty} a_k\, e^{2\pi i k t} ,$$

with $a_0 \in T^{2n}$ and $a_k \in \mathbb{C}^n$. We shall view S^1 as the boundary ∂D of the closed disc $D = \{z \in \mathbb{C} : |z| \leq 1\}$ and extend the map $x : S^1 = \partial D \to \mathbb{C}^n$ to a map $u : D \to \mathbb{C}$ by defining

$$(6.70) \qquad u(z) = a_0 + \sum_{k=1}^{\infty} a_k\, z^k , \quad z \in D .$$

Clearly $u(e^{2\pi i t}) = x(t)$ and since this loop belongs to $H^{1/2}(S^1)$, the series converges on every disc $|z| \leq \rho < 1$ uniformly and hence defines a holomorphic map from the open disc $\mathring{D} = \{z \in \mathbb{C} : |z| < 1\}$ into T^{2n}. Setting $z = s + it$ we have $u_s + iu_t = 0$ on $\mathring{D}$. We claim that

$$(6.71) \qquad \frac{1}{2}\int_D \left(|u_s|^2 + |u_t|^2\right) ds\, dt = \int_D u^*\omega_0 = \frac{1}{2}\|x\|^2_{H^{1/2}(S^1)} ,$$

provided $a_0 = 0$. We conclude that an element $x \in T^{2n} \times E^+$ can be considered as the boundary $x = u|\partial D$ of a holomorphic disc $u : D \to T^{2n}$ of finite area, which is positive if $x \neq 0$

$$0 < \int_D u^*\omega_0 < \infty .$$

The proof of (6.71) is a simple computation. Setting $z = s + it$, we find

$$\tfrac{1}{2}\int_D \left(|u_s|^2 + |u_t|^2\right) ds\, dt = \int_D \omega(u_s, u_t)\, ds\, dt$$

$$= \int_D \omega\left(\sum_{k=1}^\infty a_k\, k\, z^{k-1}, \sum_{k=1}^\infty a_k\, i\, k\, z^{k-1}\right) ds\, dt$$

$$= \int_0^1 \left(\int_0^{2\pi} \langle \sum_{k=1}^\infty a_k\, k\, r^{k-1}\, e^{2\pi i(k-1)\vartheta}, \sum a_k\, k\, r^{k-1}\, e^{2\pi(k-1)i\vartheta}\rangle\, 2\pi r\, d\vartheta\right) dr$$

$$= 2\pi \int_0^1 \sum_{k=1}^\infty |a_k|^2\, k^2\, r^{2k-1}\, dr = \pi \sum_{k=1}^\infty |a_k|^2\, |k| = \tfrac{1}{2}||x||^2_{H^{1/2}} < \infty,$$

proving the claim. Similarly $x \in T^{2n} \times E^-$ can be considered as boundary of an anti-holomorphic disc. It is represented by

$$x(t) = b_0 + \sum_{k=1}^\infty b_k\, e^{-2\pi i k t},$$

and the extension $v : D \to \mathbb{C}^n$ is defined by

$$v(z) = b_0 + \sum_{k=1}^\infty b_k \bar{z}^k, \quad z \in D,$$

so that $v(e^{2\pi i t}) = x(t)$. The map v is anti-holomorphic i.e., it satisfies $v_s - iv_t = 0$ on $\mathring{D}$. Moreover, the area is finite and negative:

(6.72) $$-\infty < \int_D v^*\omega_0 = -\tfrac{1}{2}\int_D \left(|v_s|^2 + |v_t|^2\right) ds\, dt$$
$$= -\tfrac{1}{2}||x||^2_{H^{1/2}(S^1)} \leq 0,$$

provided $b_0 = 0$. We now can rewrite the boundary value problem for X_T studied in Section 6.2 (however replacing the negative gradient flow by the positive one) as follows

(6.73) $$\begin{aligned} a_s + J a_t &= 0 & &\text{on } \mathring{D} \\ u_s - u^+ + u^- + j^* H'(t, u) &= 0 & &\text{on } (-T, T) \times S^1 \\ b_s - J b_t &= 0 & &\text{on } \mathring{D}, \end{aligned}$$

with the matching boundary conditions

(6.74) $$\begin{aligned} u(-T, t) &= a(e^{2\pi i t}) & &\text{for } t \in S^1 \\ u(T, t) &= b(e^{2\pi i t}) & &\text{for } t \in S^1, \end{aligned}$$

where $S^1 = \mathbb{R}/\mathbb{Z}$. The solution triplet (a, u, b) can be visualized as follows:

6.4 Elliptic methods and symplectic fixed points

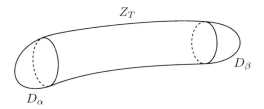

Fig. 6.8

Not regularizing the L^2-gradient in the model equation we are led to

$$
\begin{aligned}
a_s + Ja_t &= 0 & &\text{on } \mathring{D} \\
(6.75) \qquad u_s + Ju_t + H'(t,u) &= 0 & &\text{on } (-T,T) \times S^1 \\
b_s - Jb_t &= 0 & &\text{on } \mathring{D},
\end{aligned}
$$

together with the matching boundary conditions (6.74). We shall rewrite the system (6.75) together with the boundary conditions as a single equation. We define the Riemannian sphere S_T by adding two caps at the ends of the cylinder $Z_T = [-T,T] \times S^1$ such that

$$(6.76) \qquad S_T \;=\; D \bigcup_\alpha Z_T \bigcup_\beta D \,.$$

Here α and β denote the identifications

$$
\begin{aligned}
\alpha &: & e^{2\pi i t} &\sim (-T, t) \\
\beta &: & e^{-2\pi i t} &\sim (T, t) \,.
\end{aligned}
$$

Next we define the structure of a complex manifold on S_T by introducing the charts

$$\psi : (-T,T) \times S^1 \longrightarrow \mathbb{C} \;:\; (s,t) \mapsto e^{2\pi(s+it)}$$

$$(6.77) \qquad \varphi_1 : D \bigcup_\alpha [-T,0) \times S^1 \longrightarrow \mathbb{C} \;:\; \begin{cases} z \in D \mapsto z \\ (s,t) \mapsto e^{2\pi(s+T+it)} \end{cases}$$

$$\varphi_2 : (0,T] \times S^1 \bigcup_\beta D \longrightarrow \mathbb{C} \;:\; \begin{cases} z \in D \mapsto \bar z \\ (s,t) \mapsto e^{2\pi(s-T+it)} \end{cases}.$$

We observe that the transition maps $\varphi_1 \circ \psi^{-1}$ and $\varphi_2 \circ \psi^{-1}$ are holomorphic. These charts define a complex structure j on S_T. By the Uniformization Theorem (S_T, j)

is bi-holomorphic to (S^2, i), where $S^2 = \mathbb{C} \cup \{\infty\}$ is the standard Riemannian sphere, see, e.g., O. Forster [92]. For notational convenience we denote the α-disc by D_1 and the β-disc by D_2. Consider now $v \in C^\infty(S_T, T^{2n})$ and define a map Lv, which is smooth on $\mathring{D}_1, \mathring{D}_2$ and $(-T, T) \times S^1$ as

$$(6.78) \quad Lv = Tv + J \circ (Tv) \circ j + \chi_{[-T,T] \times S^1}\Big(H'(v)ds - JH'(v)dt\Big).$$

Here (s, t) are the coordinates on $[-T, T] \times S^1$ and $\chi_{[-T,T] \times S^1}$ is the corresponding characteristic function on S_T. Moreover Tv is the tangent map of $v : S_T \to T^{2n}$. Observe that $(Lv)(z)$ is a linear map $T_z S_T \to T_{v(z)} T^{2n}$, which satisfies the relation

$$-J \circ (Lv)(z) = (Lv)(z) \circ j.$$

We denote by $A_{T^{2n}} \to S_T \times T^{2n}$ the vector bundle whose fibre over (z, m) consists of all complex anti-linear maps $\gamma : T_z S_T \to T_m T^{2n}$, satisfying $\gamma j = -J\gamma$. If v is a map $S_T \to T^{2n}$, we denote by $v^* A_{T^{2n}}$ the pullback of $A_{T^{2n}}$ by the graph $z \to (z, v(z))$ of v. Then Lv may be viewed as a section of $v^* A_{T^{2n}} \to S_T$. Assume v is continuous and smooth on $S_T \setminus \{(\pm T, t) | t \in S^1\}$ and moreover satisfies $Lv = 0$ away from the circle $\{\pm T\} \times S^1$. That precisely means that Lv is the zero section of $v^* A_{T^{2n}} \to S_T$. Consider the vector field $\frac{\partial}{\partial s}$ on $(-T, T) \times S^1$. Then

$$\begin{aligned} 0 &= (Lv)(s, t)(\tfrac{\partial}{\partial s}) \\ &= v_s + J v_t + H'(t, v). \end{aligned}$$

We see that v satisfies differential equation (6.75) on $(-T, T) \times S^1$. On D_1 and D_2 we have $v_s + Jv_t = 0$ and $v_s - Jv_t = 0$ respectively. Moreover, the boundary conditions are satisfied on the common parts of D_1, D_2 and $[-T, T] \times S^1$.

This new construction can easily be generalized to arbitrary compact symplectic manifolds. For this let us denote by j the complex structure on S_T and by J an almost complex structure compatible with ω on M. For $v : S_T \to M$ we define the section Lv of $v^* A_M$ by

$$\begin{aligned}(6.79) \quad (Lv)(z) &= Tv(z) + J(v) \circ Tv(z) \circ j \\ &+ \chi_{[-T,T] \times S^1}(z) \Big[H'\big(t, v(z)\big) ds - J\big(v(z)\big) H'\big(t, v(z)\big) dt\Big].\end{aligned}$$

In order to study the equation $Lv = 0$, we need to develop a suitable functional analytic set up. If $2 < p < \infty$ we consider the Banach manifold $W^{1,p}(S_T, M)$ consisting of maps from S_T into M belonging to the Sobolev class $W^{1,p}$. For the construction of a natural differentiable structure on $W^{1,p}(S_T, M)$ we refer the reader to the article by H. Eliasson [79]. For $v \in W^{1,p}(S_T, M)$ we consider the bundle $v^* A_M \to S_T$ and let $L^p(v^* A_M)$ be the associated space of L^p-sections. Then, we can equip

$$L^p\Big(W^{1,p}(S_T, M)^* A_M\Big) = \bigcup_{v \in W^{1,p}(S_T, M)} \{v\} \times L^p(v^* A_M)$$

6.4 Elliptic methods and symplectic fixed points

with the structure of a Banach space bundle over the Banach manifold $W^{1,p}(S_T, M)$. Summing up we have formulated the elliptic system (6.75) together with the matching boundary conditions (6.74) in such a way that solutions are to be found as the zeroes of the section L of the Banach space bundle $L^p(W^{1,p}(S_T, M)^* A_M) \to W^{1,p}(S_T, M)$. We shall denote the set of solutions by $X_T = \{v \in W^{1,p}(S_T, M) | L(v) = 0\}$.

Note that we cannot expect a solution $v : S_T \to M$ of $L(v) = 0$ to be of regularity better than $W^{1,p}$ near the boundary ∂Z_T of the cylinder Z_T. This is due to the non-smoothness of the cut-off function $s \mapsto \chi_{[-T,T]}(s)$ at the boundary components $s = \pm T$ of Z_T. Recall that these two boundary components agree with the boundaries of the two disks D glued to the cylinder. However, it follows from the results in Appendix 5, that a solution v of $L(v) = 0$ is smooth except at ∂Z_T where the derivatives in s (i.e., normal to the boundaries) do not match. To be more precise we have $W^{1,p}(S_T, M) \subset C^0(S_T, M)$ by the Sobolev embedding theorem, since $p > 2$. Hence v being continuous can be studied in local coordinates. If z is a point in the interior of the cylinder Z_T we have, in local coordinates, $v \in W^{1,p}(B_\varepsilon, \mathbb{R}^{2n})$ with a small open ball B_ε centered at z. Moreover v is a weak solution of a special system of first order elliptic equations and we conclude that $v \in C^\infty(B_\varepsilon, \mathbb{R}^{2n})$ by the "inner" regularity theory of elliptic partial differential equations. Similarly, the solution v of $L(v)$ is smooth in the interior of the disks D. It is proved in Appendix 5 that, in addition, v has smooth extensions from the interior of Z_T to its boundary ∂Z_T and also from the interior of the disks to their boundaries $\partial D_1 \cup \partial D_2 = \partial Z_T$. This does, of course, not mean that the solutions fit smoothly at the boundary, since the derivatives (in s) normal to the boundary do not agree. But the restriction of v to the boundary ∂Z_T is smooth, so that, in particular, $v(-T)$ and $v(T)$ are smooth loops belonging to $C_c^\infty(S^1, M)$.

We can now use the variational structure of our problem to derive a natural a priori estimate for solutions $v \in X_T$. Recall the assumption $\omega | \pi_2(M) = 0$ and assume that $L(v) = 0$. Let $C > 0$ be a constant bounding $|\max_{(t,x)} H(t,x)|$ and $|\min_{(t,x)} H(t,x)|$. We estimate

$$\Phi_H\big(v(-T)\big) = \int_{D_1} v^* \omega - \int_0^1 H\big(t, v(-T, t)\big)\, dt$$

$$\geq -\max_{(t,x)} H(t,x) \geq -C.$$

We have used that for holomorphic maps

$$\int_{D_1} v^* \omega = \frac{1}{2} \int_{D_1} |\nabla v|_J^2\, ds\, dt \geq 0.$$

Similarly

$$\Phi_H\big(v(T)\big) \;=\; \int_{D_2} v^*\omega \;-\; \int_0^1 H\big(t, v(T,t)\big)\, dt$$

$$\leq \; -\min_{(t,x)} H(t,x) \leq C\,,$$

using that for anti-holomorphic maps

$$\int_{D_2} v^*\omega \;=\; -\frac{1}{2}\int_{D_2} |\nabla v|_J^2 \;\leq\; 0\,.$$

Summarizing, we have proved

Lemma 3. There exists a constant $C > 0$, independent of $T > 0$ such that every solution $v : S_T \to M$ of $Lv = 0$ satisfies

$$-C \;\leq\; \Phi_H\big(v(-T)\big) \;\leq\; \Phi_H\big(v(s)\big) \;\leq\; \Phi_H\big(v(T)\big) \;\leq\; C\,,\;\text{ for }\; -T \leq s \leq T\,.$$

From this we conclude

Lemma 4. There exists a constant $C > 0$, independent of $T > 0$ such that every solution $v \in W^{1,p}(S_T, M)$ of $Lv = 0$ satisfies

$$\frac{1}{2}\int_{-T}^{T}\int_0^1 \Big[\big|\tfrac{\partial v}{\partial s}\big|_J^2 + \big|\tfrac{\partial v}{\partial t} - X_{H_t}(v)\big|_J^2\Big]\, ds\, dt \;\leq\; 2C\,.$$

Proof. We compute, using Lemma 3,

$$2C \;\geq\; \Phi_H\big(v(T)\big) - \Phi_H\big(v(-T)\big)$$

$$= \int_{-T}^{T} \tfrac{d}{ds}\Phi_H\big(v(s)\big)\, ds = \int_{-T}^{T} \|\Phi_H'\big(v(s)\big)\|_{J,L^2}^2\, ds$$

$$= \tfrac{1}{2}\int_{-T}^{T}\Big[\|\tfrac{d}{ds} v(s)\|_{J,L^2}^2 + \|\dot v(s) - X_{H_t}\big(v(s)\big)\|_{J,L^2}^2\Big]\, ds$$

$$= \tfrac{1}{2}\int_{-T}^{T}\int_0^1 \Big[\big|\tfrac{\partial v}{\partial s}(s,t)\big|_J^2 + \big|\tfrac{\partial v}{\partial t} - X_{H_t}\big(v(s,t)\big)\big|_J^2\Big]\, ds\, dt\,.$$

Here $\|\;\|_{J,L^2}$ is the L^2-norm for sections associated to the Riemannian metric g_J. ∎

Our next aim is to study compactness properties of the space X_∞ of "bounded orbits" defined by

(6.80) $$X_\infty \;=\; \Big\{u \in C^\infty(\mathbb{R}\times S^1, M)\,\Big|\, u_s + J(u)u_t + H'(t,u) = 0,\; u(s) \in C_c^\infty(S^1, M)\text{ and } \big(\Phi_H(u(s))\big)_{s\in\mathbb{R}} \text{ is bounded in } \mathbb{R}\Big\}\,.$$

6.4 Elliptic methods and symplectic fixed points

We note that for the estimates it is convenient in the following to view M as a compact submanifold of some $\mathbb{R}^N$ via Whitney's embedding theorem, see, e.g., M. Hirsch [113]. Since $C^\infty(\mathbb{R} \times S^1, \mathbb{R}^N)$ is a Frechet space with the usual translation invariant metric d we may view $C^\infty(\mathbb{R} \times S^1, M)$ as a closed subspace of $C^\infty(\mathbb{R} \times S^1, \mathbb{R}^N)$. The induced metric we denote also by d. We shall see that X_∞ is a compact subspace of $C^\infty(\mathbb{R} \times S^1, \mathbb{R}^N)$. We begin with

Proposition 10. There exists a constant $C > 0$ such that for every $u \in X_\infty$ we have
$$\Phi_H\bigl(u(s)\bigr) \in [-C, C], \quad s \in \mathbb{R}.$$

Proof. By assumption $\Phi_H(u(\mathbb{R}))$ is bounded in $\mathbb{R}$. Hence, abbreviating $\|\ \| = \|\ \|_{J,L^2}$,
$$\int_{-\infty}^{+\infty} \|\Phi'_H\bigl(u(s)\bigr)\|^2 \, ds = \int_{-\infty}^{+\infty} \left(\frac{d}{ds}\Phi_H(u(s))\right) ds < +\infty.$$

Consequently there exists a sequence $(s_n) \subset \mathbb{R}$, with $s_n \to +\infty$, such that
$$\Phi_H\bigl(u(s_n)\bigr) \longrightarrow \alpha \in \mathbb{R} \text{ and } \|\Phi'_H\bigl(u(s_n)\bigr)\| \longrightarrow 0.$$

Defining $x_n = u(s_n) \in C^\infty(S^1, M)$ we obtain
$$\tag{6.81} \|\dot{x}_n - X_H(x_n)\| \longrightarrow 0.$$

Considering x_n as a map into $\mathbb{R}^N$ we deduce by the compactness of M that $\|\dot{x}_n\|_{L^2(S^1, \mathbb{R}^N)} \leq$ const, where we used the usual L^2-inner product. Hence we find a subsequence satisfying

$$\tag{6.82} \begin{aligned} x_n \rightharpoonup x &\quad \text{weakly in } H^{1,2}(S^1, \mathbb{R}^N) \\ x_n \to x &\quad \text{strongly in } C^0(S^1, \mathbb{R}^N). \end{aligned}$$

The latter convergence follows using the Arzelà-Ascoli Theorem. From (6.81) and (6.82) we conclude that $\dot{x}_n$ is a Cauchy-sequence in L^2 so that
$$\tag{6.83} x_n \to x \text{ in } H^{1,2}(S^1, M) \text{ and } \dot{x} = X_H(x).$$

Consequently $x \in C^\infty(S^1, M)$ and x is a 1-periodic solution of the Hamiltonian vector field. Now $\Phi_H(x_n) \to \Phi_H(x)$ and hence $\alpha = \Phi_H(x)$. We can argue similarly for sequences $x_n = u(s_n)$ where $s_n \to -\infty$. By the Arzelà-Ascoli Theorem the set of all 1-periodic solutions $\dot{x} = X_H(x)$ is compact in $C^\infty(S^1, \mathbb{R}^N)$, so that the set of critical levels
$$\mathcal{C} := \bigl\{\Phi_H(x) \,|\, \Phi'_H(x) = 0\bigr\}$$
is compact in $\mathbb{R}$. We choose $C > 0$ so large that the interval $[-C, C]$ contains $\mathcal{C}$. Since $s \mapsto \Phi_H(u(s))$ is monotone and since, as we have seen, the bounded solution u connects critical points we conclude that $-C \leq \Phi_H(u(s)) \leq C$ for all $s \in \mathbb{R}$ completing the proof of the proposition. ∎

Corollary 1. X_∞ is closed in $C^\infty(\mathbb{R} \times S^1, M)$.

Proof. Pick any sequence $(u_n) \subset X_\infty$. We know from Proposition 10 that $\Phi_H(u_n(\mathbb{R})) \subset [-C, C]$. Assume $u_n \to u$ for the C^∞-metric d. We obtain for every $r > 0$

$$C \geq \Phi_H\big(u_n(r)\big) \geq \Phi_H\big(u_n(-r)\big) \geq -C.$$

Hence, as $n \to \infty$ the estimate implies that $\Phi_H(u(\mathbb{R})) \subset \mathbb{R}$ is bounded. Moreover, since $\frac{\partial u_n}{\partial s} + J(u_n)\frac{\partial u_n}{\partial t} + H'(t, u_n) = 0$, we may pass to the limit and see that u satisfies the differential equation as well. This shows that indeed $u \in X_\infty$ as claimed. ∎

The geometric picture is as in Section 6.2: a "bounded orbit" $u \in X_\infty$ connects 1-periodic solutions of the Hamiltonian equation $\dot{x} = X_H(x)$ on M. We next prove that X_∞ is a compact subset of $C^\infty(\mathbb{R} \times S^1, M)$. Note that the metric space $C^\infty(\mathbb{R} \times S^1, M)$ has the Heine-Borel property; this is a consequence of the Arzelà-Ascoli Theorem applied to all the derivatives of the functions. Since, by the Corollary, X_∞ is a closed set we have to show that $X_\infty \subset C^\infty(\mathbb{R} \times S^1, M)$ is bounded. This requires that for every compact $K \subset \mathbb{R} \times S^1$ and every integer $n \in \mathbb{N}$ there is a constant $C_{K,n} > 0$, such that

(6.84) $$\sup_{\substack{(s,t) \in K \\ |\alpha| \leq n}} |D^\alpha u(s,t)| \leq C_{K,n} \text{ for all } u \in X_\infty.$$

The important ingredient for the proof of (6.84) is a gradient estimate uniform in $u \in X_\infty$. Here we shall use the assumption $\omega|\pi_2(M) = 0$ in a crucial way. Note that in order to work in local coordinates we have to make sure that, for a whole sequence of mappings, a fixed local neighborhood of a point is mapped into the same chart of the image space.

Theorem 8. Suppose $\omega|\pi_2(M) = 0$. Then there exists a constant $A > 0$ such that

$$|\nabla u(s,t)|_J \leq A$$

for all $u \in X_\infty$ and $(s,t) \in \mathbb{R} \times S^1$.

Proof. Arguing indirectly we find sequences $(s_k, t_k) \in \mathbb{R} \times S^1$ and $(u_k) \subset X_\infty$ such that

$$|\nabla u_k(s_k, t_k)|_J \longrightarrow +\infty.$$

Since X_∞ is invariant under the $\mathbb{R}$-action (6.64) we may assume that $s_k = 0$ replacing u_k by $u_k \cdot s_k$. Moreover, taking a subsequence we may assume $t_k \to t_0$. Pick any sequence (ε_k) satisfying

(6.85) $$\varepsilon_k > 0, \varepsilon_k \to 0, \varepsilon_k |(\nabla u_k)(s_k, t_k)|_J \longrightarrow +\infty.$$

We wish this sequence to meet additional conditions and use a little trick. It is based on the following:

6.4 Elliptic methods and symplectic fixed points

Lemma 5. Let (X, d) be a complete metric space and $g : X \to [0, +\infty)$ a continuous map. Assume $x_0 \in X$ and $\varepsilon_0 > 0$ are given. Then there exists $x \in X$ and $\varepsilon > 0$ such that
$$0 < \varepsilon \leq \varepsilon_0$$
$$g(x)\varepsilon \geq g(x_0)\varepsilon_0$$
$$d(x, x_0) \leq 2\varepsilon_0$$
$$g(y) \leq 2g(x) \text{ for all } y \text{ satisfying } d(y, x) \leq \varepsilon .$$

Proof of Lemma 5. Assume there is no $x_1 \in X$ satisfying $d(x_1, x_0) \leq \varepsilon_0$ and $g(x_1) > 2g(x_0)$. Then the statement follows taking $x = x_0$ and $\varepsilon = \varepsilon_0$. Otherwise we can pick such an x_1 and, proceeding inductively, we can assume there are $x_0, x_1, \ldots, x_n$ satisfying
$$d(x_k, x_{k-1}) \leq \frac{\varepsilon_0}{2^{k-1}} \text{ and } g(x_k) > 2g(x_{k-1})$$
for all $1 \leq k \leq n$. Then if there does not exist an x_{n+1} with

(6.86) $\qquad d(x_{n+1}, x_n) \leq \dfrac{\varepsilon_0}{2^n}$ and $g(x_{n+1}) > 2g(x_n)$,

the statement follows choosing $x = x_n$ and $\varepsilon = \frac{\varepsilon_0}{2^n}$. Indeed x_n satisfies $d(x_n, x_0) \leq d(x_n, x_{n-1}) + \cdots + d(x_1, x_0) \leq 2\varepsilon_0$, and $g(x_n) > 2g(x_{n-1}) > 2^2 g(x_{n-2}) > 2^n g(x_0)$ and hence $g(x_n)\varepsilon > g(x_0)\varepsilon_0$. Moreover, from (6.86) we conclude that if $d(y, x_n) \leq \varepsilon$, then $g(y_n) \leq 2g(x_n)$, hence proving the claim. If, on the other hand, there exists an x_{n+1} satisfying (6.86) we continue the process. It must terminate after finitely many steps since otherwise we have a Cauchy sequence $(x_n) \subset X$ satisfying $g(x_n) \to +\infty$. This is not possible, since $x_n \to x^*$ and g is continuous at x^*. ∎

In order to simplify the notation we shall write in the following $|\cdot|$ instead of the Riemannian norm $|\cdot|_J$ on the tangent space. In view of the above lemma we can replace $x_k = (s_k, t_k)$ and ε_k by slightly modified sequences and may assume, again using the $\mathbb{R}$-action, that the above sequence u_k satisfies, in addition

(6.87) $\quad |\nabla u_k(s, t)| \leq 2|\nabla u_k(0, t_k)| , \quad \text{if } |s|^2 + |t - t_k|^2 \leq \varepsilon_k^2, \ 0 \leq t_k \leq 1 .$

Here we conveniently view the maps u as maps defined on $\mathbb{R} \times \mathbb{R}$ by a 1-periodic continuation in the t-variable. Rescaling we define a new sequence $v_k \in C^\infty(\mathbb{R}^2, M)$ by

(6.88) $\quad v_k(s, t) = u_k\Big((0, t_k) + \dfrac{1}{R_k}(s, t)\Big), \quad \text{where } R_k = |\nabla u_k(0, t_k)| .$

Denote by D_k the "large" discs $D_k = \{x = (s,t) \in \mathbb{R}^2 | |x| \leq \varepsilon_k R_k\} \subset \mathbb{R}^2$. Then the sequence v_k has the following properties:

(6.89)
$$v_k \in C^\infty(D_k, M)$$
$$|\nabla v_k(x)| \leq 2 \text{ for } x \in D_k$$
$$|\nabla v_k(0)| = 1$$
$$\varepsilon_k R_k \longrightarrow +\infty .$$

Moreover v_k satisfies the partial differential equation

(6.90)
$$\frac{\partial}{\partial s} v_k + J(v_k) \frac{\partial}{\partial t} v_k = -\frac{1}{R_k} H'(t_k + \frac{t}{R_k}, v_k)$$

on D_k. Since $u_k \in X_\infty$ we find in view of Lemma 4

(6.91)
$$\int_{D_k} |\nabla v_k|^2 \, ds \, dt \leq 13C ,$$

for k large, with the universal constant $C > 0$ of Lemma 4. Indeed, introducing the "small" discs $B_k = B_{\varepsilon_k}(0, t_k) \subset \mathbb{R}^2$ we find

$$\int_{D_k} |\nabla v_k|^2 \, ds \, dt = \int_{B_k} |\nabla u_k|^2 \, ds \, dt$$
$$= \int_{B_k} \left\{ \left|\frac{\partial u_k}{\partial s}\right|^2 + \left|\frac{\partial u_k}{\partial t} - X_H(u_k) + X_H(u_k)\right|^2 \right\} ds \, dt$$
$$\leq 3 \int_{B_k} \left|\frac{\partial u_k}{\partial s}\right|^2 ds \, dt + 2 \int_{B_k} |X_H(u_k)|^2 \, ds \, dt$$
$$\leq 3 \int_{\mathbb{R} \times S^1} \left|\frac{\partial u_k}{\partial s}\right|^2 ds \, dt + l_k \leq 12C + l_k ,$$

where $l_k \to 0$ as $k \to \infty$. This proves the claim (6.91). Since the gradients of v_k are uniformly bounded, we shall conclude later on by the elliptic estimates for the equations (6.90) the following crucial:

Lemma 6. Assume the sequence v_k satisfies (6.89) and (6.90). Then there exists $v \in C^\infty(\mathbb{R}^2, M)$ and a subsequence such that

$$v_k \to v \text{ in } C^\infty(\mathbb{R}^2, M) .$$

Postponing the proof of this lemma, we show next how Theorem 8 follows from our topological assumption that $[\omega]|\pi_2(M) = 0$. We observe that the smooth map v guaranteed by the lemma is a holomorphic map of finite and positive area, hence not a constant map. Indeed from (6.89), (6.90) and (6.91) we deduce the following

6.4 Elliptic methods and symplectic fixed points

properties of v:

(6.92)
$$|\nabla v(0)| = 1$$
$$|\nabla v(x)| \leq 2, \quad x \in \mathbb{R}^2$$
$$v_s + J(v)v_t = 0$$
$$0 < \left|\int_{\mathbb{R}^2} v^*\omega\right| = \frac{1}{2}\int_{\mathbb{R}^2} |\nabla v|^2 < \infty.$$

Define the monotone function $\Phi(R)$ for $R > 0$ by

(6.93)
$$\Phi(R) := \left|\int_{D_R} v^*\omega\right| = \frac{1}{2}\int_{D_R} |\nabla v|^2 \nearrow \frac{1}{2}\int_{\mathbb{R}^2} |\nabla v|^2.$$

The restriction to the boundary of the holomorphic disc $v|D_R : D_R \to M$ defines the contractible loop $x_R : S^1 \to M$ by $x_R(\tau) = v(Re^{2\pi i\tau})$. In view of the differential equation in (6.92) this loop satisfies $|\dot{x}_R(\tau)| = 2\pi R|\frac{\partial v}{\partial s}(Re^{2\pi i\tau})|$. Computing in polar coordinates we find that

$$\int_{\mathbb{R}^2} |\frac{\partial v}{\partial s}|^2 = \int_0^\infty \frac{1}{2\pi R}\int_0^1 |\dot{x}_R(\tau)|^2 \, d\tau \, dR,$$

which is finite in view of (6.92). Consequently, there is a sequence of holomorphic discs $v|D_{R_j}$ such that the arc lengths $l(x_{R_j})$ of their boundaries tend to zero as $j \to \infty$. In order to estimate the area of these discs by the arc length, we cover the compact manifold M by finitely many open sets $U_j \subset M$, $1 \leq j \leq N$ such that each $\bar{U}_j$ is diffeomorphic to a closed ball. There is a constant $d > 0$ such that every loop x on M having diameter $\leq d$ is contained in one of the sets U_j. By Poincaré's lemma we conclude that $\omega|\bar{U}_j = d\lambda_j$ for a 1-form λ_j on $\bar{U}_j$. Therefore we can estimate the action of a loop $x(t) \subset U_j$ for $t \in [0,1]$ by the arc length $l(x)$,

(6.94)
$$\left|\int_{[0,1]} x^*\lambda_j\right| \leq l(x) \, ||\lambda_j||,$$

where $||\lambda_J|| = \max\{||\lambda_J(m)||, m \in \bar{U}_j\}$, with the standard norm of the linear map $\lambda_j(m) \in (T_m M)^*$. Consequently, if $\gamma = \max ||\lambda_j||$, $1 \leq j \leq N$, we conclude for every loop x on M having diameter $\leq d$ that

(6.95)
$$\left|\int_x \lambda_j\right| \leq l(x) \cdot \gamma.$$

Consider now the small loop $x_R(\tau)$ which is the boundary of the holomorphic disc v_R. Then it is contained in one of the open sets, say U_j. Since it is contractible we can choose a different disc $w : D_R \to M$ also satisfying $w|\partial D = x_R$ but which, in addition, is contained in U_j so that $w(D_R) \subset U_j$. If, for example $\alpha : \bar U_j \to \overline{B_1(0)} \subset \mathbb{R}^{2n}$ is a local chart we define such a disc by $w(re^{2\pi i\vartheta}) = \alpha^{-1}(\frac{r}{R}\alpha \circ v(Re^{2\pi i\vartheta}))$.

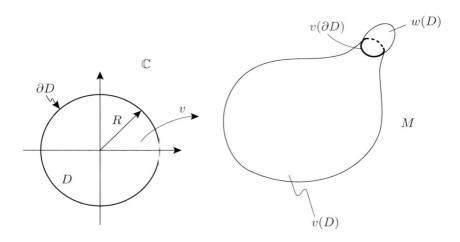

Fig. 6.9

Finally we play the Joker. Since $[\omega]|\pi_2(M) = 0$ we conclude for the discs $v|D_R$ and w

(6.96) $$\int_{D_R} v^*\omega = \int_{D_R} w^*\omega .$$

Hence applying Stokes' theorem in U_j, we find in view of (6.95) that

$$\Phi(R) = |\int_{D_R} v^*\omega| = |\int_{[0,1]} x_R^* \lambda_j| \leq l(x_R) \cdot \gamma .$$

However, there is a sequence of holomorphic discs with $l(x_{R_j}) \to 0$ as we have seen above, and hence we arrive at a contradiction to the monotonicity (6.93) of $\Phi(R)$. This contradiction shows that our assumption, namely that the gradients of $u \in X_\infty$ are not uniformly bounded, is false. Hence Theorem 8 is true.

The above techniques are usually referred to as bubbling off analysis. In order to complete the proof of Theorem 8, it remains to prove Lemma 6.

Proof of Lemma 6. In order to simplify the procedure we first get rid of the Hamiltonian term in (6.90), which tends to zero as $R_k \to \infty$. Using a trick due to M. Gromov [107] we put $\tilde u = (s, t, m) \in \mathbb{R}^2 \times M$ and define $J_k : T(\mathbb{R}^2 \times M) \to T(\mathbb{R}^2 \times M)$

6.4 Elliptic methods and symplectic fixed points

by

$$(6.97) \quad J_k(\tilde{u})(\alpha, \beta, h) = \Big(-\beta, \alpha, J(m)h + \beta H'_k(t,m) + \alpha J(m)H'_k(t,m)\Big),$$

where $H'_k(t,m) = \frac{1}{R_k} H'(t_k + \frac{t}{R_k}, m)$. One verifies readily that $J_k^2 = -id$. Hence J_k is a sequence of almost complex structures on $\mathbb{R}^2 \times M$ converging together with all its derivatives uniformly on $\mathbb{R}^2 \times M$ to the almost complex structure $i \oplus J$, if we identify $\mathbb{R}^2$ with $\mathbb{C}$ in the usual way. Define the sequence $\tilde{v}_k \in C^\infty(\mathbb{R}^2, \mathbb{R}^2 \times M)$ by

$$(6.98) \quad \tilde{v}_k(s,t) = \Big(s, t, v_k(s,t)\Big).$$

Then, on $D_k = B_{\varepsilon_k R_k}(0) \subset \mathbb{R}^2$,

$$\frac{\partial}{\partial s}\tilde{v}_k + J_k(\tilde{v}_k)\frac{\partial}{\partial t}\tilde{v}_k = 0$$

$$(6.99) \quad |\nabla \tilde{v}_k(s,t)| \leq 4$$

$$|\nabla \tilde{v}_k(0,0)| \geq 1$$

$$\varepsilon_k R_k \to \infty.$$

The aim is to find a subsequence such that $\tilde{v}_k \to \tilde{v}$ in $C^\infty(\mathbb{R}^2, \mathbb{R}^2 \times M)$ for a map $\tilde{v} \in C^\infty(\mathbb{R}^2, \mathbb{R}^2 \times M)$. In view of the Arzelà-Ascoli Theorem it is sufficient to find C^∞_{loc}-bounds for $\tilde{v}_k$, i.e., for every compact set K we have to find uniform (in k) bounds for v_k together with all the derivatives. Choosing $p > 2$ it is sufficient, in view of Sobolev's embedding theorem to establish on every compact subset $K \subset \mathbb{R}^2$ uniform bounds in $W^{l,p}(K)$ instead of $C^l(K)$ for all $l \geq 1$. In order to establish such $W^{l,p}_{loc}$-bounds, we proceed by induction. By assumption the gradients of $\tilde{v}_k$ are uniformly bounded so that the $W^{1,p}_{loc}$-bounds are guaranteed. Arguing indirectly we shall assume now that there are $W^{l,p}_{loc}$-bounds but not $W^{l+1,p}_{loc}$-bounds, for some $l \geq 1$. Hence possibly taking a subsequence of $\tilde{v}_k$, we find a sequence $x_k \in \mathbb{R}^2$ and a sequence $\varepsilon_k > 0$ satisfying $\varepsilon_k \to 0$ such that

$$(6.100) \quad x_k \to x_0 \text{ and } |\tilde{v}_k|_{W^{l+1,p}(B_{\varepsilon_k}(x_k))} \to +\infty,$$

where, viewing M as a subset of $\mathbb{R}^N$, we consider $\tilde{v}_k$ as a map from $\mathbb{R}^2$ into $\mathbb{R}^N$. Since, by assumption, $|\nabla \tilde{v}_k|$ is uniformly bounded we conclude, by the Arzelà-Ascoli theorem, for a subsequence, that

$$(6.101) \quad \tilde{v}_k(x_0) \to (x_0, m) \text{ and } \tilde{v}_k \to \tilde{v} \text{ in } C^0_{loc}.$$

This allows us to localize the problem in a coordinate chart. Around $(x_0, m) \in \mathbb{R}^2 \times M$ there is a chart given by $\alpha: U \to \mathbb{R}^{2+2n}$, satisfying $\alpha(x_0, m) = 0$ and

$$(6.102) \quad \tilde{v}_k\Big(\overline{B_{\varepsilon_0}(x_0)}\Big) \subset V \subset \overline{V} \subset U,$$

for some $\varepsilon_0 > 0$ and for a suitable open neighborhood V of (x_0, m) and for all k sufficiently large. Hence defining the smooth functions $u_k := \alpha \circ \tilde{v}_k : \overline{B_{\varepsilon_0}}(x,0) \to \mathbb{R}^{2+2n}$, and replacing J_k by $\widehat{J}_k$ defined by

$$\widehat{J}_k(y) = T\alpha\left(\alpha^{-1}(y)\right) \cdot J_k\left(\alpha^{-1}(y)\right) T\alpha^{-1}(y) , \tag{6.103}$$

$y \in \alpha(U) \subset \mathbb{R}^{2+2n}$ we have, on $B_{\varepsilon_0}(x_0)$,

$$\begin{aligned}
\frac{\partial u_k}{\partial s} + \widehat{J}_k(u_k)\frac{\partial u_k}{\partial t} &= 0 \\
u_k(x_0) &\to 0 \\
|\nabla u_k(x)| &\leq C \\
|u_k|_{W^{l+1,p}(B_{\varepsilon_0}(x_k))} &\to +\infty .
\end{aligned} \tag{6.104}$$

Moreover, $\widehat{J}_k \to \widehat{J}_\infty$ in C^∞ for some almost complex structure $\widehat{J}_\infty$ in $\mathbb{R}^{2+2n}$ near 0. We shall denote by J_0 the following constant complex structure on $\mathbb{R}^{2+2n}$

$$J_0 := \widehat{J}_\infty(0) , \quad J_0^2 = -\mathbb{1} .$$

Now we make use of the following well-known elliptic estimate for the Cauchy-Riemann operator. For every $1 < p < \infty$, m and $n \geq 0$ there exists a constant $C > 0$ such that

$$C|u|_{m+1,p} \leq |u_s + J_0 u_t|_{m,p} , \tag{6.105}$$

for every smooth function $u : \mathbb{R}^2 \to \mathbb{R}^{2n+2}$ having compact support in the unit ball. This is the classical estimate for the Cauchy-Riemann operator (if written in complex notation), J_0 being constant. For a proof we refer to the monograph by E. Stein [205], p. 60. In order to apply this estimate, we take a smooth function $\beta : \mathbb{R} \to [0, 1]$ such that $\beta(s) = 1$ for $|s| \leq \frac{1}{2}$ and $\beta(s) = 0$ for $|s| \geq 1$. For $\lambda > 0$ we define the function $\alpha_\lambda(x) := \beta(\frac{x-x_0}{\lambda})$. Then $\alpha_\lambda(x) = 0$ if $|x - x_0| \geq \lambda$. To simplify the formulas we shall use the abbreviating notation

$$|u|_l := |u|_{W^{l,p}(B_{\varepsilon_0}(x_0))} ,$$

where $p > 2$ is fixed. Moreover, we shall abbreviate $\alpha \equiv \alpha_\lambda$. We then conclude from (6.105) that

$$C|\alpha u_k|_{l+1} \leq |(\alpha u_k)_s + J_0(\alpha u_k)_t|_l .$$

By the Cauchy-Riemann equation, $(u_k)_s = -\widehat{J}_k(u_k)(u_k)_t$, and

$$C|\alpha u_k|_{l+1} \leq C_1(\lambda)|u_k|_l + |\alpha_t \widehat{J}_k(u_k) u_k|_l + |[(J_0 - \widehat{J}_k(u_k)](\alpha u_k)_t|_l .$$

Using now the Leibniz rule for differentiation and the inductive assumption, namely that $|u_k|_l \leq C_l$, we conclude

$$C|\alpha u_k|_{l+1} \leq |J_0 - \widehat{J}_k(u_k)|_{L^\infty(B_\lambda(x_0))} \cdot |\alpha u_k|_{l+1} + C_3(\lambda) .$$

6.4 ELLIPTIC METHODS AND SYMPLECTIC FIXED POINTS   243

Since $u_k(x_0) \to 0$, $u_k \to u_\infty$ uniformly on $B_{\varepsilon_0}(x_0)$ and $J_0 = \lim J_k(0)$, we see, that the factor in front of $|\alpha u_k|_{l+1}$ on the right hand side is smaller than $\frac{C}{2}$, if λ is sufficiently small, and if k is sufficiently large. Since $\alpha \equiv 1$ on $B_{\varepsilon_k}(x_k)$ as soon as k is sufficiently large, we finally conclude

$$|u_k|_{W^{l+1,p}(B_{\varepsilon_k}(x_k))} \leq C_4 < \infty ,$$

if k is sufficiently large. This obviously contradicts the last statement in (6.104) and hence the proof of Lemma 6 is finished. ∎

With Lemma 6, the proof of Theorem 8 is completed: we have uniform gradient estimates for $u \in X_\infty$. Therefore, we can apply the same arguments as in Lemma 6 and obtain C^∞_{loc}-bounds, uniform for $u \in X_\infty$. In addition, using the $\mathbb{R}$-action (6.95) on X_∞ we can get much more, namely C^l_b bounds on $\mathbb{R} \times S^1$, with the norms defined by $|u|_{C^l_b} = \sup |D^\alpha u(s,t)|$, where the supremum is taken over $|\alpha| \leq l$ and $(s,t) \in \mathbb{R} \times S^1$. We shall show that

$$\sup_{u \in X_\infty} |u|_{C^l_b(\mathbb{R} \times S^1, \mathbb{R}^N)} < \infty ,$$

for every $l \geq 0$, so that X_∞ is uniformly bounded in $C^\infty(\mathbb{R} \times S^1, M)$; here we view M again as a subset of $\mathbb{R}^N$. The easiest way to see this is the following. Since the uniform gradient estimate for X_∞ holds on all of $\mathbb{R} \times S^1$, we can take the $\mathbb{R}$-action on X_∞, given by $\tau \cdot u(s,t) = u(s+\tau, t)$ and only have to prove the uniform C^l estimates for X_∞ on a finite covering of the interval $s^* \times [0,1] \subset \mathbb{R} \times S^1$ by compact sets for some fixed $s^* \in \mathbb{R}$. But this we have just done in the proof of the previous lemma and the claim is proved. Since X_∞ is closed and bounded in $C^\infty(\mathbb{R} \times S^1, M)$ we have demonstrated

Proposition 11. The set X_∞ of bounded orbits is a compact subset of the metric space $C^\infty(\mathbb{R} \times S^1, M)$.

On X_∞ there is as we have seen a continuous flow which is gradient-like and we can estimate the number of rest points by Ljusternik-Schnirelman (Theorem 6), provided of course that $X_\infty \neq \emptyset$. In order to get an insight into the topology of this set X_∞ of bounded orbits we shall proceed as in Section 6.2 and first prove that the sets X_T are close to X_∞ if T is large. Recall that the space X_T is defined as the set of solutions

(6.106) $\qquad X_T := \{ u \in W^{1,p}(S_T, M) \mid Lu = 0 \} ,$

where the nonlinear operator L is defined in (6.79). Recall also that we cannot expect better regularity properties for these solutions, since L contains the non-smooth term $\chi_{[-T,T] \times S^1} H^1$. Since $p > 2$ we have $W^{1,p}(S_T, M) \subset C^0(S_T, M)$ in view of the Sobolev embedding theorem. However, we shall restrict now the solutions u to the cylinder $Z_T \subset S_T$. By elliptic regularity theory, these restrictions are smooth as we have pointed out above: $u|_{Z_T} \in C^\infty(Z_T, M)$. Moreover, by Lemma

4, there exists a constant $C > 0$ such that for every solutions $v \in X_T$, the following estimate (on Z_T) holds

$$(6.107) \quad \frac{1}{2} \int_{-T}^{T} \int_0^1 \left[|\frac{\partial v}{\partial s}|_J^2 + |\frac{\partial v}{\partial t} - X_{H_t}(v)|_J^2 \right] ds\, dt \leq 2C,$$

where C is independent of $T > 0$. Copying now the above C^∞-estimates word for word we find, for every given multi-index α, a constant $C_\alpha > 0$ such that for every $T > 2$ and for every $v \in X_T$, the following estimate holds:

$$(6.108) \quad \max \{|D^\alpha v(s,t)| \mid (s,t) \in [-T+2, T-2] \times [0,1]\} \leq C_\alpha.$$

From this we shall conclude that X_T is close to X_∞ provided T is large enough. To make this statement more precise we take a smooth function $\beta : \mathbb{R} \to [0,1]$ satisfying $\beta(s) = 1$ for $s \leq 1$, $\beta'(s) > 0$ for $1 < s < 2$ and $\beta(s) = 0$ for $s \geq 2$ and define the family of cut off functions $\beta_T : \mathbb{R} \to [0,1]$ for $T \geq 5$ by

$$\beta_T(s) = 1 \quad \text{for} \quad |s| \leq T - 3$$

$$\beta_T(s) = \beta(s - T + 3) \quad \text{for} \quad s \geq T - 3$$

$$\beta_T(s) = \beta_T(-s) \quad \text{for } s \geq 0.$$

Given $T_0 > 5$ we define for $T \geq T_0$ a map

$$(6.109) \quad \begin{aligned} \sigma_T : X_T &\to C^\infty(\mathbb{R} \times S^1, M) \\ \sigma_T(u)(s,t) &= u\Big(s\beta_T(s), t\Big). \end{aligned}$$

By (6.108), we have uniform C^∞-estimates on $[-T+1, T-1] \times [0,1]$. Assume we are given an open neighbourhood U of X_∞ in $C^\infty(\mathbb{R} \times S^1, M)$. This means that there exists $R > 0$, $\varepsilon > 0$ and $k \in \mathbb{N}$ such that every $w \in C^\infty(\mathbb{R} \times S^1, M)$ for which there exists an $u \in X_\infty$ satisfying

$$| D^\alpha(u(s,t) - w(s,t)) | < \varepsilon, \text{ for all } (s,t) \in [-R, R] \times S^1, \text{ and for all } |\alpha| \leq k,$$

belongs to U. We conclude that for T large

$$\sigma_T(X_T) \subset U.$$

Indeed, arguing indirectly we find a sequence $T_n \to \infty$ and a sequence $u_n \in X_{T_n}$ such that $v_n := \sigma_{T_n}(u_n) \notin U$. However, in view of (6.108) we find that (v_n) has a subsequence converging with all its derivative on bounded subsets of $\mathbb{R} \times S^1$ to some smooth map $v : \mathbb{R} \times S^1 \to M$. Hence we may assume without loss of generality that (v_n) converges to some $v \notin U$. We observe that for every $R > 0$

$$(6.110) \quad \frac{1}{2} \int_{-R}^{R} \int_0^1 \left[|\frac{\partial v}{\partial s}|_J^2 + |\frac{\partial v}{\partial t} - X_{H_t}(v)|_J^2 \right] ds\, dt \leq 2C,$$

6.4 Elliptic methods and symplectic fixed points

in view of (6.107). If $R \to \infty$, we obtain

$$\frac{1}{2}\int_{-\infty}^{+\infty}\int_0^1 \left[|\frac{\partial v}{\partial s}|_J^2 + |\frac{\partial v}{\partial t} - X_{H_t}(v)|_J^2\right] ds\, dt \leq 2C.$$

Clearly v satisfies the partial differential equation

$$\frac{\partial v}{\partial s} + J(v)\frac{\partial v}{\partial s} + H'(t, v) = 0.$$

This implies that $v \in X_\infty$ hence contradicting the fact that $v \notin U$.

Summing up we have shown that for a given open neighbourhood U of X_∞ in $C^\infty(\mathbb{R} \times S^1, M)$, there exists T_0 such that, for $T \geq T_0$, the set $\sigma_T(X_T)$ belongs to U. Consequently, we have the following commutative diagram

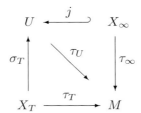

where the maps τ_T, τ_∞ and τ_U are the evaluation maps at the point $(s, t) = (0, 0)$.

Assume now that τ_T^* is injective in cohomology. Then τ_U^* is also injective and we consider the following diagram for the Alexander-Spanier cohomology:

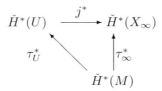

In view of the continuity property of the Alexander-Spanier cohomology, we can take the direct limit over all open neighborhoods U of X_∞ in $C^\infty(\mathbb{R} \times S^1, M)$ and conclude that the map τ_∞^* is injective. We see that the set of bounded solutions contains the cohomology of the underlying manifold M, in particular cup-length $(X_\infty) \geq$ cup-length (M). Applying the Ljusternik-Schnirelman Theorem (Theorem 6) to the continuous gradient-like flow on X_∞, we conclude the desired statement

Theorem 9. (A. Floer, H. Hofer) Assume (M, ω) is a compact symplectic manifold satisfying $[\omega]|\pi_2(M) = 0$. Then every smooth time-periodic Hamiltonian vector field $\dot{x} = X_H(x)$ on M possesses at least cup-length $(M) + 1$ contractible periodic solutions of period 1.

In order to complete the proof of Theorem 9, we still have to verify that the map τ_T^* is injective. Proceeding as in Section 6.2, we shall use Fredholm theory in order to approximate X_T by smooth compact manifolds of dimension $2n$, whose cohomology can be estimated using degree theory. For the following we choose a fixed number $T > 5$ and consider the operator $L = L^T$ as a smooth section of the bundle

$$L^p(W^{1,p}(S_T, M)^* A_M) = \bigcup_{v \in W^{1,p}(S_T, M)} \{v\} \times L^p(v^* A_M).$$

over $W^{1,p}(S_T, M)$. We abbreviate this bundle by

$$\mathcal{E} \to \mathcal{B}.$$

The idea now is to homotope the given section L to the Cauchy-Riemann type section $\frac{\partial}{\partial s} + J\frac{\partial}{\partial t}$ by considering the homotopy $\rho \mapsto L_\rho$, $0 \leq \rho \leq 1$, defined by

$$L_\rho(v) = Tv + J(v)(Tv)j + \rho\chi(s)[H'(t,v)ds - J(v)H'(t,v)dt].$$

Introducing the evaluation map $\tau : \mathcal{B} \to M$ by $v \mapsto v(0,0)$ we define the associated homotopy $\hat{L}_\rho$, $0 \leq \rho \leq 1$ by

$$\hat{L}_\rho(v) = \Big(L_\rho(v), \tau(v)\Big), 0 \leq \rho \leq 1.$$

This is, by definition, a map from $\mathcal{B}$ into $\mathcal{E} \times M$, for every $0 \leq \rho \leq 1$. We may view $\mathcal{E} \times M$ as a Banach manifold bundle over $\mathcal{B}$. For $\rho = 0$, the equation $\hat{L}_0(v) = (0, m)$, for given $m \in M$, only has the constant solution $v_0 \equiv m$, since there are no nontrivial holomorphic spheres because of our assumption $[\omega]|\pi_2(M) = 0$ on the manifold M. The linearization at this particular section v_0 gives in the first component the linear Cauchy-Riemann operator, and in the second component the evaluation at 0. To be more precise, denoting by $D = \hat{L}'_0(v_0)$ the linearization at the particular point $v_0 \equiv m$, one computes that

$$Dh = (Th + J(m)(Th)j, h(0)).$$

In view of the results in Appendix 4, the operator $D : W^{1,p}(S_T, T_m M) \to L^p(v_0^* A_M) \times T_m M$ is an isomorphism. Now we use the $\mathbb{Z}_2$-degree for Fredholm sections having Fredholm index zero. Given the necessary compactness conditions we have just demonstrated that this degree is equal to 1. Let us assume for the moment that the set of solutions of $L_\rho(v) = 0, 0 \leq \rho \leq 1$ is compact. Then there exists an open neighborhood of this solution set on which the Fredholm map is proper, see S. Smale [201]. Consequently, by the homotopy property of the $\mathbb{Z}_2$-index, the original map for $\rho = 1$ given by $v \mapsto (L_1(v), \tau(v))$ has also $\mathbb{Z}_2$-degree equal to 1. To continue our line of thought, let us also assume for the moment that the image of the map $v \mapsto L_1(v)$ is transversal to the zero section $0_\mathcal{E}$ of the bundle $\mathcal{E}$. Then

6.4 ELLIPTIC METHODS AND SYMPLECTIC FIXED POINTS   247

the preimage of $0_\mathcal{E}$ is a smooth compact submanifold Σ of $\mathcal{B}$ with dimension $2n$. Moreover, the evaluation map $\tau := \tau|\Sigma : \Sigma \to M$, defined by $v \mapsto v(0,0)$ will have $\mathbb{Z}_2$-mapping degree equal to 1. This allows us to argue precisely as in Proposition 5 of Section 6.2 in order to prove that in Alexander-Spanier cohomology the map τ_T^* is injective as desired, hence completing the proof of Theorem 9. To reach this conclusion, we assumed all the necessary compactness properties and, in addition, the transversality of the image of $v \mapsto L_1(v)$ to the zero section $0_\mathcal{E}$. The transversality can be established by an arbitrary C^∞-small perturbation of the Hamiltonian H. However, the argument is quite intricate and we refer the reader to A. Floer, H. Hofer and D. Salamon [87] for a proof. Hence with this perturbation argument together with the continuity property of the Alexander-Spanier cohomology, we are left with the compactness question, and it remains to prove the following

Proposition 12. Assume $2 < p < \infty$ and $[\omega]|\pi_2(M) = 0$. Then the set of solutions

$$\hat{X}_T := \{(\rho, v) \in [0,1] \times W^{1,p}(S_T, M) \,|\, L_\rho(v) = 0\}$$

is compact in $[0,1] \times W^{1,p}(S_T, M)$.

Proof of Proposition 12. If $v \in \hat{X}_T$ we define the number $\varepsilon(v) \geq 0$ by

(6.111) $$\varepsilon(v) = \inf\left\{\varepsilon > 0 \,\bigg|\, \text{there exists } z \in S_T\right.$$

$$\left. \text{satisfying } |\nabla v|_{p, B_\varepsilon(z)} = \varepsilon^{\frac{2-p}{p}} \right\}.$$

Here $B_\varepsilon(z)$ denotes the ε-ball around z for some fixed metric on S_T. (The choice of the metric is irrelevant since S_T is compact.) Note that the function $\varepsilon \mapsto \varepsilon^{\frac{p-2}{p}}|\nabla v|_{p,B_\varepsilon(z)}$ is nondecreasing. Assume now that

$$\inf_{v \in \hat{X}_T} \varepsilon(v) \equiv \varepsilon > 0.$$

Then

$$|\nabla v|_{p, B_\varepsilon(v)}(z) \leq \varepsilon^{\frac{2-p}{p}}$$

for every $z \in S_T$, and hence we have a $W^{1,p}$-bound for $\hat{X}_T$. This bound implies the compactness of $\hat{X}_T$ in $[0,1] \times W^{1,p}(S_T, M)$ as we shall show next. Viewing M as a subset of some $\mathbb{R}^N$, we can consider $W^{1,p}(S_T, M)$ as a subset of $W^{1,p}(S_T, \mathbb{R}^N)$. Pick a sequence $(\rho_k, v_k) \in \hat{X}_T$; then we have to find a convergent subsequence. By taking a suitable subsequence we can assume that $\rho_k \to \rho$, and $v_k \to v$ in $C^0(S_T, \mathbb{R}^N)$ (in view of the compact embedding of $W^{1,p}(S_T, \mathbb{R}^N) \subset C^0(S_T, \mathbb{R}^N)$) and $v_k \rightharpoonup v$ weakly in $W^{1,p}(S_T, \mathbb{R}^N)$ for some $v \in W^{1,p}(S_T, \mathbb{R}^N)$ (in view of the reflexivity of the space $W^{1,p}(S_T, \mathbb{R}^N)$). We shall show that v_k is a Cauchy sequence in $W^{1,p}(S_T, \mathbb{R}^N)$. Since we have convergence in $C^0(S_T, \mathbb{R}^N)$ we now can work locally in charts of S_T and of M as well. Any difficulty that may arise occurs near the boundary $\partial Z_T \subset S_T$ of cylinder $Z_T \subset S_T$, since away from this boundary, the maps $u \in \hat{X}_T$ are smooth (by elliptic regularity theory) and we can therefore establish C^∞-bounds arguing as above. Hence we carry out the local argument for the worst case only focussing on the lower cap of S_T. In order to find local coordinates on S_T near the boundary ∂Z_T belonging to the lower cap we take a biholomorphic embedding φ of $[-1,1] \times S^1$ into S_T satisfying $\varphi(s,t) = (s-T, t)$ for all $0 \leq s \leq 1$. Choosing now symplectic coordinates on M we are confronted with the following situation. We have a sequence of mappings $v_k \in W^{1,p}(B_\varepsilon(0), \mathbb{R}^{2n})$ and a sequence of numbers $\rho_k \in [0,1]$ satisfying

$$v_k \longrightarrow v \quad \text{in} \quad C^0\left(B_\varepsilon(0), \mathbb{R}^{2n}\right)$$

$$v_k \rightharpoonup v \quad \text{in} \quad W^{1,p}\left(B_\varepsilon(0), \mathbb{R}^{2n}\right)$$

$$\rho_k \longrightarrow \rho \,,$$

where $v \in W^{1,p}(B_\varepsilon(0), \mathbb{R}^{2n})$ and $v(0) = 0$, and where $B_\varepsilon(0) \subset \mathbb{R}^2$ is the open disk centered at $(s,t) = (0,0)$. Moreover, these maps satisfy the following system of first order elliptic partial differential equations

$$\frac{\partial v_k}{\partial s} + J(v_k)\frac{\partial v_k}{\partial t} + \rho_k \chi(s) H'(t, v_k) = 0$$

on $B_\varepsilon(0)$, where χ is the characteristic function of the positive half line in s. We conclude that v solves the equation

$$\frac{\partial v}{\partial s} + J(v)\frac{\partial v}{\partial t} + \rho \chi(s) H'(t, v) = 0 \,.$$

We shall prove that v_k is a Cauchy sequence in $W^{1,p}(B_\rho(0), \mathbb{R}^{2n})$ for some small $\rho > 0$. Let $\beta : \mathbb{R}^2 \to [0,1]$ be a smooth function having its support in the unit disk and taking on the value 1 on the smaller disk of radius $1/2$. For $\tau > 0$ we then define the function β_τ on $\mathbb{R}^2$ by $\beta_\tau(z) = \beta(z/\tau)$. By J_0 we denote the constant almost complex structure $J(0)$ on $\mathbb{R}^{2n}$. Finally, we write $|.|_j$ for the $W^{j,p}$-norm on the ε-disk in $\mathbb{R}^2$, recalling that $2 < p < \infty$ and fixed. If $0 < \tau < \varepsilon$ we now compute, using the classical estimate (6.105) for the Cauchy-Riemann operator,

$$c|\beta_\tau(v_k - v)|_1 \leq |\bar{\partial}(\beta_\tau(v_k - v))|_0$$

$$\leq |\frac{\partial}{\partial s}(\beta_\tau(v_k - v)) + J(v)\frac{\partial}{\partial t}(\beta_\tau(v_k - v))|_0$$

$$+ |J_0 - J(v)|_{L^\infty(B_\tau)} |\beta_\tau(v_k - v)|_1$$

$$\leq |\beta_\tau(\frac{\partial}{\partial s}(v_k - v) + J(v)\frac{\partial}{\partial t}(v_k - v))|_0 + c_1(\tau)|v_k - v|_0$$

$$+ \sigma(\tau)|\beta_\tau(v_k - v)|_1.$$

6.4 Elliptic methods and symplectic fixed points

Here $\sigma(\tau) \to 0$ for $\tau \to 0$, and $c_1(\tau)$ is a constant depending on τ. Hence, if $\tau > 0$ is small enough,

$$\frac{c}{2}|\beta_\tau(v_k - v)|_1 \leq c_1(\tau)|v_k - v|_0 + |J(v) - J(v_k)|_{L^\infty(B_\tau)}|v_k|_1$$
$$+ |\rho_k \chi H'(t, v_k) - \rho \chi H'(t, v)|_0 .$$

By assumption, $|v_k|_1$ is a bounded sequence, $|v_k - v|_0 \to 0$, $J(v_k) \to J(v)$ uniformly on $B_\varepsilon(0)$ and $|\rho_k H'(t, v_k) - \rho \chi H'(t, v)|_0 \to 0$. Consequently $v_k \to v$ in $W^{1,p}(B_\delta(0), \mathbb{R}^{2n})$ for some $\delta > 0$ sufficiently small. Covering S_T by finitely many charts we have demonstrated the $v_k \to v$ in $W^{1,p}(S_T, M)$, so that $\hat{X}_T$ is compact, as claimed in the proposition.

To finish our argument, it remains to verify our assumption that the numbers $\varepsilon(v)$ are uniformly bounded away from 0 on the set $\hat{X}_T$. We argue by contradiction, assuming that we have no uniform $W^{1,p}$-bound on $\hat{X}_T$, that we find a sequence $(\rho_k, v_k) \in \hat{X}_T$ satisfying $\varepsilon(v_k) \to 0$. By the definition of $\varepsilon(v)$ we find a sequence $z_k \in S_T$ such that with $\varepsilon_k := \varepsilon(v_k)$ we have

$$|\nabla v_k|_{0, B_{\varepsilon_k}(z)} \leq |\nabla v_k|_{0, B_{\varepsilon_k}(z_k)}$$

for all $z \in S_T$. Moreover,

$$1 = \varepsilon_k^{\frac{p-2}{p}} |\nabla v_k|_{0, B_{\varepsilon_k}(z_k)} .$$

Since $\varepsilon_k \to 0$, and since we have C^∞ bounds away from ∂Z_T, we deduce that dist $(z_k, \partial Z_T) \to 0$. Hence, after taking a subsequence, we can assume that $z_k \to z_0 \in \partial Z_T$. The argument now is similar to the one used previously in the bubbling off analysis. Namely we rescale near z_k and obtain in this way a new sequence for which we have a $W^{1,p}_{loc}$-bound. Therefore, arguing as above, the new sequence, u_k, will be $W^{1,p}_{loc}$-precompact. Hence taking a convenient subsequence we can pass to the limit and obtain a solution $v \in W^{1,p}(\mathbb{C}, M)$ of $v_s + J(v)v_t = 0$ on $\mathbb{C}$, which is nonzero. By ellitpic regularity theory $v \in C^\infty(\mathbb{C}, M)$. Moreover, $\int_\mathbb{R} |\nabla v|_J^2 ds dt < \infty$; this follows from (6.110) applied to our sequence u_k, together with the following a priori estimate for the caps:

$$\left| \int_D u_k^* \omega \right| \leq 2C ,$$

which follows immediately from the estimate before Lemma 3. Arguing now as in the proof of Theorem 8, we arrive at a contradiction to our topological assumption $[\omega]|\pi_2(M) = 0$. The proof of Proposition 12, and therefore the proof of Theorem 9 is complete. ∎

Remarks: We note that the torus (T^{2n}, ω) meets the assumption of Theorem 9 for every symplectic structure ω.

Theorem 9 is due to A. Floer [86] and H. Hofer [115], 1987. They both derive the result from a more general intersection result for compact Lagrangian manifolds. M. Chaperon, in his Bourbaki lecture [42] 1982–83, had already noticed that the variational approach is applicable to the global intersection problem of Lagrangian manifolds suggested by V.I. Arnold in [9].

The cohomology in the above statement uses $\mathbb{Z}_2$ coefficients and we would like to point out that the statement actually holds for every coefficient field. The proof, however, requires more intricate tools from Fredholm theory and algebraic topology. The crucial point is to resolve a number of orientation questions for a distinguished class of Fredholm operators. For the technology involved, we refer to A. Floer and H. Hofer [89]. It turns out that X_T has a natural orientation inherited from the orientation of (M, ω) provided it is a manifold.

The above proof of the Arnold conjecture for a general symplectic manifold (M, ω) crucially uses the restrictive assumption requiring $[\omega]\|\pi_2(M) = 0$, and we point out that without this assumption the conjecture is still open. However, more can be said if one (a priori) requires that all the 1-periodic solutions of the Hamiltonian vector field under consideration are nondegenerate, as we shall see in the next section.

6.5 Floer's approach to Morse theory for the action functional

We shall sketch Floer's seminal approach to solving the Arnold conjecture in the nondegenerate case where one assumes that all the 1-periodic solutions of the Hamiltonian system on (M, ω) are nondegenerate. Since M is compact, there are only finitely many such 1-periodic solutions. Floer studies the set of bounded orbits of the unregularized gradient equation defined by the action functional on the space of contractible loops of M. The bounded orbits are special solutions of a distinguished system of first order elliptic equations of Cauchy-Riemann type, as we have shown above. But unlike the general case studied in the previous section, in the nondegenerate case the bounded orbits tend in a strong sense asymptotically to the 1-periodic solutions of the Hamiltonian equation. This additional structure of the solution set is used by Floer in order to construct by means of Fredholm theory a chain complex generated by the finitely many 1-periodic solutions. The homology of this complex is called Floer homology; it is an invariant of the underlying manifold. By continuing the complex to the classical Morse complex defined by the gradient system $\dot{y} = -\nabla H(y)$ of the Hamiltonian function on M, Floer finally shows that his homology is a model of the singular homology of the manifold. This then finishes the proof of the Arnold conjecture.

6.5 Floer's Appraoch to Morse theory ...

We consider a compact symplectic manifold (M, ω) of dimension $2n$ restricted by the requirement that $[\omega]|\pi_2(M) = 0$. This means that the integral of the symplectic structure vanishes over every sphere

$$\int_{S^2} u^*\omega = 0, \quad u : S^2 \to M. \tag{6.112}$$

To the smooth time-dependent Hamiltonian function $H : S^1 \times M \to \mathbb{R}$, periodic in time, $H(t, x) = H(t + 1, x)$ for $t \in \mathbb{R}$ and $x \in M$ we associate as usual the Hamiltonian vector field $X_H : \mathbb{R} \times M \to TM$ by $\omega(X_H, \cdot) = -d_x H(t, x)$. The solutions $x(t)$ of the time-dependent Hamiltonian equation

$$\dot{x}(t) = X_H\Big(t, x(t)\Big) \text{ on } M \tag{6.113}$$

define the flow ψ^t by $\psi^t(x(0)) = x(t)$. We are looking for periodic solutions $x(t + 1) = x(t)$ which are contractible loops on M, $x \simeq 0$. We shall assume that all the contractible periodic solutions $x(t)$ are nondegenerate, requiring for the flow map $\psi = \psi^1$ that

$$\det\Big(1 - d\psi(x(0))\Big) \neq 0. \tag{6.114}$$

By $\mathcal{P}_H$ we shall denote the set of these periodic solutions:

$$\mathcal{P}_H = \{x : S^1 \to M \mid x(t+1) = x(t), x \text{ solves} \tag{6.115}$$
$$(6.113) \text{ and } (6.114), \text{ and } x \simeq 0\}.$$

Note that this is a finite set, since M is compact and the nondegenerate 1-periodic solutions are isolated in M. By Ω we shall denote the set of smooth contractible loops γ of M, where, as usual, $S^1 = \mathbb{R}/\mathbb{Z}$. A loop is represented by its cover $\gamma : \mathbb{R} \to M$ satisfying $\gamma(t + 1) = \gamma(t)$. The required 1-periodic solutions are distinguished in Ω as the critical points of the action functional $\varphi_H : \Omega \to \mathbb{R}$ defined by

$$\varphi_H(\gamma) = -\int_D \bar{\gamma}^*\omega + \int_0^1 H\Big(t, \gamma(t)\Big) dt. \tag{6.116}$$

Here $D \subset \mathbb{C}$ denotes the closed unit disc and $\bar{\gamma} : D \to M$ is a smooth function extending γ and hence satisfying $\bar{\gamma}(e^{2\pi i t}) = \gamma(t)$. Such an extension exists since γ is assumed to be contractible and it follows from the assumption (6.112) that the first integral on the right hand side of (6.116) does not depend on the choice of the extension and hence depends only on the loop γ. With $T_\gamma\Omega$ we shall denote the tangent space at $\gamma \in \Omega$ consisting of the smooth vector fields $\xi \in C^\infty(\gamma^* TM)$ along γ satisfying $\xi(t+1) = \xi(t)$. Computing the derivative of φ_H at γ in the direction of ξ, one finds readily that

$$d\varphi_H(\gamma)\xi = \int_0^1 \{\omega(\dot\gamma, \xi) + dH(t,\gamma)\xi\}\, dt, \tag{6.117}$$

which indeed vanishes for every $\xi \in T_\gamma \Omega$ if and only if the loop $\gamma \in \Omega$ is a solution of the Hamiltonian equation (6.113). It is convenient to choose an almost complex structure J on M compatible with ω (see Chapter 1). This is an endomorphism $J \in C^\infty(\operatorname{End} TM)$, i.e., $J_x \in \mathcal{L}(T_x M, T_x M)$ satisfying $J^2 = -1$ such that

$$g(\xi,\eta) = \omega\big(\xi, J(x)\eta\big), \quad \xi,\eta \in T_x M \tag{6.118}$$

defines a Riemannian metric g on M. The Hamiltonian vector field is then represented by

$$X_H(t,x) = J(x) \nabla H(t,x), \tag{6.119}$$

where ∇ denotes the gradient of a function on M with respect to the x-variable in the metric g above. For the derivative we obtain

$$d\varphi_H(\gamma)\xi = \int_0^1 g\Big(J(\gamma)\dot\gamma + \nabla H(t,\gamma), \xi\Big)\, dt, \tag{6.120}$$

so that the gradient of φ_H with respect to an induced L^2 metric on Ω is given by

$$\operatorname{grad} \varphi_H(\gamma) = J(\gamma)\dot\gamma + \nabla H(\gamma) \in T_\gamma \Omega. \tag{6.121}$$

However, the gradient equation

$$\frac{d}{ds} x = -\operatorname{grad} \varphi_H(x), \quad x \in \Omega \tag{6.122}$$

does not determine a flow. As we have pointed out in the previous section, due to a loss of derivatives there is no Banach space setup in which $\operatorname{grad} \varphi_H(x)$ can be considered as a vector field on a Banach manifold. Indeed the initial value problem defined by (6.122) on Ω is not well-posed. Floer views a solution x of (6.122) as a map

$$u : \mathbb{R} \times S^1 \longrightarrow M, \quad (s,t) \mapsto u(s,t) = u(s,t+1) \tag{6.123}$$

which solves the partial differential equation

$$\frac{\partial u}{\partial s} + J(u) \frac{\partial u}{\partial t} + \nabla H(t,u) = 0. \tag{6.124}$$

6.5 Floer's Appraoch to Morse theory ...

The crucial objects from now on, are therefore the solutions $u : \mathbb{R} \times S^1 \to M$ of the nonlinear partial differential equation (6.124), which are contractible so that $u(s, \cdot) \in \Omega$. These solutions correspond to the flow lines of the gradient equation (6.122). The equation (6.124) is a perturbation of a Cauchy-Riemann equation. A smooth map $u : S^2 \to M$ solving the Cauchy-Riemann equation

$$(6.125) \qquad \frac{\partial u}{\partial s} + J(u) \frac{\partial u}{\partial t} = 0$$

is called a holomorphic sphere. It satisfies

$$(6.126) \qquad \int_{S^2} u^* \omega = \frac{1}{2} \int_{S^2} |\nabla u|^2 ,$$

so that our assumption (6.112) excludes, in particular, nonconstant holomorphic spheres.

From the torus case we know that not all solutions of (6.124) are relevant and we are interested only in the set of "bounded orbits" which we denote by $\mathcal{M} \equiv \mathcal{M}(H, J)$. This set is defined as the set of smooth mappings $u : \mathbb{R} \times S^1 \to M$ which are contractible, solve the equation (6.124) and have, in addition, finite "energy"

$$(6.127) \, E(u) := \frac{1}{2} \int_{-\infty}^{\infty} \int_0^1 \left\{ \left|\frac{\partial u}{\partial s}\right|^2 + \left|\frac{\partial u}{\partial t} - X_H(t, u)\right|^2 \right\} ds\, dt < \infty .$$

Equivalently $\mathcal{M}$ is the space of contractible solutions u along which the decreasing function $\varphi_H u(s)$ remains bounded for all $s \in \mathbb{R}$, where $u(s)(t) = u(s, t) \in \Omega$.

Floer proves in [85] that the space $\mathcal{M}$ of bounded solutions of the partial differential equation (6.124) has a structure which looks like that of a Morse-Smale system on a smooth, compact, and finite dimensional manifold, as we now describe.

Since M is compact and since $[\omega]$ vanishes over $\pi_2(M)$ according to the assumption (6.112) it follows using elliptic regularity theory that the set of solutions $\mathcal{M}$ is compact in the topology of uniform convergence with all derivatives on compact sets. This is not easy to prove and we refer to A. Floer [85], D. Salamon [188], and H. Hofer and D. Salamon [119]. Moreover, for every bounded orbit $u \in \mathcal{M}$ there exists a pair $x, y \in \mathcal{P}_H$ of periodic solutions, such that u is a connecting orbit from y to x, i.e.,

$$(6.128) \qquad \lim_{s \to -\infty} u(s, t) = y(t) , \quad \lim_{s \to +\infty} u(s, t) = x(t) ,$$

the convergence being uniformly in t as $|s| \to \infty$ and $\frac{\partial u}{\partial s}$ converging to zero again uniformly in t. Given two periodic solutions $x, y \in \mathcal{P}_H$ we abbreviate by

$$(6.129) \qquad \mathcal{M}(y, x) = \mathcal{M}(y, x; H, J)$$

the set of solutions $u \in \mathcal{M}$ satisfying the asymptotic boundary conditions (6.128). For the energy one computes

$$(6.130) \qquad E(u) = \varphi_H(y) - \varphi_H(x), \quad \text{if } u \in \mathcal{M}(y,x).$$

The energy is positive for nonconstant orbits. From the previous section we know that the set $\mathcal{M}$ of solutions consists of the periodic solutions and the connecting orbits

$$(6.131) \qquad \mathcal{M} = \bigcup_{y,x \in \mathcal{P}_H} \mathcal{M}(y,x).$$

The group $\mathbb{R}$ acts naturally on $\mathcal{M}$ by shifting $u(s,t)$ in the s direction, hence defining a continuous flow on the space $\mathcal{M}$. Under this action the sets $\mathcal{M}(y,x)$ are invariant subspaces. Their compactness properties can be formulated analogously to the finite dimensional Morse theory as represented by M. Schwarz in his book [191]:

Proposition 12. A sequence $u_\nu \in \mathcal{M}(y,x)$ possesses a subsequence with the following property: there is a sequence $s_\nu^j \in \mathbb{R}$ of times, $j = 1, 2, \ldots, m$, such that $u_\nu(s + s_\nu^j, t)$ converges together with all its derivatives uniformly on compact sets to solutions $u^j \in \mathcal{M}(x^j, x^{j-1})$ where $x^j \in \mathcal{P}_H$ for $j = 0, 1, \ldots m$, with $x^0 = x$ and $x^m = y$.

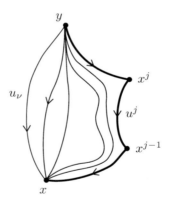

Fig. 6.10

Using of Fredholm theory in the appropriate functional analytic setting, one proves that for a generic choice of the pair (H, J) the sets $\mathcal{M}(y,x)$ are smooth, finite dimensional manifolds. Linearizing the equation (6.124) in the direction of a vector field $\xi \in C^\infty(u^*TM)$ along u leads to the linear first order differential operator

$$(6.132) \qquad F(u)\xi := \nabla_s \xi + J(u) \nabla_t \xi + \nabla_\xi J(u) \frac{\partial u}{\partial t} + \nabla_\xi \nabla H(t, u),$$

6.5 Floer's Appraoch to Morse theory ...

where $\nabla_s, \nabla_t, \nabla_\xi$ denote the covariant derivatives with respect to the metric g associated with J. If $u : \mathbb{R} \times S^1 \to M$ is any smooth map connecting y with x, hence satisfying (6.128) for two nondegenerate solutions $y, x \in \mathcal{P}_H$, then $F(u)$ is a Fredholm operator between appropriate Sobolev spaces. A pair (H, J) with an almost complex structure J satisfying (6.118) is called regular, if $F(u)$ is a surjective linear operator for every $u \in \mathcal{M}$. It can be proved that the set of regular pairs is dense with respect to the C^∞-topology, using of the Sard-Smale theorem. It then follows by the implicit function theorem that $\mathcal{M}(y, x)$ is indeed a smooth manifold whose local dimension near $u \in \mathcal{M}(y, x)$ is

$$\text{(6.133)} \qquad \dim \mathcal{M}(y, x) = \text{Index } F(u) .$$

This index depends only on the boundary condition on u, i.e., on $y, x \in \mathcal{P}_H$, if we require, in addition to (6.112), that the first Chern class of (TM, J) vanishes over $\pi_2(M)$. Note that ω determines an almost complex structure J satisfying (6.118) uniquely up to homotopy and we denote by $c_1 = c_1(TM) \in H^2(M)$ the first Chern class of TM. We shall assume that $[c_1]|\pi_2(M) = 0$ requiring that the integral of the 2-form c_1 vanishes over every sphere

$$\text{(6.134)} \qquad \int_{S^2} u^* c_1 = 0 , \quad u : S^2 \to M .$$

Consider a nondegenerate periodic orbit $x \in \mathcal{P}_H$. Since $x : S^1 \to M$ is contractible we can extend it to a map $u : D \to M$ of the closed disk D satisfying $u|\partial D = x$. Now we take a trivialization of $u^*(TM)$; it determines a trivialization of $x^*(TM)$. Note that if we take another extension $v : D \to M$ of x, we find another trivialization of $x^*(TM)$ which, however, is homotopic in view of the assumption $[c_1]|\pi_2(M) = 0$. In the trivialized tangent bundle $x^*(TM)$, the linearized flow of the Hamiltonian vector field X_H along x determines a unique Maslov-type index $\mu(H, x) \in \mathbb{Z}$ associated with x. (The integers are in one-to-one correspondence with the homotopy classes of linear Hamiltonian systems on $(\mathbb{R}^{2n}, \omega_0)$ which are periodic in time and nondegenerate). This gives

$$\text{(6.135)} \qquad \mu_H : \mathcal{P}_H \longrightarrow \mathbb{Z} , \quad x \mapsto \mu(H, x) .$$

If a smooth map $u : \mathbb{R} \times S^1 \to M$ satisfies (6.128) for two nondegenerate periodic solutions $y, x \in \mathcal{P}_H$ of the Hamiltonian vector field, the Fredholm index of $F(u)$ is given by

$$\text{(6.136)} \qquad \text{Index } F(u) = \mu(H, y) - \mu(H, x) .$$

We conclude for a generic pair (H, J) that $\dim \mathcal{M}(y, x) = \mu_H(y) - \mu_H(x)$. For the definition of the index μ_H and also for a proof of the above index formula, we refer to D. Salamon and E. Zehnder [190]. We remark that if $[c_1]|\pi_2(M) \neq 0$, the index (6.135) would depend on the homotopy class of the chosen extension. The ambiguity would be the integer $2N$, where $N\mathbb{Z} = \{(u^*c_1)[S^2] \,|\, u \in \pi_2(M)\}$. Hence $\mu(H)$ is only well-defined as a map $\mu_H : \mathcal{P}_H \to \mathbb{Z}/2N\mathbb{Z}$.

We are now prepared to define the Floer homology groups associated to a regular pair (H, J) on (M, ω). With $C = C(M, H)$ we shall denote the vector space over $\mathbb{Z}_2$ generated by the finitely many elements of the set $\mathcal{P}_H$ of nondegenerate contractible periodic orbits of the Hamiltonian vector field. This vector space comes with a grading defined by the index map $\mu : \mathcal{P}_H \to \mathbb{Z}$:

$$(6.137) \qquad \begin{aligned} C &= \bigoplus_{k \in \mathbb{Z}} C_k \\ C_k &= C_k(M, H) = \operatorname{span}_{\mathbb{Z}_2} \{x \in \mathcal{P}_H \,|\, \mu(x, H) = k\}. \end{aligned}$$

Consider now $y, x \in \mathcal{P}_H$ satisfying $\mu(y) - \mu(x) = 1$. Then the manifold $\mathcal{M}(y, x)$ is one-dimensional $\dim \mathcal{M}(y, x) = 1$. In view of the compactness, combined with the manifold structure, it can be shown [85] that this manifold has finitely many components, each of which consists of a connecting orbit together with all its translates by the time $-s$ shift. This was used by Floer to construct a linear boundary operator $\partial : C \to C$ respecting the grading by counting the connecting orbits modulo 2. The map $\partial_k = \partial_k(M, H, J) : C_k \to C_{k-1}$ is defined by

$$(6.138) \qquad \partial_k y = \sum_{\{x \in \mathcal{P}_H \,|\, \mu(x) = k-1\}} \langle \partial y, x \rangle x,$$

for $y \in \mathcal{P}_H$ satisfying $\mu(y) = k$. The matrix element $\langle \partial y, x \rangle \in \mathbb{Z}_2$ is the number of components of $\mathcal{M}(y, x)$ counted modulo 2. Floer proves [85] that $\partial \circ \partial = 0$, so that the pair (C, ∂) defines a chain complex. Its homology

$$(6.139) \qquad HF_k(M, H, J) := \frac{\ker(\partial : C_k \to C_{k-1})}{\operatorname{im}(\partial : C_{k+1} \to C_k)}$$

is called the Floer homology of the pair (H, J) on (M, ω).

In order to explain the relation $\partial_k \circ \partial_{k+1} = 0$ for the boundary operators $\partial_{k+1} : C_{k+1} \to C_k$, we pick $x \in C_{k+1}$, $y \in C_k$ and $z \in C_{k-1}$. Assume that there exist connections $u \in \mathcal{M}(x, y)$ and $v \in \mathcal{M}(y, z)$. Then we may view the pair (u, v) as a "broken trajectory" connecting x with z. By a perturbation argument, called the gluing method, Floer established in [85] a unique 1-parameter family of associated true connections in $\mathcal{M}(x, z)$. Taking the quotient by the $\mathbb{R}$-action of the time-s shift, one finds this way a connected 1-dimensional manifold without boundaries of unparameterized orbits which represents one component of $\mathcal{M}(x, z)$. Such 1-dimensional manifolds are either circles or intervals with two ends. The compactness argument in Proposition 12 shows that each end converges in a suitable sense to a well-defined broken trajectory $(u', v') \in \mathcal{M}(x, y') \times \mathcal{M}(y', z)$ for some $y' \in C_k$. The gluing construction shows in a reverse process that the broken trajectories correspond uniquely to such an end. Thus, the 1-dimensional manifold $\mathcal{M}(x, z)/\mathbb{R}$

6.5 Floer's approach to Morse theory ...

has an even number of ends.

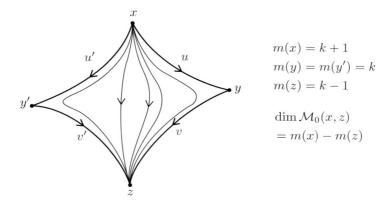

$$m(x) = k+1$$
$$m(y) = m(y') = k$$
$$m(z) = k-1$$

$$\dim \mathcal{M}_0(x,z) = m(x) - m(z)$$

Consequently the "broken trajectories" (u,v) between x and z occur in pairs. Since we use $\mathbb{Z}_2$ coefficients, we find for every $x \in C_{k+1}$ that

$$\sum_{m(z)=k-1} \left(\sum_{m(y)=k} \langle \partial x, y \rangle \langle \partial y, z \rangle \right) z = 0.$$

For the intricate details of this gluing construction we refer to A. Floer [85] and C. Taubes [210, 211] and in particular to the recent book by M. Schwarz [191].

The chain complex is defined by the one dimensional components of the solution space $\mathcal{M}$ of bounded orbits. It is obviously constructed analogously to the familiar Morse complex for Morse-Smale gradient flows on finite dimensional compact manifolds. The transversality conditions of the Morse-Smale flow correspond to the assumption that the Fredholm operator $F(u)$ is a surjective map which is met only "generically" as in the finite dimensional situation. The role of the Morse index of a nondegenerate critical point of the Morse-Smale flow is played by the integer $\mu(x, H)$ associated with the nondegenerate contractible periodic orbit $x \in \mathcal{P}_H$. The Morse index of $x \in \mathcal{P}_H$ is infinite!

Up to now it is conceivable that the space $\mathcal{M}$ might be empty and the Floer homology groups all trivial. At this point it is useful to recall a crucial aspect of C. Conley's index theory for flows on compact spaces: it is designed in such a way that it is invariant under continuation. Prompted by this theory, Floer established an analogous homotopy invariance property for his homology groups, which can be used to show that the groups are independent of (H, J) used for its construction. A special choice of H and J then allowed him to compute the homology:

Theorem 10. (Floer continuation) For every two regular pairs (H^α, J^α) and (H^β, J^β) there exists a natural isomorphism of the Floer homology respecting

the grading
$$\psi^{\beta\alpha} : HF_*(M, H^\alpha, J^\alpha) \longrightarrow HF_*(M, H^\beta, J^\beta).$$

Moreover, if (H^γ, J^γ) is another regular pair, then
$$\psi^{\alpha\beta} \circ \psi^{\beta\gamma} = \psi^{\alpha\gamma} \text{ and } \psi^{\alpha\alpha} = id.$$

We see that the Floer homology $FH_*(M, H, J)$ is independent of the regular pair (H, J) used in the construction of the connecting orbits of $\mathcal{P}_H$; hence it is an invariant of the underlying manifold (M, ω). We have a category whose objects are the Floer homology groups associated with regular pairs (H, J) and whose morphisms are the natural isomorphisms $\psi^{\beta\alpha}$ induced by regular homotopies as explained below. In the terminology of C. Conley [54] such a category with unique morphisms is called a connected simple system.

We shall sketch the ideas of the proof which is again based on special solutions of a P.D.E. and Fredholm theory. We follow the presentation in D. Salamon and E. Zehnder [190]. In order to construct a homomorphism $\varphi^{\beta\alpha} : C(M, H^\alpha, J^\alpha) \to C(M, H^\beta, J^\beta)$, we choose a smooth homotopy $H^{\alpha\beta} : \mathbb{R} \times S^1 \times M \to \mathbb{R}$ of Hamiltonian functions together with a smooth homotopy of almost complex structures $J^{\alpha\beta} : \mathbb{R} \times M \to \text{End}(TM)$ such that $H^{\alpha\beta}(s, t, x)$ and $J^{\alpha\beta}(s, x)$ are independent of s for $|s|$ sufficiently large and satisfy

(6.140)
$$\lim_{s \to -\infty} H^{\alpha\beta}(s, t, x) = H^\alpha(t, x)$$
$$\lim_{s \to +\infty} H^{\alpha\beta}(s, t, x) = H^\beta(t, x)$$

and

(6.141)
$$\lim_{s \to -\infty} J^{\alpha\beta}(s, x) = J^\alpha(x)$$
$$\lim_{s \to +\infty} J^{\alpha\beta}(s, t, x) = J^\beta(x).$$

Recall that the space of almost complex structures compatible with ω is contractible. Given such a pair $(H^{\alpha\beta}, J^{\alpha\beta})$ of homotopies connecting the data one studies the smooth solutions $u(s, t), u : \mathbb{R} \times S^1 \to M$ of the P.D.E.

(6.142)
$$\frac{\partial u}{\partial s} + J(s, u)\frac{\partial u}{\partial t} + \nabla H(s, t, u) = 0$$

which satisfy the asymptotic boundary conditions

(6.143)
$$\lim_{s \to -\infty} u(s, t) = x^\alpha(t), \quad \lim_{s \to +\infty} u(s, t) = x^\beta(t)$$

where $x^\alpha \in \mathcal{P}_{H^\alpha}$ and $x^\beta \in \mathcal{P}_{H^\beta}$.

The solutions can be considered as flow lines of a time-dependent gradient flow defined by the action functional φ_{H_s}, $H_s(t, x) = H^{\alpha\beta}(s, t, x)$ with respect to the time-dependent metric determined by the almost complex structure J_s. As

6.5 Floer's Appraoch to Morse theory ...

before, a solution of bounded energy satisfies the boundary conditions. We shall abbreviate by

(6.144) $$\mathcal{M}(x^\alpha, x^\beta, H^{\alpha\beta}, J^{\alpha\beta})$$

the set of smooth solutions $u : \mathbb{R} \times S^1 \to M$ of (6.142) and (6.143). This set is analyzed locally by linearizing the equation in the direction of a vector field $\xi \in C^\infty(u^*TM)$ along u which leads to a linear operator $F^{\beta\alpha}(u) : W^{k,p}(u^*TM) \to W^{k-1,p}(u^*TM)$ between appropriate Sobolev spaces. For every smooth map $u : \mathbb{R} \times S^1 \to M$ satisfying the boundary conditions (6.143) with x^α, x^β nondegenerate, the operator $F^{\beta\alpha}(u)$ is Fredholm and its index is computed as

(6.145) $$\text{Index } F^{\beta\alpha}(u) = \mu(x^\alpha, H^\alpha) - \mu(x^\beta, H^\beta) .$$

Moreover, for a regular homotopy from (H^α, J^α) to (H^β, J^β) the operator is surjective. Here a pair $(H^{\alpha\beta}, J^{\alpha\beta})$ is called regular, if both pairs (H^α, J^α) and (H^β, J^β) are regular and if the linear operator $F^{\beta\alpha}(u)$ is surjective for every solution u of (6.142) and (6.143) for every $x^\alpha \in \mathcal{P}_{H^\alpha}$ and $x^\beta \in \mathcal{P}_{H^\beta}$. Floer proved [84, 85] that the set of regular homotopies connecting regular pairs is dense in the set of all smooth homotopies connecting (H^α, J^α) to (H^β, J^α) with respect to the C^∞ topology. (Floer carried out his proof for the analogous geometric problem of Lagrangian intersections.) It then follows that the space $\mathcal{M}(x^\alpha, x^\beta, H^{\alpha\beta}, J^{\alpha\beta})$ of connecting orbits is a finite dimensional manifold and

(6.146) $$\dim \mathcal{M}(x^\alpha, x^\beta, H^{\alpha\beta}, J^{\alpha\beta}) = \mu(x^\alpha, H^\alpha) - \mu(x^\beta, H^\beta) .$$

This allows us to construct a chain homomorphism

$$\varphi^{\beta\alpha} = \varphi(H^{\alpha\beta}, J^{\alpha\beta}) : C_k(M, H^\alpha) \to C_k(M, H^\beta)$$

by counting the connecting orbits modulo 2 between periodic solutions having the same index

(6.147) $$\varphi^{\beta\alpha}(x^\alpha) = \sum_{\mu(x^\beta, H^\beta)=k} \langle \varphi^{\beta\alpha}(x^\alpha), x^\beta \rangle x^\beta ,$$

where $\mu(x^\alpha, H^\alpha) = k$ and where $\langle \varphi^{\beta\alpha} x^\alpha, x^\beta \rangle$ is the number of solutions modulo 2.

One can show that for every regular homotopy this linear map $\varphi^{\beta\alpha}$ is a chain homomorphisms, i.e., satisfies $\partial^\beta \varphi^{\beta\alpha} = \varphi^{\beta\alpha} \partial^\alpha$. Moreover, it respects the grading. It, therefore, induces a natural homomorphism $\psi^{\beta\alpha}$ of Floer homology, which turns out to be independent of the choice of the homotopy $(H^{\alpha\beta}, J^{\alpha\beta})$. More precisely one shows that for two regular homotopies $(H_0^{\alpha\beta}, J_0^{\alpha\beta})$ and $(H_1^{\alpha\beta}, J_1^{\alpha\beta})$ from (H^α, J^α) to (H^β, J^β), the associated chain homomorphisms $\varphi_0^{\beta\alpha}$ and $\varphi_1^{\beta\alpha}$ are chain homotopy equivalent. This finishes our sketch of Theorem 10.

Finally, in order to compute the homology, Floer used the continuation theorem and connected a regular pair (H^α, J^α) with a very special pair (H, J), whose Floer homology agrees with the homology of the Morse complex defined by the gradient flow of the function $H : M \to \mathbb{R}$ on the underlying manifold M.

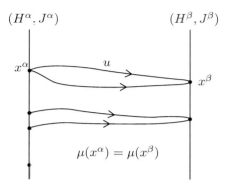

Fig. 6.11

Theorem 11. (Computation of the homology) For every regular pair (H^α, J^α) there exists a natural isomorphism $\psi^\alpha : FH_*(M, H^\alpha, J^\alpha) \to H_*(M, \mathbb{Z}_2)$ respecting the grading

$$\psi_k^\alpha : FH_k(M, H^\alpha, J^\alpha) \to H_{k+n}(M, \mathbb{Z}_2)$$

between the Floer homology of the pair (H^α, J^α) and the singular homology of M with $\mathbb{Z}_2$ coefficients. Moreover, if (H^β, J^β) is another regular pair, then

$$\psi^\alpha = \psi^\beta \circ \psi^{\beta\alpha}.$$

In particular, the Floer homology vanishes for $|k| > n$.

We outline the idea of the proof of Theorem 11 following the presentation in [190]. We choose a special Hamiltonian function $H : M \to \mathbb{R}$ which is independent of time t, and which is a Morse-function, such that all its critical points on M are nondegenerate. Moreover, we assume that H is sufficiently small in the C^2 topology. Then, as we have seen above, the periodic solutions of period 1 of the Hamiltonian system $\dot{x} = X_H(x)$ on M are constant and agree with the critical points of H such that in this case

(6.148) $$\mathcal{P}_H = \{\, x(t) = x \in M \mid dH(x) = 0 \,\}.$$

In this situation of a "small" Hamiltonian H, the Morse index $\text{ind}_H(x)$ of the critical point x of H is related to the index $\mu(x, H)$ of the constant 1-periodic solution $x(t) = x$ of $\dot{x} = X_H(x)$ by

(6.149) $$\mu(x, H) = \text{ind}_H(x) - n, \quad x \in \mathcal{P}_H,$$

see [190]. We can choose an almost complex structure J such that (H, J) is a regular pair. One then shows that the solutions $u \in \mathcal{M}(y, x)$ for $x, y \in \mathcal{P}_H$ satisfying $\mu(y, H) - \mu(x, H) = 1$ of the partial differential equation (6.124) and (6.128)

6.5 Floer's Appraoch to Morse theory ...

are also independent of the time t. Hence setting $u(s,t) = \gamma(s)$, they satisfy the gradient equation

$$\text{(6.150)} \qquad \frac{d\gamma}{ds} + \nabla H\bigl(\gamma(s)\bigr) = 0 \text{ on } M,$$

and the boundary conditions $\gamma(s) \to y$ as $s \to -\infty$ and $\gamma(s) \to x$ as $s \to +\infty$. These solutions agree with the 1-dimensional connecting orbits of the Morse-Smale gradient flow $\dot{x} = -\nabla H(x)$ for which

$$\text{(6.151)} \qquad \mathcal{M}(y,x) = W^u(y) \cap W^s(x),$$

whenever ind $F(u) = \text{ind}_H(y) - \text{ind}_H(x) = 1$. Hence, the chain complex $C(M,H)$ agrees with the familiar Morse complex, in which $C_k(H) = \{x \in M | dH(x) = 0 \text{ and } \text{ind}_H(x) = k\}$ and where the boundary operator $\partial_k = \partial_k(H,J) : C_k \to C_{k-1}$ defined above counts the connecting orbits (mod 2) of the gradient flow on M. It is well-known, that its homology groups

$$\text{(6.152)} \qquad H\,M_k(M,H,J) = \frac{\ker \partial_k}{\operatorname{im} \partial_{k+1}}$$

agree with the singular homology of M

$$\text{(6.153)} \qquad H\,M_k(M,H,J) \cong H_k(M,\mathbb{Z}_2).$$

For a proof we refer to D. Salamon [188] and to the recent monograph by M. Schwarz [191] in which he develops a Morse homology theory on finite dimensional manifolds based on the trajectory spaces of Morse functions. We see in view of (6.149) that Floer's chain complex for our special symplectic manifold (M,ω) agrees with the familiar Morse complex of a gradient flow of a function, the grading being shifted by n:

$$\text{(6.154)} \qquad H\,F_k(M,H,J) \cong H\,M_{k+n}(M,H,J) \cong H_{n+k}(M,\mathbb{Z}_2).$$

Consequently, Floer's connecting orbits of nondegenerate periodic solutions can serve as a model for the homology of the underlying manifold. We conclude, in particular, that

$$\text{(6.155)} \qquad \sum_{x \in \mathcal{P}_H} t^{\mu(x,H)} = P(t,M) t^{-n} + (1+t) Q(t).$$

This Morse equation relates the nondegenerate contractible periodic solutions $x \in \mathcal{P}_H$ having indices $\mu(x,H) \in \mathbb{Z}$ to the topology of M described by the Poincaré polynomial $P(t,M)$. Recall that we posed the assumptions $\omega|\pi_2 = 0 = c_1|\pi_2$ on (M,ω). In case of a time-independent and small Morse function H, the equation (6.155) becomes the familiar Morse equation

$$\text{(6.156)} \qquad \sum_{\nabla H(x)=0} t^{m(x)} = P(t,M) + (1+t) Q(t)$$

which relates the critical points of the function H to the Poincaré polynomial $P(t,M)$. In particular, we conclude from (6.155)

Theorem 12. (Floer) Assume the compact symplectic manifold (M, ω) satisfies $\omega|\pi_2(M) = 0$ and $c_1|\pi_2(M) = 0$, assume the time periodic Hamiltonian vector field $\dot{x} = X_H(x)$ on M possesses nondegenerate 1-periodic solutions only. Then their number is at least the sum of the Betti numbers $SB(M, \mathbb{Z}_2)$ of M.

For example $\pi_2(T^{2n}) = 0$ so that every symplectic torus (T^{2n}, ω) meets the assumption of the theorem. Actually Floer proved in [85] a more general result, assuming merely that (M, ω) a monotone in the sense that

$$\text{(6.157)} \qquad \int_{S^2} u^* c_1 = \lambda \int_{S^2} u^* \omega ,$$

for every sphere $u : S^2 \to M$, with a constant $\lambda > 0$.

More recently, H. Hofer and D. Salamon succeeded, [119], in defining the Floer homology groups under the assumption that

$$\text{(6.158)} \qquad \omega(A) \leq 0 \text{ for every } A \in \pi_2(M) \text{ satisfying } 3 - n \leq c_1(A) < 0 ,$$

and to compute them in the cases $c_1|\pi_2(M) = 0$ and $N \geq \frac{1}{2}\dim(M)$, where $N\mathbb{Z} = c_1(\pi_2(M)) \subset \mathbb{Z}$, proving the Arnold conjecture in these cases. They demonstrate that the Floer groups form a module over Novikov's ring of generalized Laurent series. Such a ring was introduced by S. Novikov in his generalization of Morse theory for closed one-forms [170]. Under the assumption (6.158) the obtained theory is a Novikov type variant of Floer homology. Modifying the ideas of Hofer and Salamon, K. Ono [172] gave a modified construction of Floer's homology theory, again under the assumption (6.158), which can be computed without additional hypothesis. It turns out that for a suitable coefficient field these Floer groups are isomorphic to the singular homology groups of the manifold establishing, in particular, the Arnold conjecture for the dimension 2, 4 and 6 in which (6.158) is vacuous. Manifolds meeting the condition (6.158) have the property used in the proof that, for a generic almost complex structure, there is no holomorphic sphere of negative Chern number.

The Arnold conjecture for a general compact symplectic manifold with dim $M \geq 8$ is still open.

It should be mentioned that the Floer homology groups HF_* can be defined for integer coefficients $\mathbb{Z}$ (instead of $\mathbb{Z}_2$) and, therefore, for every Abelian group. In order to do this one has to assign an integer $+1$ or -1 to every connecting orbit $u \in \mathcal{M}(y, x)$ whenever $\mu(y) - \mu(x) = 1$. However, the determination of such an orientation is quite subtle and requires a consistent orientation of the moduli spaces $\mathcal{M}(y, x)$ for which we refer to A. Floer and H. Hofer [89]. Thus all the statements described above are true not only for $\mathbb{Z}_2$ coefficients but for $\mathbb{Z}$ coefficients as well.

The 1-periodic solutions of $\dot{x} = X_H(x)$ on M correspond in a one-to-one way to the fixed points of the Hamiltonian map ψ^1 which is the time-1 map of the flow generated by the Hamiltonian vector field. Since M is compact one should

6.5 Floer's Appraoch to Morse theory ...

expect, in view of Poincaré's recurrence theorem, in addition to these fixed points an abundance of periodic points having large periods. Indeed, if one of the fixed points is of general elliptic type as explained in the introduction, then it is a cluster point of other periodic points whose prime periods tend to infinity. This follows simply by the Birkhoff-Lewis theorem, a local theorem for the proof of which we refer to J. Moser [164]. Unfortunately, our global approach to the existence of the fixed points gives only little information about their nature and, in particular, no information about their nonlinear Birkhoff invariants required to deduce the periodic points nearby. We shall show next under additional assumptions on the 1-periodic solutions that infinitely many periodic solutions are deduced from the Floer homology. A 1-periodic solution $x(t) = x(t+1)$ can be considered as a τ-periodic solutions $x(t) = x(t+\tau)$ for every integer $\tau \in \mathbb{N}$, which is then called the τ-iterated solution of x and denoted by x^τ. We call a 1-periodic solution x strongly nondegenerate if all its iterates x^τ are nondegenerate, i.e., if

$$\det\left(1 - \psi^\tau(x(0))\right) \neq 0$$

for all positive integers τ, not only for $\tau = 1$.

Theorem 13. Assume the compact symplectic manifold (M, ω) satisfies $\omega|\pi_2 = 0$ and $c_1|\pi_2 = 0$ and assume $H: S^1 \times M \to \mathbb{R}$ is a smooth Hamiltonian so that all the contractible 1-periodic solutions of the Hamiltonian system are strongly nondegenerate. Then there are infinitely many contractible periodic solutions having large integer periods.

This is a special case of a more general statement proved by D. Salamon and E. Zehnder in [190] which requires merely that the 1-periodic solutions $x(t) = x(t+1)$ are weakly nondegenerate, i.e., they possess at least one Floquet multiplier not equal to 1 so that $\sigma(d\psi^1(x(0))) \neq \{1\}$. The condition on $\pi_2(M)$ cannot be removed as is demonstrated by the example of a rotation on $M = S^2$ with irrational frequency (Fig. 6.12).

This is a Hamiltonian map, but has only two periodic points, namely the strongly nondegenerate fixed points at the north and the south pole.

In sharp contrast to this example, an area preserving diffeomorphism of S^2 having ≥ 3 fixed points, has automatically infinitely many periodic points. This striking phenomenon has been discovered by J.M. Gamboudo and P. Le Calvez [103] and J. Franks [96]. One is tempted to conjecture that every Hamiltonian map on a compact symplectic manifold (M, ω) possessing more fixed points than necessarily required by the V. Arnold conjecture possesses always infinitely many periodic points. Similarly one could expect, that a compact and strictly convex hypersurface $S \subset \mathbb{R}^4$, carries either 2 or ∞ many closed characteristics.

Theorem 13 is an easy consequence of the Floer homology. We make use of the fact proved in [190], that for the index $\mu(x, H)$ of a strongly nondegenerate

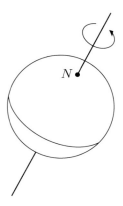

Fig. 6.12

1-periodic solution $x(t) = x(t+1)$, the following iteration formula holds

$$\mu(x^\tau) = \frac{1}{\pi}\left(\tau \triangle(x) + r_\tau(x)\right).$$

Here $\triangle(x) \in \mathbb{R}$ is a mean winding number associated with the 1-periodic solution and $|r_\tau(x)| < \pi n$ for every $\tau \in \mathbb{N}$. We see that either $|\mu(x^\tau)| \to \infty$ as $\tau \to \infty$ or $|\mu(x^\tau)| < n$ for every τ. Assume now, by contradiction, that there are only finitely many periodic solutions of integer periods. Then they have a common period, say $\tau_0 \in \mathbb{N}$, and we pick a prime number $\tau > \tau_0$. Then all the τ-periodic solutions of $\dot{x} = X_H(x)$ are iterated 1-periodic solutions and hence, are nondegenerate by assumption. Considering the Floer complex for the τ periodic solutions $x \in \mathcal{P}_H(\tau)$ we conclude that

$$|\mu(x)| \neq n \text{ for all } x \in \mathcal{P}_H(\tau)$$

if only τ is sufficiently large. Hence we miss the highest and the lowest homology groups, in contradiction to the Morse equation (6.155). Consequently there are infinitely many periodic solutions. Modifying this idea it is shown in [190] that if there exists only finitely many contractible 1-periodic solutions x perhaps even degenerate but satisfying $\triangle(x) \neq 0$ then, for every $\tau \in \mathbb{N}$ large and prime there exists a τ-periodic solution which is contractible and has prime period τ.

Floer's approach to the Morse theory in the loop space of a compact symplectic manifold (M, ω) leads to an invariant of the underlying manifold which turned out to be the familiar singular homology of M. However, Floer carried out his construction also for the Chern-Simons functional on the space of $SU(2)$ connections on a homology 3 sphere X, see [83]. Here the connecting orbits can be interpreted as self-dual Yang-Mills connections on the manifold $X \times \mathbb{R}$ with finite Yang-Mills action. For this functional, the Floer homology groups are surprising new invariants of the underlying homology 3-spheres refining the Casson invariant. Indeed it turns out that the Euler characteristic of Floer's instanton homology of X agrees with double the Casson invariant. We shall not discuss these developments in topology here and refer instead to J.C. Sikorav's Séminaire Bourbaki [199] in 1990. Floer's instanton homology was extended to general oriented 3-manifolds by K. Fukaya in [102]. Recent developments on instanton homology are presented in the Floer memorial volumes [120].

6.6 Symplectic homology

In Chapter 3 we introduced the special symplectic capacity $c_0(M,\omega)$ dynamically in terms of Hamiltonian systems. The existence proof was based on a distinguished critical point of the action functional. This critical point represents a particular periodic orbit and the question arises whether the structure of all critical points together with their connecting orbits leads to a more subtle symplectic invariant which is not necessarily a number anymore. Indeed, combining the construction of the capacity c_0 with Floer's homology construction in the loop space H. Hofer, and A. Floer established in [90] a symplectic homology theory. Our aim is to sketch, without proofs, this theory which is not yet in its final form.

We first recall the setting from the previous section and consider a symplectic manifold (M,ω) satisfying $\pi_2(M) = 0$, for simplicity. By $\Omega(M)$ we denote the set of parameterized contractible loops $x : S^1 \to M$. If $H : S^1 \times M \to \mathbb{R}$ is a time-periodic Hamiltonian, the action functional $\varphi_H : \Omega(M) \to \mathbb{R}$ is defined by

$$\varphi_H(x) = -\int_D \bar{x}^*\omega + \int_0^1 H\big(t, x(t)\big)\, dt \ .$$

Here $\bar{x} : D \to M$ is an extension of x to the closed unit disc D satisfying $\bar{x}|\partial D = x$. The critical points of φ_H are the contractible 1-periodic solutions of the Hamiltonian vector field $\dot{x} = X_H(x)$ on M. We shall assume that all of them are nondegenerate and denote this finite set by $\mathcal{P}_H$. By $C(M, H)$ we denote the vector space over $\mathbb{Z}_2$ generated by $\mathcal{P}_H$ to be

$$C = \oplus C_k \ , \ C_k(M, H) = \mathrm{span}_{\mathbb{Z}_2}\left\{x \in \mathcal{P}_H \,|\, \mu(H, x) = k\right\} \ .$$

The grading is determined by the index $\mu(H, x) \in \mathbb{Z}$ intrinsically associated to the nondegenerate periodic solutions $x \in \mathcal{P}_H$. For a generic almost complex structure J associated with ω there exists a boundary operator $\partial_k = \partial_k(M, H, J) : C_k \to C_{k-1}$. It is defined by counting (mod 2) the connecting orbits $\mathcal{M}(x^-, x^+, H, J)$ which connect periodic solutions of index difference $\mu(H, x^-) - \mu(H, x^+) = 1$. The pair (C, ∂) is a chain complex and we denote by

$$F_*(M, H, J) = \frac{\ker \partial}{\mathrm{im}\, \partial}$$

the Floer homology of the generic pair (H, J). Turning now to the construction of the symplectic homology of (M,ω), we first make the observation that for a fixed generic pair (H, J) there exists an additional filtration of the Floer complex defined by the values of the functional φ_H on $\Omega(M)$. We define for $a \in \mathbb{R} \cup \{\infty\}$ the graded free groups

$$C^a(M, H) = \bigoplus_{k \in \mathbb{Z}} C_k^a$$
$$C_k^a = \mathrm{span}_{\mathbb{Z}_2}\ \{x \in \mathcal{P}_H \,|\, \mu(H, x) = k \text{ and } \varphi_H(x) < a\}\ .$$

Assume now that u is a connecting orbit in $\mathcal{M}(x^-, x^+, H, J)$ and denote by $u_s \in \Omega$ the loops $u_s(t) = u(s, t)$. By the variational structure, $\varphi_H(x^+) \leq \varphi_H(u_s) \leq \varphi_H(x^-)$, hence the periodic orbits lying between two different levels together with their connecting orbits constitute a subcomplex of the Floer complex. More precisely $\partial = \partial(H, J)$ satisfies

$$\partial: C_k^a \to C_{k-1}^a.$$

Therefore, defining for $-\infty < a \leq b \leq \infty$, the quotient group by

(6.159) $$C^{[a,b)}(H, J) = \frac{C^b(H, J)}{C^a(H, J)},$$

then the boundary map induces a boundary operator $\partial_k : C_k^{[a,b)} \to C_{k-1}^{[a,b)}$, and we can define

(6.160) $$F_k^{[a,b)}(H, J) = \frac{\ker(\partial_k)}{\mathrm{im}\ (\partial_{k+1})}.$$

Yet, this Floer homology group depends on the choice of a generic pair (H, J) and is not an invariant of M. The aim is to construct a directed set of very special generic pairs, so that the directed limit of the corresponding groups defines a symplectic invariant of M. The class of admissible Hamiltonians in a generic pair is prompted by the class used in the construction of $c_0(M, \omega)$ in Chapter 3.

We sketch this construction for compact symplectic manifolds (M, ω) possibly having a boundary ∂M. In the case that $\partial M \neq \emptyset$ we shall assume that ∂M is of contact type as defined above. Equivalently, this requires that there exists a 1-form λ on ∂M satisfying $d\lambda = \omega$ on ∂M and $\lambda(x, \xi) \neq 0$ for all nonvanishing $\xi \in T_x(\partial M)$ for which $\omega(\xi, \eta) = 0$ for all $\eta \in T_x(\partial M)$. Moreover, we assume for simplicity, as above, that $\pi_2(M) = 0$. By $\mathcal{H}$ we denote the collection of all smooth and time-periodic Hamiltonian functions $H : S^1 \times M \to \mathbb{R}$ satisfying

(6.161) $\mathrm{supp}(H)$ is compact in $S^1 \times (M \backslash \partial M)$ and $H(S^1 \times M) \subset (-\infty, 0]$.

Clearly this condition implies that there are many periodic solutions x of X_H on M which are degenerate and satisfy $\varphi_H(x) = 0$. In addition to (6.161) we, therefore, require that all contractible periodic solutions x of X_H satisfying $\varphi_H(x) < 0$ are nondegenerate. Consequently, if $-\infty \leq a \leq b \leq 0$ we can define as above the relative Floer groups $F^{[a,b)}(H, J)$. Here, however, not every J is admissible.

6.6 Symplectic homology

The J has to be chosen in such a way that every solution $u : \mathbb{R} \times S^1 \to M$ in $\mathcal{M}(x^-, x^+, H, J)$ stays away from the boundary, i.e., $u(\mathbb{R} \times S^1) \subset M \setminus \partial M$, provided x^- and x^+ are contained in $M \setminus \partial M$. Such admissible J's do exist; the boundary ∂M is then called J-pseudo convex in the sense of D. McDuff [154]. We shall call a pair (H, J) satisfying the above conditions admissible and introduce in the set of admissible pairs a partial ordering, by defining (this is not a typing error)

$$(H, J) \leq (K, \tilde{J}) \iff H(t, x) \geq K(t, x)$$

for all $(t, x) \in S^1 \times M$. For such pairs one can define a monotone homotopy. It consists of a pair $(L, \hat{J})$, where $L : \mathbb{R} \times S^1 \times M \to \mathbb{R}$ and where the almost complex structure $\hat{J}(s, t, x)$ is adapted to the symplectic form ω. Moreover, $L = H, \hat{J} = J$ for $s \to -\infty$ while $L = K, \hat{J} = \tilde{J}$ for $s \to +\infty$, and $\frac{\partial L}{\partial s} \leq 0$. Given a monotone homotopy for $(H, J) \leq (K, \tilde{J})$ one considers the partial differential equations for $u : \mathbb{R} \times S^1 \to M$:

$$\frac{\partial u}{\partial s} + \hat{J}(s, t, u) \frac{\partial u}{\partial t} + (\nabla_j L)(s, t, u) = 0$$
$$u(s, \cdot) \to x \in \mathcal{P}_H \quad (s \to -\infty)$$
$$u(s, \cdot) \to y \in \mathcal{P}_K \quad (s \to +\infty).$$

Using the monotonicity assumption we conclude that $\varphi_K(y) \leq \varphi_H(x)$. This implies that the boundary operator ∂ preserves the real filtration; this was the reason for the monotonicity assumption. Proceeding as in Floer's construction, the combinatorics of the set of solutions is used to construct a natural homomorphism

$$F^{[a,b)}(H, J) \to F^{[a,b)}(K, \tilde{J})$$

of $\mathbb{Z}_2$ vector spaces, which is independent of the chosen generic monotone homotopy. The partial ordering $\leq$ turns the set of admissible pairs (H, J) into a directed set, since given (H_1, J_1) and (H_2, J_2), there exists an admissible pair (H, J) satisfying

$$(H_j, J_j) \leq (H, J) \text{ for } j = 1, 2.$$

Hence we may pass to the limit obtaining the symplectic homology group

$$S^M_{[a,b)} = \text{dir. lim } F^{[a,b)}(H, J).$$

In addition, if $U \subset M$ is an open subset, and if $\mathcal{H}_U$ consists of those admissible pairs (H, J) which are restricted by the requirement that $\text{supp}(H) \subset S^1 \times (U \setminus \partial M)$, we may take the directed limit over $\mathcal{H}_U$ and obtain a symplectic homology group of U defined by

$$S^M_{[a,b)}(U) := \text{dir. lim}_{\mathcal{H}_U} F^{[a,b)}(H, J).$$

We next list some of the properties of these groups known so far. It can be shown that there are natural homomorphisms of $\mathbb{Z}_2$ vector spaces

$$S^M_{[a,b)}(U) \to S^M_{[a,b)}.$$

Moreover, the short exact sequence $0 \to F^{[a,b)}(H,J) \to F^{[a,c)}(H,J) \to F^{[b,c)}(H,J) \to 0$, for $-\infty \leq a \leq b \leq c \leq 0$, gives rise to an exact triangle $\triangle_{a,b,c}(U)$ of symplectic homology groups:

$$\begin{array}{ccc} S^{[a,b)}(U) & \longrightarrow & S^{[a,c)}(U) \\ & \partial \nwarrow \quad \swarrow & \\ & S^{[b,c)}(U). & \end{array}$$

Given a symplectic diffeomorphism $\psi : M \to M$ we can define for any admissible pair (H, J), with $H \in \mathcal{H}_U$, a new admissible pair (H_ψ, J_ψ) as follows:

$$H_\psi(t, x) = H\left(t, \psi^{-1}(x)\right)$$
$$J_\psi(x) = T\psi\left(\psi^{-1}(x)\right) J\left(\psi^{-1}(x)\right) T\psi^{-1}(x).$$

This is used to show that ψ induces a natural isomorphism

$$\psi_\# : S^M_{[a,b)}(U) \xrightarrow{\cong} S^M_{[a,b)}\left(\psi(U)\right).$$

If $U \subset V$ then the inclusion map induces a homomorphism

$$\sigma_{V,U} : S^M_{[a,b)}(U) \to S^M_{[a,b)}(V).$$

If $\psi : M \to M$ is a symplectic diffeomorphism satisfying $\psi(U) \subset V$, we define the induced map ψ_* by

$$\psi_* = \sigma_{V,\psi(U)} \circ \psi_\# : S^M_{[a,b)}(U) \to S^M_{[a,b)}(V).$$

Then if $\varphi : M \to M$ is another symplectic diffeomorphism satisfying $\varphi(V) \subset W$, one can prove

$$\begin{aligned} (\varphi \circ \psi)_* &= \sigma_{W, \varphi \circ \psi(U)} \circ (\varphi \circ \psi)_\# \\ &= \left(\sigma_{W, \varphi(V)} \circ \varphi_\#\right) \circ \left(\sigma_{V, \psi(U)} \circ \psi_\#\right) \\ &= \varphi_* \circ \psi_*. \end{aligned}$$

Moreover, the identity map of M induces the identity map $\mathrm{id} : S^M_{[a,b)}(U) \to S^M_{[a,b)}(U)$. The most useful property is the isotopy invariance formulated as follows: if $\psi_s : M \to M$ for $0 \leq s \leq 1$ is a smooth family of symplectic diffeomorphisms satisfying $\psi_s(U) \subset V$ for all s, then $(\psi_s)_* : S^M_{[a,b)}(U) \to S^M_{[a,b)}(V)$ is independent of s.

It is rather difficult to compute the symplectic homology groups sketched above and we refer to [51] for computations and applications. It is not surprising anymore that the homology groups are related to distinguished periodic orbits, which sit on or near the boundary of a domain. This is illustrated by the theorem below.

6.6 Symplectic homology

We consider a compact symplectic manifold (M, ω) with boundary $\partial M \neq \emptyset$ which is exact symplectic, i.e., $\omega = d\lambda$ for a 1-form λ. We assume the boundary to be of contact type with respect to λ. Then we can define the action spectrum $\sigma(\partial M) \subset \mathbb{R}$ as follows: $\sigma(\partial M) = \{|\int_x \lambda|$, where x is a periodic solution of the Reeb vector field on $\partial M\}$, note that iterated solutions are also considered, in particular, $k \cdot \sigma(\partial M) \subset \sigma(\partial M)$ for $k \in \mathbb{N}$.

Theorem 14. (K. Cieliebak, A. Floer, H. Hofer and K. Wysocki [52]) Consider two exact symplectic and compact manifolds $(M, d\lambda)$ and $(N, d\tau)$ with boundaries ∂M and ∂N of contact type with respect to λ and τ. Assume there exists an exact symplectic diffeomorphism

$$\varphi : \overset{\circ}{M} \to \overset{\circ}{N}$$

(i.e., $\varphi^* \tau - \lambda = dF$ for a smooth function $F : \overset{\circ}{M} \to \mathbb{R}$). If the periodic solutions on ∂M and ∂N and all their iterates are nondegenerate, then

$$\sigma(\partial M) = \sigma(\partial N).$$

Examples by Ya. Eliashberg and H. Hofer [75] demonstrate that under the assumptions of the theorem, the flows on ∂M and ∂N are not necessarily conjugated, so that φ cannot be extended to a symplectic diffeomorphism $M \to N$.

The proof of Theorem 14 is based on the following nontrivial fact [52]. Let $\alpha < 0$ and define

$$d_M(\alpha) = \frac{1}{2} \lim_{\varepsilon \to 0} \dim S_M^{[\alpha-\varepsilon, \alpha+\varepsilon)}(\overset{\circ}{M}).$$

Then $d_M(\alpha) = 0$ if $-\alpha \notin \sigma(\partial M)$ and

$$d_M(\alpha) = \#\left\{ \text{periodic solutions } x \text{ on } \partial M \text{ with action } \left|\int_x \lambda\right| = -\alpha \right\}.$$

Since $S_M^{[a,b)}(\overset{\circ}{M}) = S_N^{[a,b)}(\overset{\circ}{N})$ for all $a < b$, we conclude $d_M(\alpha) = d_N(\alpha)$ so that the theorem follows. For the intricate details of the proof and for applications of this result, we refer to the forthcoming paper [52].

By computing the homology groups one also proves a conjecture by Gromov on the symplectic classification of open polydiscs. Define for $r = (r_1, r_2, \ldots, r_n)$, where $0 < r_1 \leq r_2 \leq \cdots \leq r_n$, the polydisc $D(r) = B^2(r_1) \times \cdots \times B^2(r_n)$.

Theorem 15. The open polydiscs $D(r)$ and $D(r')$ are symplectomorphic if and only if $r = r'$, assuming r and r' are ordered.

We recall that we have proved this statement in Chapter 2 for the special case $n = 2$. The proof was based on two symplectic invariants: the volume and a capacity. The proof for $n \geq 2$ follows with other applications, in [52]. The details of the above construction of the symplectic homology are carried out in K. Cieliebak,

A. Floer and H. Hofer [51] where also possible extensions of this theory are outlined. The proof is based on the action principle in the loop space of symplectic manifolds and combines the construction of the dynamical capacity of Chapter 3 and the construction of the Floer homology, with an additional real filtration. An alternate more recent construction of a symplectic homology theory for open subsets of $(\mathbb{R}^{2n}, \omega_0)$ can be found in L. Traynor [212]; the construction is based on the generating function approach in a finite dimensional setting introduced in [218] by C. Viterbo.

Concluding remarks. As already pointed out in the introduction we have selected only a few and rather elementary topics in symplectic geometry and explained them in detail. We would, therefore, like to mention some additional literature. There is a whole sequence of relative symplectic capacities for subsets in $(\mathbb{R}^{2n}, \omega_0)$ in the framework of symplectic diffeomorphisms defined on all of $\mathbb{R}^{2n}$ constructed by I. Ekeland and H. Hofer in [68]. These capacities have an outer regularity property, in contrast to the intrinsically defined dynamical capacity of Chapter 3 which has an inner regularity property. Quite recently S. Kuksin extended in [130] the capacity concept to some special infinite dimensional Hamiltonian systems defined by partial differential equations, e.g. nonlinear string equation on S^1, some nonlinear Schrödinger equations on T^n. He proved, in particular, a non squeezing result for the flows of such systems. The construction of the capacity is based on a Galerkin approximation using the finite dimensional capacities introduced in Chapter 3. Kuksin's result prompted a non squeezing result for a very special nonlinear Schrödinger equation on S^1, whose flow maps are not compact perturbations of linear maps in the symplectic Hilbert space, see J. Bourgain [31]. Extending the embedding capacity, called the Gromov-width in Chapter 2, one can define, for every integer k, a new capacity as follows: one takes k symplectic embeddings of equal balls $B(r)$ having disjoint images in (M, ω) and determines an upper bound of the radius r. This leads to the symplectic packing problem related to algebraic geometry studied by D. McDuff and L. Polterovich in [155]; for interesting packing constructions we point out L. Traynor [213]. The intrinsic L^∞-geometry on the group of compactly supported Hamiltonian mappings of $(\mathbb{R}^{2n}, \omega_0)$ described in Chapter 5 has been extended to general symplectic manifolds by F. Lalonde and D. McDuff in [134] and [135]. It turns out that the corner stone of this geometry, namely the energy-capacity inequality, is equivalent to the non squeezing theorem. Conjugate points on geodesics in the Hofer metric have in the meantime been investigated by I. Ustilovsky in [214]. For a survey on these interesting developments we point out F. Lalonde [133].

As already emphasized, the V.I. Arnold conjecture about fixed points of Hamiltonian mappings is open for general compact symplectic manifolds of dim $M \geq 8$. For a survey on Floer's contributions to this problem we refer to D. McDuff [153] and to the Floer memorial volume [120]. Recently an analogue of the Arnold conjecture for symplectic mappings φ which are not necessarily Hamiltonian but isotopic to the identity through symplectic diffeomorphisms has been studied by

Lê Hông Vân and Kaoru Ono [137]. They prove for compact manifolds satisfying $c_1|\pi_2(M) = \lambda\omega|\pi_2(M)\,(\lambda \geq 0)$ assuming that all the fixed points are nondegenerate, that the number of fixed points of φ is at least the sum of the Betti-numbers of the Novikov homology over $\mathbb{Z}_2$ associated to the Calabi-invariant α of φ, Nov (α). In the special case of the torus T^{2n} considered in Chapter 6, Nov $(\alpha) = 0$ if $\alpha \neq 0$, so that one obtains nothing new for the torus.

Finally, for an abundance of exciting developments and discoveries in symplectic geometry not treated in our book, e.g. pseudo-holomorphic spheres, filling techniques, Lagrangian intersections, contact geometry, examples of symplectic manifolds, and so on, we refer to the forthcoming books by D. McDuff and D. Salamon [156], M. Audin, J. Lafontaine [13], B. Aebischer, M. Borer, Ch. Leuenberger, M. Kälin, H.M. Reimann [3] and C. Abbas, H. Hofer [1].

Appendix

A.1 Generating functions of symplectic mappings in $\mathbb{R}^{2n}$

We shall start with the very useful observation that every symplectic map can locally be represented in terms of a single scalar function, a so-called generating function. This will be used in Appendix 2 to prove the Arnold-Jost theorem concerning the construction of so-called action-angle coordinates in integrable systems. Such Hamiltonian systems are characterized by the property of having sufficiently many integrals so that, as we shall show, the task of solving (or "integrating") the differential equations becomes trivial, after a global symplectic transformation.

All the considerations in section A.1 are local in nature. We consider a symplectic mapping $u : (\xi, \eta) \mapsto (x, y)$ in $(\mathbb{R}^{2n}, \omega_0)$ represented by

(A.1)
$$x = a(\xi, \eta)$$
$$y = b(\xi, \eta) \, .$$

If we assume

(A.2)
$$\det(a_\xi) \neq 0 \, ,$$

then, locally, the first equation in (A.1) can be solved for $\xi = \alpha(x, \eta)$ and inserting this solution into the second equation in (A.1) we find the following equivalent representation of the map u:

$$\xi = \alpha(x, \eta)$$
$$y = \beta(x, \eta) \, ,$$

where $\beta(x, \eta) = b(\alpha(x, \eta), \eta)$ and

$$\det(\alpha_x) \cdot \det(a_\xi) = 1 \, .$$

The advantage of scrambling the variables in this seemingly ugly manner is that the condition for u to be symplectic is much easier to express in terms of α and β than in terms of a and b. We shall show that under the condition (A.2) the map u is symplectic if and only if there exists a scalar function $W = W(x, \eta)$ with

(A.3)
$$\xi = \alpha(x, \eta) = \frac{\partial}{\partial \eta} W(x, \eta)$$
$$y = \beta(x, \eta) = \frac{\partial}{\partial x} W(x, \eta) \, .$$

Hence (β, α) is the gradient of a function. To prove this statement we consider on $\mathbb{R}^{4n}$, having the coordinates $(\xi, \eta, x, y) \in \mathbb{R}^{4n}$, the one-form,

$$\sigma = \sum_{j=1}^{n} y_j dx_j + \xi_j d\eta_j,$$

whose exterior derivative is the two-form

$$d\sigma = \sum_{j=1}^{n} dy_j \wedge dx_j - \sum_{j=1}^{n} d\eta_j \wedge d\xi_j .$$

Introducing the embeddings i and j: $\mathbb{R}^{2n} \to \mathbb{R}^{4n}$ by

$$i : (x, \eta) \mapsto \big(\alpha(x,\eta),\ \eta,\ x,\ \beta(x,\eta)\big)$$

$$j : (\xi, \eta) \mapsto \big(\xi,\ \eta,\ a(\xi,\eta),\ b(\xi,\eta)\big)$$

and defining the local diffeomorphism $\varphi : \mathbb{R}^{2n} \to \mathbb{R}^{2n}$ by $(x, \eta) \mapsto (\xi, \eta) = (\alpha(x, \eta), \eta)$, we have $i = j \circ \varphi$. Consequently,

$$\begin{aligned} d(i^*\sigma) &= i^* d\sigma = \varphi^*(j^* d\sigma) \\ &= \varphi^*(u^*\omega_0 - \omega_0) . \end{aligned}$$

If u is symplectic, that is $u^*\omega_0 - \omega_0 = 0$, then $i^*\sigma$ is closed and, therefore, locally exact, such that

$$\begin{aligned} i^*\sigma(x, \eta) &= \sum_{j=1}^{n} \beta_j(x, \eta) dx_j + \alpha_j(x, \eta) d\eta_j \\ &= dW(x, \eta) = \sum_{j=1}^{n} W_{x_j}(x, \eta) dx_j + W_{\eta_j}(x, \eta) d\eta_j \end{aligned}$$

for a function $W = W(x, \eta)$. From this identity we read off (A.3). We can reverse all the steps and hence have proved:

Proposition 1. *Every symplectic mapping u as in (A.1) satisfying (A.2) can locally be represented in the implicit form*

(A.4)
$$\begin{aligned} \xi &= W_\eta(x, \eta) \\ y &= W_x(x, \eta) , \end{aligned}$$

where $\det(W_{x\eta}) \neq 0$. *Conversely, any smooth function* $W = W(x, \eta)$ *satisfying this inequality defines implicitly by (A.4) a symplectic mapping u near a point.*

The function $W(x, \eta)$ is called a generating function of the symplectic map u. The proposition shows that a symplectic map u satisfying (A.2) can locally be

A.1 Generating functions of symplectic mappings in $\mathbb{R}^{2n}$

described in terms of a single function. For example, the identity map corresponds to $W(x,\eta) = \langle x,\eta \rangle$. Thus every function W,

$$W(x,\eta) = \langle x,\eta \rangle + w(x,\eta),$$

w being small together with its first two derivatives, defines a symplectic mapping near the identity, and all such mappings can, locally, be parameterized by functions. Of course, other generating functions can be constructed. If we single out (x,ξ) as independent variables, where we assume

(A.5) $$\det(a_\eta) \neq 0,$$

instead of (A.2) we can represent the map u in (A.1) locally in the implicit form

(A.6) $$\begin{aligned} \eta &= - V_\xi(\xi,x) \\ y &= V_x(\xi,x), \quad \det(V_{x\xi}) \neq 0 \end{aligned}$$

with a generating function $V(\xi,x)$. The proof proceeds as above, but starts with the one-form on $\mathbb{R}^{4n}$ given by $\sigma = y\,dx - \eta\,d\xi$, in short notation. For example, the identity map violates conditions (A.5) but satisfies (A.2) while the symplectic rotation

$$x = \eta, \quad y = -\xi$$

satisfies (A.5) but violates (A.2). However, for every symplectic mapping there is some generating function representing it, as we shall prove next. The so-called group of elementary symplectic transformations of $(\mathbb{R}^{2n}, \omega_0)$ is generated by the rotation

$$x_1 = \eta_1, \qquad x_k = \xi_k, \qquad k \geq 2.$$
$$y_1 = -\xi_1 \qquad y_k = \eta_k$$

and the mappings

$$x_j = \xi_{\pi(j)}, y_j = \eta_{\pi(j)}, \quad j = 1, 2, \ldots, n,$$

with a permutation π of the integers $(1, 2, \ldots, n)$. All these mappings are symplectic and can be represented by symplectic matrices which we abbreviate by E.

Proposition 2. Let U be a linear symplectic map of $(\mathbb{R}^{2n}, \omega_0)$. Then there exists an elementary symplectic transformation E satisfying

$$U \cdot E = \begin{pmatrix} A & B \\ C & D \end{pmatrix} \quad \text{and} \quad \det A \neq 0.$$

With U and U^T is a symplectic matrix, hence Proposition 2 follows immediately from

Lemma 1. Let A respectively B be two $n \times n$ matrices having column vectors a_j respectively $b_j \in \mathbb{R}^n$ for $1 \leq j \leq n$. Assume that the rank of the $2n \times n$ matrix (A, B) is equal to r. If $AB^T = BA^T$ then there are r different indices $j_p \in \{1, 2, \ldots, n\}$ such that there are column vectors $v_p = a_{j_p}$ or $v_p = b_{j_p}$, which are linearly independent.

Proof. The following argument is due to C. de Boor. Choose $I \subset \{1, 2, \ldots, n\}$ such that $\{a_j\}_{j \in I}$ are linearly independent and $Im(A) = \text{span}\{a_j | j \in I\}$. It is sufficient to show that

(A.7) $$Im(B) \subset Im(A) + \text{span}\{b_s \mid s \notin I\}.$$

Pick a basis c_j of $\mathbb{R}^n$ satisfying $\langle a_j, c_k \rangle = \delta_{jk}$ for $j, k \in I$. Then one computes, for $j \in I$, that

$$BA^T c_j = b_j + \sum_{s \notin I} d_j(s) b_s,$$

where $d_j(s)$ is the s-th component of the vector $A^T c_j$. Since $Im(BA^T) \subset Im(A)$, due to the assumption $BA^T = AB^T$, we conclude that $b_j, j \in I$, belongs to $Im(A) + \text{span}\{b_s, s \notin I\}$ and so the claim (A.7) follows and the lemma is proved. ∎

Applying Proposition 2 to a symplectic map u with Jacobian U at some point, we achieve the condition (A.2) to hold for $u \circ E$ with some elementary symplectic transformation E and thus find a local representation in terms of a generating function near this point. To illustrate the use of generating functions we shall determine the most general symplectic map u in (A.1) for which

(A.8) $$x = a(\xi)$$

is prescribed, satisfies $\det(a_\xi) \neq 0$, and is independent of η. Representing the desired map u by a generating function $V(\xi, y)$ in the form

$$x = V_y(\xi, y)$$
$$\eta = V_\xi(\xi, y),$$

we find $V_y(\xi, y) = a(\xi)$ and hence $V(\xi, y) = \langle a(\xi), y \rangle - v(\xi)$ or $\eta = V_\xi = a_\xi^T y - v_\xi$ and the explicit representation of u is

(A.9) $$u : \begin{aligned} x &= a(\xi) \\ y &= (a_\xi^T)^{-1}\left(\eta + v_\xi(\xi)\right) \end{aligned}$$

where $v = v(\xi)$ is an arbitrary scalar function. This is the most general symplectic map u compatible with (A.8). For the special case that $a = id$ we find

(A.10) $$u : \begin{aligned} x &= \xi \\ y &= \eta + v_\xi(\xi). \end{aligned}$$

A.1 Generating functions of symplectic mappings in $\mathbb{R}^{2n}$

In order to generalize these considerations we first recall a definition. If (M, ω) is any symplectic manifold, then the Poisson bracket of two functions $F, G : M \to \mathbb{R}$ is defined to be the function

(A.11) $$\{F, G\} = -\omega(X_F, X_G)$$

where X_F is the Hamiltonian vector field associated to the function F. Clearly $\{F, G\} = -\{G, F\}$, and, since $\omega(X_F, \cdot) = -dF$, we have

(A.12) $$\{F, G\} = dF(X_G) = -dG(X_F).$$

A symplectic map u on M, i.e., $u^*\omega = \omega$, leaves the Poisson-bracket invariant:

(A.13) $$\{F, G\} \circ u = \{F \circ u, G \circ u\}.$$

In order to verify this, recall the transformation formula for Hamiltonian vector fields, $u^* X_G = (du)^{-1} X_G \circ u = X_{G \circ u}$ if $u^*\omega = \omega$. Therefore $\{F, G\} \circ u = dF \circ u(X_G \circ u) = d(F \circ u)(u^* X_G) = d(F \circ u)(X_{G \circ u}) = \{F \circ u, G \circ u\}$ proving the claim (A.13). Conversely, one verifies readily that a map u satisfying (A.13) for all functions F and G is necessarily a symplectic map. One verifies, moreover, easily that for all functions F, G and H on M

(A.14) $$[X_F, X_G] = X_{\{G,F\}}$$

and

(A.15) $$\{F, \{G, H\}\} + \{H, \{F, G\}\} + \{G, \{H, F\}\} = 0.$$

Actually, it can be shown for a nondegenerate 2-form ω on M that the condition $d\omega = 0$ is equivalent to each of the conditions (A.14) and (A.15), see, e.g., [167].

Locally in a Darboux chart of M i.e., in $(\mathbb{R}^{2n}, \omega_0)$ the Poisson-bracket of two functions $F = F(x, y)$ and $G = G(x, y)$ is expressed as

(A.16) $$\{F, G\} = \langle F_x, G_y \rangle - \langle F_y, G_x \rangle.$$

The special coordinate functions $(x, y) \mapsto x_j$, for example, satisfy $\{x_i, x_j\} = 0$. Hence, if $u : (\xi, \eta) \mapsto (x, y) = (a(\xi, y), b(\xi, y))$ is a symplectic map we conclude from the invariance of the Possion brackets that $\{a_i, a_j\} = 0$, where $x_j = a_j(\xi, \eta)$, $1 \leq j \leq n$. Our aim is to prove the converse:

Theorem 1. (Liouville) Consider n functions $a_j = a_j(\xi, \eta)$, $1 \leq j \leq n$ satisfying

(i) $$\{a_i, a_j\} = 0$$

(ii) $$\text{rank}(a_\xi, a_\eta) = n.$$

Then there is a local symplectic diffeomorphism $u = (a, b)$,

$$x = a(\xi, \eta)$$
$$y = b(\xi, \eta),$$

extending the relation $x = a(\xi, \eta)$.

Proof. By applying an elementary symplectic transformation we may assume that $\det(a_\xi) \neq 0$ and solve the equation $x = a(\xi, \eta)$ for $\xi = \alpha(x, \eta)$. We shall prove that a_η is symmetric as a consequence of the condition (i). Indeed this condition can be written as $a_\xi a_\eta^T - a_\eta a_\xi^T = 0$. Multiplying this equation by a_ξ^{-1} from the left and its transpose from the right we find $a_\eta^T (a_\xi^T)^{-1} - a_\xi^{-1} a_\eta = 0$, i.e., $a_\xi^{-1} a_\eta$ is symmetric. On the other hand, differentiating the identity $\xi = \alpha(a(\xi,\eta), \eta)$ yields

$$\mathbb{1} = \alpha_x \, a_\xi \quad \text{and} \quad 0 = \alpha_x \, a_\eta + \alpha_\eta \,,$$

and consequently $\alpha_\eta = -c_\xi^{-1} a_\eta$ which is symmetric, as claimed. Therefore, locally, there is a function $W = W(x, \eta)$ with

$$\alpha(x, \eta) = W_\eta(x, \eta) \,.$$

Moreover, $\det(W_{x\eta}) = \det(\alpha_x) = (\det a_\xi)^{-1} \neq 0$, and hence W is a generating function of the desired symplectic map u extending $x = a(\xi, \eta)$. ∎

Corollary. If $a_0, a_1, \ldots a_n$ are $(n + 1)$ functions for which

(i) $\quad \{a_j, a_k\} = 0, \quad j, k = 0, 1, \ldots n$

(ii) $\quad \operatorname{rank}\left(a_{j\xi}, a_{j\eta}\right)_{j=1,2,\ldots n} = n\,,$

then a_0 can be expressed as a function of $a_1, \ldots a_n$. In particular, there are at most n independent functions satifying (i).

Proof. Extending the relation $x = a(\xi, \eta)$ by the theorem, we find a symplectic map u for which $a \circ u(x, y) = x$. Now set $a_0 \circ u = f(x, y)$. Since the Poisson bracket is preserved under u we have $0 = \{a_j, a_0\} = \{x_j, f\} = f_{y_j}$ so that f is independent of y. Hence $a_0 = f(a_1, \ldots a_n)$ proving the corollary. ∎

For further examples of generating functions we refer to the Notes on Dynamical Systems by J. Moser and E. Zehnder [167]. We turn now to a global version of Liouville's theorem.

A.2 Action-angle coordinates, the Theorem of Arnold and Jost

We consider a symplectic manifold (M, ω) of $\dim M = 2n$.

Definition. A Hamiltonian vector field X_H on M is called integrable (in the sense of Liouville) if there exist n functions $F_j : M \to \mathbb{R}$, $1 \leq j \leq n$ satisfying at every point of M:

(i) $\quad dF_1, \ldots dF_n$ are linearly independent

(ii) $\quad \{F_i, F_j\} = 0 \quad \text{for all } i, j\,.$

(iii) $\quad \{H, F_j\} = 0 \quad \text{for all } j\,.$

A.2 ACTION-ANGLE COORDINATES, THE THEOREM OF ARNOLD AND JOST

Introducing the map

(A.17) $$F : M \longrightarrow \mathbb{R}^n, \quad F(x) = \left(F_1(x), \ldots F_n(x)\right)$$

we conclude from (i) that the level sets

(A.18) $$N_c = \{x \in M \mid F(x) = c\}, \quad c \in \mathbb{R}^n$$

are n-dimensional submanifolds of M. The Hamiltonian vector fields X_{F_j} are linearly independent and, since $dF_i(X_{F_j}) = 0$ in view of (ii), they span the tangent space of N_c:
$$T_x N_c = \text{span}\{X_{F_j}(x) \mid 1 \leq j \leq n\},$$
$x \in N_c$. We remark, moreover, that in view of $\omega(X_{F_j}, X_{F_i}) = 0$ we have $\omega(v, w) = 0$ for all pairs $v, w \in T_x N_c$. Hence N_c is a Lagrangian submanifold of M. By (iii), the functions F_j are all integrals of X_H which is therefore tangent to the level sets N_c so that the flow of X_H leaves N_c invariant. Under an additional topological assumption on N_c, the flow can be determined explicitly for all times in view of the following theorem which goes back to V. Arnold [12] in the special case $M = \mathbb{R}^{2n}$. However, Arnold required an additional assumption which was removed by R. Jost in [126] who proved it in the general case.

Theorem 2. (Arnold-Jost) Let (M, ω) be a symplectic manifold of dim $M = 2n$. Assume there are n functions F_j satisfying (i) and (ii) at every point. Assume, moreover, that one of the level sets, say $N = F^{-1}(0) \subset M$ is compact and connected. Then

1) N is an embedded n-dimensional torus T^n.

2) There is an open neighborhood U of N in M which can be described by the action and angle variables (x, y) in the following manner. If $x = (x_1, \ldots, x_n)$ are the variables on the torus $T^n = \mathbb{R}^n / \mathbb{Z}^n$ and $y = (y_1, \ldots, y_n) \in D_1$, where D_1 and D_2 are some domains in $\mathbb{R}^n$ containing $y = 0$, then there exists a diffeomorphism
$$\psi : T^n \times D_1 \longrightarrow U = \bigcup_{c \in D_2} \left(F^{-1}(c) \cap U\right)$$
and a diffeomorphism $\mu : D_2 \to D_1$, with $\mu(0) = 0$, such that

(A.19) $$\psi^* \omega = \sum_{j=1}^{n} dy_j \wedge dx_j = \omega_0$$

(A.20) $$\mu \circ F \circ \psi(x, y) = y.$$

In particular, ψ maps the torus $T^n \times \{0\}$ diffeomorphically onto $F^{-1}(0) = N$ and the torus $T^n \times \{y\}$ onto $N_c \cap U$ where $y = \mu(c)$.

Corollary: Any integrable Hamiltonian system given by H having the integrals F_j is by the symplectic diffeomorphism ψ transformed into the following system on $(T^n \times D_1, \omega_0)$:

(A.21) $$H \circ \psi(x,y) = h(y),$$

where the Hamiltonian h depends on the integrals only and not on the angle variables x.

Proof of the Corollary: In view of (A.13) the functions $F_j \circ \psi$ and therefore, by (A.20) also the coordinate functions y_j, are integrals of $H \circ \psi$, and hence

$$0 = \{H \circ \psi, y_j\} = \{h, y_j\} = \frac{\partial}{\partial x_j} h$$

so that $h(x,y) = H \circ \psi(x,y)$ does indeed not depend on the x variables. ∎

Consequently, on $\psi^{-1}(U)$ the transformed Hamiltonian system looks extremely simple:

$$X_h : \dot{x} = \frac{\partial h}{\partial y}(y), \quad \dot{y} = 0.$$

It is easily "integrated" with the solutions

$$x(t) = x(0) + t\frac{\partial h}{\partial y}\left(y(0)\right), \quad y(t) = y(0)$$

hence the name "integrable systems". Geometrically, every torus $T^n \times \{y\}$ in $T^n \times D_2$ is invariant under the flow φ^t of X_h and the restriction onto such a torus is linear:

$$\varphi^t \mid T^n \times \{y\} : (x,y) \mapsto (x + t\omega, y)$$

where $\omega = \frac{\partial h}{\partial y}(y) \in \mathbb{R}^n$ are the frequencies of this particular torus. We see that in the open set $U \subset M$ all the solutions of $\dot{x} = X_H(x)$ are quasi-periodic.

Proof of Theorem 2. The proof proceeds in several steps. First we show that N is a torus. Second we introduce convenient coordinates (x,y) locally near a point p on N by means of Liouville's theorem. Thirdly, these coordinates are extended to an open neighborhood of N by means of the flows of the Hamiltonian vector fields X_{F_j}. Finally, the normalization of the periods requires an additional symplectic change of variables to conclude the proof.

a) In order to prove the first statement of the theorem, we denote by $\varphi_j^{s_j}$ the flow of X_{F_j}. Since $[X_{F_j}, X_{F_i}]$ is a Hamiltonian vector field with Hamiltonian function $\{F_i, F_j\} = 0$, the vector fields X_{F_j} and hence also their flows do commute and we abbreviate

(A.22) $$\varphi^s = \varphi_1^{s_1} \circ \varphi_2^{s_2} \circ \ldots \varphi_n^{s_n}, \quad s = (s_1, \ldots, s_n) \in \mathbb{R}^n,$$

wherever it is defined. Since the F_j are integrals of X_{F_i}

$$F\Big(\varphi^s(x)\Big) = F(x), \quad x \in M$$

and hence φ^s leaves all the level sets $N_c \subset M$ invariant. Fix now a point $p \in N$. Then the flow $\varphi^s(p)$ exists for all $s \in \mathbb{R}^n$, since N is compact, and we can define the action of $\mathbb{R}^n$ on N by

(A.23) $\qquad A : \mathbb{R}^n \longrightarrow N, \quad s \mapsto \varphi^s(p) .$

Clearly $\varphi^s \circ \varphi^t = \varphi^{s+t}$. The map A is an immersion, since the tangent map at t maps $e_j \in \mathbb{R}^n$ onto $X_{F_j}(\varphi^t(p)) \in T_{\varphi^t(p)} N$ and hence these vectors span the tangent space. Consequently, A is a local diffeomorphisms and its image is both open and closed and hence equal to N, since N is assumed to be connected. The map A, however, is not injective. The isotropy group $\Gamma = \{s \in \mathbb{R}^n | \varphi^s(p) = p\}$ is a discrete subgroup of $\mathbb{R}^n$, i.e., a lattice. Therefore Γ is generated by vectors $\gamma_1, \ldots, \gamma_d \in \mathbb{R}^n$ which are linearly independent over $\mathbb{R}$ and

(A.24) $\qquad \Gamma = \Big\{ \gamma = \sum_{j=1}^{d} n_j \gamma_j \mid n_j \in \mathbb{Z} \Big\} .$

In view of $\varphi^{s+\gamma}(p) = \varphi^s(p)$ for all $s \in \mathbb{R}^n$, the map A induces a diffeomorphism between $\mathbb{R}^n / \Gamma$ and N. Since N and hence $\mathbb{R}^n / \Gamma$ are compact, we necessarily have $d = n$ so that N is an n-dimensional torus as claimed.

b) We next introduce convenient local coordinates near the point $p \in N$. By the Darboux and Liouville theorems and its corollary we have

Lemma 2. (Liouville) If (M, ω) is a symplectic manifold of dim $M = 2n$, and if F_j are n functions with $\{F_j, F_i\} = 0$ and dF_j linearly independent everywhere, then every point $q \in M$ has an open neighborhood W and a diffeomorphism $\psi : V \to W$ where V is an open neighborhood of the origin in $\mathbb{R}^{2n}$ such that

(i) $\qquad \psi(0) = q$

(ii) $\qquad \psi^* \omega = \sum_{j=1}^{n} dy_j \wedge dx_j .$

(iii) $\qquad F \circ \psi(x, y) = y ,$

where (x, y) are the symplectic coordinates of $(\mathbb{R}^{2n}, \omega_0)$.

Corollary. In the above "Liouville"-coordinates the flow φ^s defined in (A.22) looks very simple, namely

(A.25) $\qquad \varphi^s \circ \psi(x, y) = \psi(x + s, y), \quad s \in \mathbb{R}^n$

if (x, y) and $(x + s, y) \in V$.

Indeed, $\psi^{-1} \circ \varphi_j^{s_j} \circ \psi$ is the flow of the Hamiltonian vector field $X_{F_j \circ \psi}$ on V for every $1 \leq j \leq n$. Since $F_j \circ \psi(x,y) = y_j$, it is indeed linear and given by $(x,y) \mapsto (x + s_j e_j, y)$.

c) Going back to our point $p \in N$, we choose by Lemma 2 the Liouville coordinates $\psi : V \to U \subset M$, where V is an open neighborhood of 0 in $\mathbb{R}^{2n}$ and $\psi(0) = p$. In order to extend these coordinates to $(x,y) \in \mathbb{R}^{2n}$ we shall define a map

(A.26)
$$\vartheta : \quad \mathbb{R}^n \times D_2 \to M$$
$$(x,y) \mapsto \vartheta(x,y) = \varphi^x \circ \psi(0,y)$$

where φ^x is the flow of the integrals X_{F_j} defined in (A.22), and where $D_2 \subset \mathbb{R}^n$ is an open (and small) neighborhood of $y = 0$. For small x we have, in view of the corollary $\vartheta(x,y) = \psi(x,y)$, for $(x,y) \in V$. Geometrically, the definition (A.26) says that we apply the flow φ^x to the local section $S = \{\psi(0,y) \in M : y \in D_2\}$ which is, in view of the corollary, transversal to the vector fields X_{F_j}.

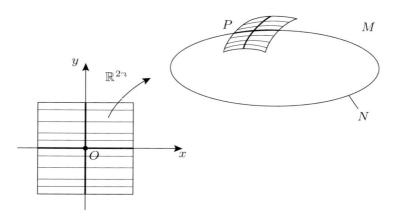

Fig. A2.1

Moreover, since $F \circ \varphi^x = F$ we conclude from Lemma 2 (iii) that $F(\vartheta(x,y)) = F(\varphi^x \circ \psi(0,y)) = F(\psi(0,y)) = y$ and hence

(A.27) $$\vartheta : \mathbb{R}^n \times \{y\} \longrightarrow N_y = F^{-1}(y)$$

for $y \in D_2$, wherever the map is defined. If $y = 0$ then by $\psi(0,0) = p$, we have $\vartheta(x,0) = \varphi^x(p)$, and therefore by the considerations in a) the map is defined for all $x \in \mathbb{R}^n$ and maps $\mathbb{R}^n \times \{0\}$ onto $N = F^{-1}(0)$. It follows that ϑ is defined on $K \times D_2$ for a compact set $K \subset \mathbb{R}^n$ containing the lattice Γ in its interior provided D_2 is small enough. We shall prove next that ϑ is in fact defined on all of $\mathbb{R}^n \times D_2$. Recalling that if $y = 0$, then $\vartheta(x + \gamma, 0) = \vartheta(x,0)$ for $x \in \mathbb{R}^n$ and $\gamma \in \Gamma$ we shall first construct, for y in a small neighborhood of 0, an n-dimensional

A.2 ACTION-ANGLE COORDINATES, THE THEOREM OF ARNOLD AND JOST

lattice $\Gamma(y) \subset \mathbb{R}^n$ close to $\Gamma(0) = \Gamma$ such that $\vartheta(x + \gamma(y), y) = \vartheta(x, y)$ for every $\gamma(y) \in \Gamma(y)$.

Let $\gamma = \gamma_j \in \Gamma$ be an arbitrary basis vector of the lattice defined in (A.24). Then $\varphi^\gamma(p) = p$ and $p = \psi(0)$. Hence, in the Liouville coordinates $V \subset \mathbb{R}^{2n}$

(A.28) $$\psi^{-1} \circ \varphi^\gamma \circ \psi : (x, y) \mapsto (\xi, \eta)$$

defines a symplectic diffeomorphism locally near the fixed point 0 in V.

Lemma 3. *The symplectic map* (A.28) *in V is of the form*

(A.29) $$\begin{aligned} \xi &= x - v(y), \quad v(y) = \tfrac{\partial}{\partial y} Q(y), \\ \eta &= y \end{aligned}$$

for a smooth scalar function Q satisfying $Q(0) = 0$. Moreover, $v(0) = 0$.

Proof. Since, by definition, $\varphi^\gamma \circ \psi(x, y) = \psi(\xi, \eta)$ we conclude from $F \circ \varphi^x = F$, in view of Lemma 2 (ii), that

$$y = F \circ \psi(x, \eta) = F \circ \varphi^\gamma \circ \psi(x, \eta) = F \circ \psi(\xi, \eta) = \eta .$$

Every symplectic map extending the relation $\eta = y$ is necessarily of the announced form as we have seen in the previous section. That $v(0) = 0$ follows from $\psi^{-1} \circ \varphi^\gamma \circ \psi(0) = 0$. ∎

By definition (A.28), $\varphi^\gamma \circ \psi(x, y) = \psi(\xi, \eta)$ and we conclude from Lemma 3 and Lemma 2 that

(A.30) $$\psi(x, y) = \varphi^{\gamma + v(y)} \circ \psi(x, y)$$

for $(x, y) \in V' \subset V$. We can do this for every basis vector $\gamma_j \in \Gamma$ and find $v_j(y) = \tfrac{\partial}{\partial y} Q_j(y)$ such that (A.30) holds true for $\gamma_j + v_j(y)$ replacing $\gamma + v(y)$. We, therefore, introduce the lattice $\Gamma(y) \subset \mathbb{R}^n$ by

(A.31) $$\Gamma(y) = \left\{ \gamma(y) = \sum_{j=1}^n n_j \gamma_j(y) \,\Big|\, n_j \in \mathbb{Z} \right\}$$

where, in view of Lemma 3,

$$\gamma_j(y) = \gamma_j + v_j(y) = \frac{\partial}{\partial y} W_j(y), \quad 1 \leq j \leq n .$$

Here $\gamma_j \in \Gamma$ and W_j are scalar functions satisfying $W_j(0) = 0$. Moreover, since $v_j(0) = 0$ we have $\gamma_j(0) = \gamma_j$ and hence

(A.32) $$\Gamma(0) = \Gamma .$$

Consequently, for small y, the lattice vectors $\gamma_j(y)$ are linearly independent and satisfy, in view of (A.30)

(A.33) $$\psi(x, y) = \varphi^{\gamma_j(y)} \circ \psi(x, y), \quad 1 \leq j \leq n ,$$

if $(x, y) \in V' \subset V$.

Lemma 4. The map $\vartheta : (\mathbb{R}^n \times D_2, \omega_0) \to (M, \omega)$ defined in (A.26) exists for all $(x, y) \in \mathbb{R}^n \times D_2$ and satisfies

(i) $\qquad\qquad\qquad \vartheta^* \omega \;=\; \omega_0 \;.$

(ii) $\qquad\qquad \vartheta\Big(x + \gamma(y), y\Big) \;=\; \vartheta(x, y) \;, \quad \text{if } \gamma(y) \in \Gamma(y) \;.$

Proof. The map ϑ exists for $(x, y) \in K \times D_2$ with a compact set $K \subset \mathbb{R}^n$ containing $\Gamma(0)$ in its interior, if D_2 is small enough. Since, for x small, we have, in view of (A.33),

$$\vartheta\Big(x + \tilde{\gamma}_j(y), y\Big) \;=\; \varphi^{x+\gamma_j(y)} \circ \psi(0, y)$$
$$= \varphi^x \circ \varphi^{\gamma_j(y)} \circ \psi(0, y)$$
$$= \varphi^x \circ \psi(0, y) \;=\; \vartheta(x, y)$$

for every $\gamma_j(y) \in \Gamma(y)$, the flow $x \mapsto \varphi^x \circ \psi(0, y)$ does indeed exist for all $x \in \mathbb{R}^n$ by the standard continuation theorem of solutions of ordinary differential equations. It remains to prove (i). If $(x, y) \in \mathbb{R}^n \times D_2$ we choose x_0 close to x such that $(x - x_0, y) \in V$ and represent the map ϑ in the form

$$\vartheta(x, y) \;=\; \varphi^{x_0 + (x - x_0)} \circ \psi(0, y)$$
$$= \varphi^{x_0} \circ \psi(x - x_0, y) \;=\; \varphi^{x_0} \circ \psi \circ \sigma_{x_0}(x, y)$$

where $\sigma_{x_0}(x, y) = (x - x_0, y)$ and where we have used the corollary to Lemma 2. Hence, ϑ being locally a composition of symplectic mappings is indeed symplectic and the lemma is proved. ∎

Recall that $\vartheta(\mathbb{R}^n \times \{y\}) \subset F^{-1}(y)$. We claim that, for every $y \in D_2$, the map ϑ induces a diffeomorphism

(A.34) $\qquad\qquad\qquad \vartheta_0 : \mathbb{R}^n / \Gamma(y) \longrightarrow F^{-1}(y)$

onto its image in $F^{-1}(y)$, which then is a torus T^n. We only have to prove that ϑ_0 is injective provided y is sufficiently small. Arguing by contradiction there are sequences (x_j, y_j) and (x'_j, y_j) satisfying

(A.35)
$$\vartheta(x_j, y_j) \;=\; \vartheta(x'_j, y_j) \;, \quad y_j \to 0$$
$$x_j - x'_j \notin \Gamma(y_j) \;.$$

Taking a subsequence we may assume $x_j \to x^*$ and $x'_j \to x'^*$. Hence $\vartheta(x^*, 0) = \vartheta(x'^*, 0)$ and, therefore $x^* - x'^* \in \Gamma(0) = \Gamma$. Consequently, $x_j - x'_j$ is close to a point in $\Gamma(y_j)$. Since $\Gamma(y)$ is discrete with distances between points greater than a positive constant uniformly in y, we conclude that $x_j - x'_j \in \Gamma(y_j)$. This contradicts (A.35) and hence proves the claim (A.34).

A.2 Action-angle coordinates, the Theorem of Arnold and Jost

d) Finally, we normalize the lattices $\Gamma(y)$ by means of another symplectic diffeomorphism

$$\sigma : \mathbb{R}^n \times D_1 \longrightarrow \mathbb{R}^n \times D_2$$
$$(\xi, \eta) \longmapsto (x, y) .$$

Recalling the definition of the lattice $\Gamma(y) \subset \mathbb{R}^n$ in (A.31) the symplectic map σ is implicitly defined by the generating function

(A.36)
$$V(\xi, y) = \sum_{j=1}^{n} \xi_j W_j(y) ;$$

hence,

(A.37)
$$\eta_j = \frac{\partial}{\partial \xi_j} V(\xi, y) = W_j(y)$$
$$x_j = \frac{\partial}{\partial y_j} V(\xi, y) = \sum_{s=1}^{n} \xi_s \frac{\partial}{\partial y_j} W_s(y) .$$

The first relation $\eta = W(y)$ in (A.37) defines a local diffeomorphism near the fixed point $y = 0$. In the statement of Theorem 2 it is denoted by $\mu : D_2 \to D_1$. This follows since

$$\frac{\partial}{\partial y} W(y) = \bigl(\gamma_1(y), \ldots, \gamma_n(y)\bigr)$$

is of rank n. Indeed the column vectors $\gamma_j(y)$ are the n linearly independent generators of $\Gamma(y)$. Hence $\det(V_\xi) \neq 0$. Moreover, $\sigma(e_j, \eta) = (\gamma_j(y), y)$ for $\eta = \mu(y)$ and therefore

$$\sigma : \mathbb{Z}^n \times \{\eta\} \longrightarrow \Gamma(y) \times \{y\} , \quad \eta = \mu(y) .$$

The composition

$$\Phi := \vartheta \circ \sigma : (\mathbb{R}^n \times D_1, \omega_0) \longrightarrow (M, \omega)$$

is a symplectic map satisfying $\Phi(\xi + e_j, \eta) = \Phi(\xi, \eta)$. It induces a symplectic diffeomorphism

$$\Phi : \mathbb{R}^n / \mathbb{Z}^n \times D_1 \longrightarrow U \subset M$$

which is the desired map defining the action and angle variables denoted by $(\xi, \eta) \in T^n \times D_1$. Indeed, using the definitions we compute $\mu \circ F \circ \Phi(\xi, \eta) = \mu \circ F \circ \vartheta \circ \sigma(\xi, \eta) = \mu \circ F \circ \vartheta(x, y) = \mu(y) = \eta$ as claimed in Theorem 2 whose proof is therefore completed. ∎

The symplectic manifold (M, ω) is not required to be exact (exact meaning that $\omega = d\vartheta$ for a one-form ϑ on M). But it follows from the theorem that the restriction of ω onto the image $U = \psi(T^n \times D)$ in M of the action-angle coordinates is an exact form. Indeed $\omega|U = d\sigma$ for the one-form $\sigma = (\psi^{-1})^* \lambda$, where $\lambda = \sum_{j=1}^{n} y_j dx_j$ on $T^n \times D$ satisfies $d\lambda = \omega_0$. This allows us to obtain the action variables $y = \mu \circ F$ as "action integrals" of σ over closed curves on $N_c \cap U$ whose homology classes form a basis in $H_1(N_c \cap U, \mathbb{Z})$ with $\mu(x) = y$. Hence the name action variables.

The action-angle variables (x,y) we constructed are, of course, not unique. Given a unimodular matrix $M \in GL(n, \mathbb{Z})$, a function $w(y)$ and constants $c \in \mathbb{R}^n$, we find other action-angle variables (x', y') by means of the symplectic diffeomorphism of $T^n \times D$:

$$x' = M\left(x + \frac{\partial w}{\partial y}(y)\right)$$
$$y' = (M^T)^{-1} y + c.$$

This corresponds to a change of the basis of the fundamental group of T^n by the matrix M and a phase shift $\frac{\partial w}{\partial y}(y)$ of the action variables. Straight lines on T^n are mapped into straight lines under this map.

We have constructed the action-angle coordinates near a torus and one can ask how these local action-angle charts fit together. We do not discuss this global question and refer instead to N. Nekhoroshev [168] (1972), to A.S. Mishchenko and A.T. Fomenko [158] (1978) and, in particular, to J.J. Duistermaat [64] (1980). Integrable systems constitute a very restricted even exceptional class of Hamiltonian systems. Examples are described in the Notes [167], from which this appendix is taken with minor modifications. For interesting developments in the field of integrable systems we refer to J. Moser [166].

A.3 Embeddings of $H^{1/2}(S^1)$ and smoothness of the action

In our proof of the smoothness of the action functional below the following embedding properties of the space $H^{1/2}(S^1)$ introduced in Chapter 3 will be crucial. Formally they follow from Sobolev's embedding theorem. However, for fractional derivatives a separate proof is needed.

Theorem 3. If $u \in H^{1/2}(S^1)$ then $u \in L^p(S^1)$ for every $1 \leq p < \infty$ and there are constants C_p with

$$\|u\|_{L^p(S^1)} \leq C_p \|u\|_{H^{1/2}(S^1)}, \quad u \in H^{1/2}(S^1).$$

The embeddings are, moreover, compact.

Proof. It clearly suffices to find constants C_p satisfying the estimates for smooth functions $u \in C^\infty(S^1)$ only, since they are dense in $H^{1/2}(S^1)$. Assuming for simplicity also the functions to have mean value zero we represent them as

$$u(e^{2\pi i \vartheta}) = \sum_{j \in \mathbb{Z}} c_j e^{2\pi i \vartheta j} \text{ with } c_0 = 0 \text{ and } c_{-j} = \bar{c}_j.$$

By definition,

$$\|u\|^2_{H^{1/2}(S^1)} = 2\pi \sum_{j \in \mathbb{Z}} |j| |c_j|^2.$$

A.3 Embeddings of $H^{1/2}(S^1)$ and smoothness of the action

Set $z = re^{2\pi i\vartheta} \in \mathbb{C}$. Considering $S^1 = \partial D$ as the boundary of the disc $D = \{z \in \mathbb{C} : |z| \leq 1\}$ we can extend the function u to the harmonic function $u(z)$ on $\mathring{D}$ by setting

$$u(z) = \sum_{j \neq 0} c_j r^{|j|} e^{2\pi i \vartheta j} = \frac{1}{2}\left(f(z) + \overline{f(z)}\right)$$

where

$$f(z) = 2 \sum_{j=1}^{\infty} c_j z^j \,.$$

Since $u(e^{2\pi i\vartheta})$ is in $H^{1/2}(S^1)$ the function f is holomorphic in $\mathring{D}$ and $u(z)$ attains the desired boundary values. We claim

$$\int_D |\nabla u|^2 \, dx \, dy = \|u\|^2_{H^{1/2}(\partial D)} \,.$$

Indeed using $f'(z) = \frac{\partial u}{\partial x} - i\frac{\partial u}{\partial y}$ and integrating over the angle variables first, we find

$$\int_D |\nabla u|^2 dx \, dy = \int_D |f'(z)|^2 r \, dr \, d\vartheta$$
$$= 8\pi \sum_{j=1}^{\infty} j^2 |c_j|^2 \int_0^1 r^{2j-1} dr = 8\pi \sum_{j \geq 1} |c_j|^2 \frac{j^2}{2j}$$
$$= 2\pi \sum_{j \in \mathbb{Z}} |j| |c_j|^2 = \|u\|^2_{H^{1/2}(S^1)}$$

as claimed. A similar computation shows that $\|u\|_{L^2(D)} \leq \|\nabla u\|_{L^2(D)}$ using the assumption $c_0 = 0$. The desired estimate of Theorem 3 is now an immediate consequence of

Proposition 3. For $2 \leq p < \infty$ there are constants C_p, such that

$$\|u\|_{L^p(\partial D)} \leq C_p\left(\|u\|_{L^2(D)} + \|\nabla u\|_{L^2(D)}\right).$$

for $u \in C^1(D)$

In order to prove this statement, which is a special case of a general "trace" theorem, we start with a lemma.

Lemma 5. For $2 \leq p < \infty$ there are constants C_p satisfying

$$\|u\|_{L^p(\partial D)} \leq C_p\left(\|u\|_{L^{2p-2}(D)} + \|\nabla u\|_{L^2(D)}\right)$$

for $u \in C^1(D)$.

Proof. Consider at first the function $u = u(r, \vartheta)$ on the annulus $A = \{z | 1/2 \le |z| \le 1\}$. By continuity of u we can find a radius $\frac{1}{2} < \rho < 1$ with

$$\|u\|_{L^1(A)} = \int_{\frac{1}{2}}^{1} \int_{0}^{2\pi} |u(r,\vartheta)| r \, dr \, d\vartheta = \frac{\rho}{2} \int_{0}^{2\pi} |u(\rho,\vartheta)| d\vartheta \ge \frac{1}{4} \int_{0}^{2\pi} |u(\rho,\vartheta)| d\vartheta \,.$$

Hence integrating

$$u(1,\vartheta) = u(\rho,\vartheta) + \int_{\rho}^{1} \frac{\partial u}{\partial r}(r,\vartheta) dr$$

over the angle variable, we get

$$\frac{1}{4}\|u\|_{L^1(\partial D)} \le \|u\|_{L^1(A)} + \left\|\frac{\partial u}{\partial r}\right\|_{L^1(A)} \,.$$

Replacing u by $v = |u|^{p-1} u$, with $p \ge 2$, using $\frac{\partial v}{\partial r} = p|u|^{p-1}\frac{\partial u}{\partial r}$ and the Cauchy-Schwarz inequality we find

$$\|u\|_{L^p(\partial D)} \le C_p \left(\|u\|_{L^p(A)} + \|u\|_{L^{2p-2}(A)}^{\frac{p-1}{p}} \left\|\frac{\partial u}{\partial r}\right\|_{L^2(A)}^{\frac{1}{p}} \right),$$

with a constant C_p depending on p only. Using Young's inequality ($ab \le \frac{a^p}{p} + \frac{b^q}{q}$ if $\frac{1}{p} + \frac{1}{q} = 1$), we conclude

$$\|u\|_{L^p(\partial D)} \le C_p \left(\|u\|_{L^{2p-2}(A)} + \left\|\frac{\partial u}{\partial r}\right\|_{L^2(A)} \right),$$

from which the lemma follows. ∎

In view of this lemma, Proposition 3 is a consequence of the next proposition which is a special case of a more general Sobolev embedding theorem. It states that $H^{1,2}(D)$ is continuously embedded in $L^p(D)$, for every $p < \infty$.

Proposition 4. *There are constants $C_p, 2 \le p < \infty$ such that*

$$\|u\|_{L^p(D)} \le C_p \left(\|u\|_{L^2(D)} + \|\nabla u\|_{L^2(D)} \right)$$

for all $u \in C^1(D)$.

Proof. First extend the functions $u \in C^1(D)$ to C^1-functions of compact support in a neighborhood of D. This may be achieved as follows. Choose $\chi \in C_c^\infty(\mathbb{R}^2)$ such that $\chi = 1$ on D and $\chi = 0$ outside a disc of radius $\frac{5}{4}$. Then define the extension of $u = u(r,\vartheta)$ by

$$u(1+r,\vartheta) = \{4u(1-\frac{1}{2}r,\vartheta) - 3u(1-r,\vartheta)\} \cdot \chi(1+r,\vartheta) \,.$$

A.3 Embeddings of $H^{1/2}(S^1)$ and smoothness of the action

This defines a linear extension map $E : C^1(D) \to C_c^1(Q)$ which is bounded in the $W^{1,2}$ norm, where Q is a bigger domain. In order to prove the proposition it is, therefore, sufficient to find constants C_p such that for all $u \in C_c^1(Q)$

$$\|u\|_{L^p(Q)} \leq C_p \big(\|u\|_{L^2(Q)} + \|\nabla u\|_{L^2(Q)} \big). \tag{A.38}$$

If $u \in C_c^1(Q)$, then

$$|u(x,y)| = \Big| \int_{-\infty}^{x} \frac{\partial u}{\partial x}(t,y) dt \Big| \leq \int_{-\infty}^{\infty} \Big| \frac{\partial u}{\partial x}(t,y) \Big| dt,$$

and hence, with a similar estimate for the y-derivative:

$$|u(x,y)|^2 \leq \int_{-\infty}^{\infty} \Big|\frac{\partial u}{\partial x}(t,y)\Big| dt \cdot \int_{-\infty}^{\infty} \Big|\frac{\partial u}{\partial y}(x,t)\Big| dt.$$

By integration,

$$\|u\|_{L^2(Q)}^2 \leq \Big\|\frac{\partial u}{\partial x}\Big\|_{L^1(Q)} \cdot \Big\|\frac{\partial u}{\partial y}\Big\|_{L^1(Q)}.$$

Replacing u by $v = |u|^{t-1} u$, with $t \geq 2$, we find by the Cauchy-Schwarz inequality

$$\|u\|_{L^{2t}(Q)}^{2t} \leq C_t \|u\|_{L^{2t-2}(Q)}^{2t-2} \Big\|\frac{\partial u}{\partial x}\Big\|_{L^2(Q)} \Big\|\frac{\partial u}{\partial y}\Big\|_{L^2(Q)}$$

$$\leq C_t \|u\|_{L^{2t-2}(Q)}^{2t-2} \|\nabla u\|_{L^2(Q)}^2.$$

Hence by Young's inequality,

$$\|u\|_{L^{2t}(Q)} \leq C_t \big(\|u\|_{L^{2t-2}(Q)} + \|\nabla u\|_{L^2(Q)} \big). \tag{A.39}$$

If $t = 2$, we have the $W^{1,2}$ norm on the right hand side and hence by iterating the inequality (A.39) the desired estimate (A.38) follows for every $2 \leq p < \infty$. This completes the proof of Proposition 4 and the first claim of Theorem 3.

Finally we shall show that the embeddings $H^{1/2} \to L^p$ are compact for $p < \infty$. For the special case $p = 2$, this is proved in Chapter 3. Applying the Cauchy-Schwarz inequality to $|u| \, |u|^{q-1} = |u|^q$, we find the following interpolation inequality

$$\|u\|_{L^q} \leq \|u\|_{L^2}^{\frac{1}{q}} \|u\|_{L^{2q-2}}^{1-\frac{1}{q}}, \quad q > 2.$$

Since $H^{1/2} \to L^{2q-2}$ is continuous by the first part of Theorem 3 and $H^{1/2} \to L^2$ is compact, we conclude from the above interpolation inequality that also $H^{1/2} \to L^q$ is compact. This proves the second statement of Theorem 3. ∎

The case $p = \infty$ is excluded in Theorem 3 and we give an example of an unbounded function in $H^{1/2}(S^1)$. If $0 < \varepsilon < \frac{1}{2}$ then the function

$$u(e^{i\vartheta}) := Re[\log(1 - e^{i\vartheta})]^{\frac{1}{2}-\varepsilon}$$

belongs to $H^{1/2}(S^1)$, where the branch of log is chosen so that $\log 1 = 0$. In order to verify this it suffices, to prove for $f(z) := [\log(1-z)]^{\frac{1}{2}-\varepsilon}$, that the integral

$$\int_{|z|<1} |f'(z)|^2 \, dx \, dy$$

is finite. We would like to add that a function $u \in H^{1/2}$ not only belongs to L^p for every $1 \leq p < \infty$ but even $e^{\alpha u^2}$ is in L^1 for every $\alpha \in \mathbb{R}$, and there is a constant $C > 0$ such that

$$\int_0^{2\pi} e^{[u(e^{i\vartheta})]^2} \, d\vartheta \leq C \quad \text{for every} \quad u \in H^{1/2}(S^1) \text{ satisfying}$$

$$\|u\|_{H^{1/2}(S^1)} \leq 1 \quad \text{and} \quad \int_0^{2\pi} u(e^{i\vartheta}) d\vartheta = 0 \, .$$

This result, due to D.E. Marshall, sharpens an estimate of A. Beurling and we refer to S.Y. Chang [40] and more recently W. Beckner [17].

Next we consider a smooth Hamiltonian $H \in C^\infty(\mathbb{R} \times \mathbb{R}^{2n}, \mathbb{R})$ which is periodic in time, $H(t,x) = H(t+1,x)$ for all $(t,x) \in \mathbb{R} \times \mathbb{R}^{2n}$. We assume that all the derivatives of H have polynomial growth.

Theorem 4. The action $a : H^{1/2}(S^1) \to \mathbb{R}$, defined by

$$a(x) := \int_0^1 H\big(t, x(t)\big) dt \, ,$$

belongs to $C^\infty(H^{1/2}(S^1), \mathbb{R})$.

Proof. Fixing an integer k we start with the Taylor formula for the function $x \mapsto H(t,x)$:

$$H(t, y + \eta) = \sum_{j=0}^k \frac{1}{j!} h_j(t, y) \cdot \eta^j + o_k(\eta) \, ,$$

where

$$h_j(t,y) \cdot \eta^j := \sum_{|\alpha|=j} \frac{j!}{\alpha!} D_x^\alpha H(t,y) \eta^\alpha$$

and $y, \eta \in \mathbb{R}^{2n}$. If $x, \xi \in H^{1/2}$ we insert $y = x(t)$ and $\eta = \xi(t)$ and integrating over t we find the Taylor formula for a in the form

$$a(x + \xi) = \sum_{j=0}^k \frac{1}{j!} a_j(x) \cdot \xi^j + r_k(x, \xi) \, ,$$

with
$$a_j(x) \cdot \xi^j := \int_0^1 h_j\left(t, x(t)\right) \cdot \xi(t)^j \, dt .$$

Now $a_j(x)$ is a continuous j-linear form on $H^{1/2}$ which, in addition, depends continuously on $x \in H^{1/2}$. This follows immediately from Hölder's inequality applied to the above integral, using Theorem 3 and the assumption that the derivatives of H are polynomially bounded. Moreover $|r_k(x,\xi)| \leq C_k \cdot |\xi|_{1/2}^{k+1}$. Since this holds true for every k we conclude by the converse to Taylor's theorem (see e.g R. Abraham and J. Robbin [2]) that a is smooth and, as formally expected, has the derivatives
$$D^j a(x) = a_j(x)$$
for $j \geq 1$. The proof of Theorem 4 is finished. ∎

A.4 The Cauchy-Riemann operator on the sphere

The Cauchy-Riemann operator is the linear operator defined by
$$\bar{\partial}: C^\infty(\mathbb{C}, \mathbb{C}^n) \to C^\infty(\mathbb{C}, \mathbb{C}^n) : u \to u_s + i u_t.$$

We drop for convenience the factor $\frac{1}{2}$ in the usual definition of $\bar{\partial}$. Here $s + it$ are the coordinates on $\mathbb{C}$. Our first goal is to solve the equation
$$\bar{\partial} u = g$$
for given $g \in \mathcal{D}$, where $\mathcal{D}$ denotes the vector space of compactly supported $\mathbb{C}^n$-valued smooth functions on $\mathbb{C}$. If $\varepsilon > 0$ and $u \in \mathcal{D}$ we define the function $A_\varepsilon u : \mathbb{C} \to \mathbb{C}^n$ using the Green's function of $\bar{\partial}$ by
$$(A_\varepsilon u)(z) = -\frac{1}{2\pi} \int_{|\xi| \geq \varepsilon} \frac{1}{\xi} u(z + \xi) \, ds \, dt ,$$
where $\xi = s + it$, $s, t \in \mathbb{R}$. Since u has compact support there exists a constant $M > 0$ depending on u, such that for all $z_1, z_2 \in \mathbb{C}$
$$|u(z_1) - u(z_2)| \leq M |z_1 - z_2|.$$
Using polar coordinates (r, θ) we can estimate
$$\begin{aligned}
|(A_\varepsilon u)(z_2) - (A_\varepsilon u)(z_1)| &\leq \frac{1}{2\pi} \int_{|\xi| \geq \varepsilon} \left| \frac{u(z_2+\xi) - u(z_1+\xi)}{\xi} \right| ds \, dt \\
&\leq \frac{1}{2\pi} \int_\varepsilon^R \int_0^{2\pi} M |z_1 - z_2| \, dr \, d\theta \\
&\leq MR |z_2 - z_1| ,
\end{aligned}$$

where $R >$ diameter ($\operatorname{supp} u$) $+ \max\{|z_1|, |z_2|\}$. The estimate implies that $A_\varepsilon u \in C^0(\mathbb{C}, \mathbb{C}^n)$. Similarly one finds for $\varepsilon_1, \varepsilon_2 > 0$

$$|(A_{\varepsilon_1} u)(z) - (A_{\varepsilon_2} u)(z)| \leq \|u\|_0 \, |\varepsilon_2 - \varepsilon_1| \, .$$

Here $\|.\|_0$ denotes the sup-norm. By the latter estimate the limit of $A_\varepsilon u$ for $\varepsilon \to 0$ exists uniformly and defines a continuous function $Au : \mathbb{C} \to \mathbb{C}^n$.

If $e \in \mathbb{C}$ is a vector of unit length, $h \in \mathbb{R} \setminus \{0\}$ and $u \in \mathcal{D}$ we denote by $d_h u$ the difference quotient associated to h and e:

$$(d_h u)(z) = \frac{u(z + he) - u(z)}{h} \, .$$

It follows immediately from the definition of A that

$$A d_h = d_h A.$$

From the definition we deduce that for every compact $K \subset \mathbb{R}^{2n}$ there is a constant $M = M_K$ such that $|A_\varepsilon u(z)| \leq M \|u\|_0$ for all $u \in \mathcal{D}$ having $\operatorname{supp}(u) \subset K$. Therefore, also $|Au(z)| \leq M \|u\|_0$. Consequently, denoting by D_e the directional derivative associated to the vector e, we find for fixed z, that $|(d_h Au)(z) - (AD_e u)(z)| = |A(d_h u - D_e u)(z)| \leq M \|d_h u - D_e u\|_0 \to 0$ as $h \to 0$, where M depends only on $\operatorname{supp}(u)$. Hence $AD_e = D_e A$ and repeating the arguments

$$D^\alpha Au = AD^\alpha u \, .$$

We see that A defines a linear map from $\mathcal{D}$ into C^∞, satisfying, in addition,

$$A \circ \bar{\partial} = \bar{\partial} \circ A \, .$$

In fact, even more is true and we claim that

$$\bar{\partial} \circ A = \operatorname{Id} \text{ on } \mathcal{D} \, .$$

In order to prove this we define $v(s,t) = u(z + \xi)$ for a fixed $z \in \mathbb{C}$ and $\xi = s + it$. We compute for $\xi \neq 0$

$$\begin{aligned}
\tfrac{1}{\xi}(\bar{\partial} v) ds \wedge dt &= \tfrac{1}{\xi} \left[\tfrac{\partial v}{\partial s} + i \tfrac{\partial v}{\partial t}\right] ds \wedge dt \\
&= \tfrac{1}{\xi} \left[\left(\tfrac{\partial v}{\partial s}\right) ds + \left(\tfrac{\partial v}{\partial t}\right) dt\right] \wedge (dt - ids) \\
&= \tfrac{-i}{\xi} (dv) \wedge (ds + idt) \\
&= d\left(v \tfrac{1}{\xi}(ds + idt)\right)(-i) \, ,
\end{aligned}$$

A.4 THE CAUCHY-RIEMANN OPERATOR ON THE SPHERE

where we have used that $d(\frac{1}{\xi}) = -\frac{1}{\xi^2}(ds + idt)$. Because u has compact support we infer by Stokes' theorem for large R

$$\begin{aligned}
A_\varepsilon \bar\partial u(z) &= \tfrac{-1}{2\pi} \int_{\varepsilon \le |\xi| \le R} \tfrac{1}{\xi}(\bar\partial u)(z+\xi)\, ds\, dt \\
&= \tfrac{i}{2\pi} \int_{\varepsilon \le |\xi| \le R} d\Big(u(z+\xi)(\tfrac{1}{\xi}(ds+idt))\Big) \\
&= \tfrac{1}{2\pi i} \int_{|\xi|=\varepsilon} u(z+\xi)\tfrac{1}{\xi}(ds+idt) \\
&= \tfrac{1}{2\pi i} \int_0^{2\pi} u(z+\varepsilon e^{i\theta})\tfrac{1}{\varepsilon} e^{-i\theta}(i\varepsilon)e^{i\theta}\, d\theta \\
&= \tfrac{1}{2\pi} \int_0^{2\pi} u(z+\varepsilon e^{i\theta})\, d\theta \ .
\end{aligned}$$

The last expression tends to $u(z)$ as $\varepsilon \to 0$. Thus we have proved $A\bar\partial = \mathrm{Id}$ and hence $\bar\partial \circ A = \mathrm{Id}$ as claimed. Next we study the asymptotic behaviour of $Au(z)$ as $z \to \infty$.

Proposition 5.
$$(Au)(z) \to 0 \text{ as } |z| \to \infty \text{ for } u \in \mathcal{D}\ .$$

Proof. We choose $R > 0$ be so large that $\mathrm{supp}(u) \subset B_R(0)$ and first observe that

$$\begin{aligned}
(Au)(z) &= -\lim_{\varepsilon \to 0} \tfrac{1}{2\pi} \int_{|\xi| \ge \varepsilon} u(z+\xi)\tfrac{1}{\xi}\, ds\, dt \\
&= -\lim_{\varepsilon \to 0} \tfrac{1}{2\pi} \int_\varepsilon^\infty \int_0^{2\pi} u(z+re^{i\theta})e^{-i\theta}\, dr\, d\theta \\
&= -\tfrac{1}{2\pi} \int_0^\infty \int_0^{2\pi} u(z+re^{i\theta})e^{-i\theta}\, dr\, d\theta \ .
\end{aligned}$$

For $|z| > R$ consider the function $v(r,\theta) := u(z+re^{i\theta})e^{-i\theta}$. Pick (r_0, θ_0) in such a way that $z = -r_0 e^{i\theta_0}$. Then the support of v lies in a cone with vertex at 0 having an angle σ so that

$$\frac{R}{r_0} = \sin\left(\frac{\sigma}{2}\right).$$

Moreover if (r, θ) is in this sector then $v(r, \theta) = 0$ if

$$r \in [0, r_0 - R] \cup [r_0 + R, +\infty)\ .$$

This is illustrated by the following Fig. A4.1.

As $r_0 = |z| \to \infty$ we see that $\sigma \to 0$. More precisely for $|z|$ large

$$0 \le \sigma \le \frac{4R}{|z|}\ .$$

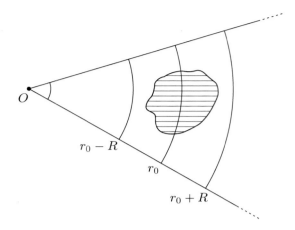

Fig. A4.1

This implies for $|z|$ large enough

$$|(Au)(z)| \leq \tfrac{1}{2\pi} \int_{\text{supp}(v)} |v(r,\theta)| dr\, d\theta$$

$$\leq \tfrac{1}{2\pi} \int_{|z|-R}^{|z|+R} \left(\int_\Gamma ||u||_\infty d\theta \right) dr ,$$

where Γ is an open set with measure $(\Gamma) \leq \sigma$. Hence

$$|(Au)(z)| \leq \tfrac{1}{2\pi} 2R||u||_\infty \sigma$$

$$\leq \tfrac{4R^2}{\pi} ||u||_\infty \tfrac{1}{|z|}$$

$$\leq \text{const}\, (u) \tfrac{1}{|z|}$$

for $|z|$ large enough. ∎

We shall use the proposition to extend the map A to the sphere and study next the Cauchy-Riemann operator on the Riemann sphere $S^2 = \mathbb{C} \cup \{\infty\}$. We equip S^2 with its structure of a complex manifold by defining two charts as follows:

(A.40)
$$U = \mathbb{C} \xrightarrow{\alpha} \mathbb{C} : z \longrightarrow z$$
$$V = (\mathbb{C}\setminus\{0\}) \cup \{+\infty\} \xrightarrow{\beta} \mathbb{C} : z \longrightarrow \begin{cases} \tfrac{1}{z} & \text{if } z \neq \infty \\ 0 & \text{if } z = \infty. \end{cases}$$

Clearly α and β are homeomorphisms and $\alpha \circ \beta^{-1} : \beta(U \cap V) \to \alpha(U \cap V)$ is the map $z \to \tfrac{1}{z}$ and hence holomorphic. We have a natural complex structure on S^2

A.4 THE CAUCHY-RIEMANN OPERATOR ON THE SPHERE

denoted by i. Consider now the vector bundles

$$S^2 \times \mathbb{C}^n \longrightarrow S^2$$
$$X \longrightarrow S^2$$

where the fibre X_z over $z \in S^2$ consists of all complex anti-linear maps $\phi : T_z S^2 \to \mathbb{C}^n$. In order to define the Cauchy-Riemann operator $\bar{\partial}$ on the C^∞-sections of these bundles, we denote by $C^\infty(S^2 \times \mathbb{C}^n)$ the space of smooth sections of $S^2 \times \mathbb{C}^n \to S^2$, i.e., $C^\infty(S^2 \times \mathbb{C}^n) = C^\infty(S^2, \mathbb{C}^n)$, and with $C^\infty(X)$ the space of smooth sections of $X \to S^2$. We define $\bar{\partial} : C^\infty(S^2 \times \mathbb{C}^n) \to C^\infty(X)$ by

$$(A.41) \qquad (\bar{\partial} f)(z) = Tu(z) + i\, Tu(z) i \,,$$

where $Tu(z) : T_z S^2 \to \mathbb{C}^n, z \in S^2$ is the tangent map of u. Note that

$$(\bar{\partial} u)(z) \circ i = -i \circ (\bar{\partial} u)(z) \,,$$

so that $\bar{\partial}$ is indeed well-defined.

We now assume that $g \in C^\infty(X)$. Using a partition of unity subordinate to our distinguished atlas (A.40), we have the splitting

$$g = g_U + g_V \,,$$

where $\operatorname{supp}(g_U) \subset U$, $\operatorname{supp}(g_V) \subset V$. The map $g_U(z) : \mathbb{C} \to \mathbb{C}^n$ is complex anti-linear for every $z \in U = \mathbb{C}$, and has compact support with respect to z. Hence

$$z \to g_U(z)(1)$$

belongs to $\mathcal{D}$. We can solve

$$u_s + i\, u_t = g_U(\cdot)(1) =: \hat{g}_U \in C^\infty(\mathbb{C}, \mathbb{C}^n)$$

for $u \in C^\infty(\mathbb{C}, \mathbb{C}^n)$ by

$$u = A \hat{g}_U \,.$$

We observe that

$$\begin{aligned}(\bar{\partial} u)(z)(1) &= \big[(Tu)(z) + i\, Tu(z) i\big](1) \\ &= u_s(z) + i\, u_t(z) \\ &= g_U(z)(1) \,.\end{aligned}$$

Because $\bar{\partial} u(z)$ is anti-linear we have $\bar{\partial} u = g_U$ over $U = \mathbb{C}$. By Proposition 5, we have $u(z) \to 0$ as $|z| \to \infty$ and, moreover, $\bar{\partial} u(z) = 0$ outside the support of $\hat{g}_U$ so that u is holomorphic for $|z|$ large. Consider the map on $V \setminus \{\infty\}$,

$$\mathbb{C}\setminus\{0\} \longrightarrow \mathbb{C}^n \ : \ z \longrightarrow u\left(\frac{1}{z}\right) = u \circ \beta^{-1}(z) \,.$$

This map is holomorphic in a punctured neighborhood of 0 and $u(\frac{1}{z}) \to 0$ for $z \to 0$. Hence 0 is a removable singularity by the well-known classical result in the theory of one complex variable. Therefore, u may be viewed as a smooth function on S^2 which vanishes at ∞. From Proposition 5 we conclude in particular

Proposition 6. The map A defines a linear operator from $\mathcal{D}$ into the subspace $\{f \in C^\infty(S^2, \mathbb{C}^n) | f(\infty) = 0\}$ of $C^\infty(S^2, \mathbb{C}^n)$, satisfying

$$\bar{\partial} \circ A = Id.$$

Back to our previous discussion, we have found $u \in C^\infty(S^2 \times \mathbb{C}^n)$ solving the equation $\bar{\partial} u = g_U$ and since $S^2 \setminus \{0\}$ is biholomorphic to $\mathbb{C}$, we find by the same reasoning a $g_v \in C^\infty(S^2 \times \mathbb{C}^n)$ satisfying $\bar{\partial} v = g_V$. Therefore, it follows that $\bar{\partial}(u+v) = g_U + g_V = g$ proving that $\bar{\partial}$ is surjective. On the other hand, if $\bar{\partial} u = 0$ we obtain a bounded holomorphic function $u : \mathbb{C} \to \mathbb{C}^n$ which by Liouville's theorem is a constant map. Summing up we have proved

Theorem 5. The map $\bar{\partial} : C^\infty(S^2 \times \mathbb{C}^n) \to C^\infty(X)$ is a surjective linear operator whose kernel consists of all constant maps. As a real (respectively complex) linear operator the kernel is of dimension $2n$ (respectively n).

Next we would like to study $\bar{\partial}$ as an operator in a Sobolev space setting based on L^p-spaces $1 < p < \infty$. We start with the following estimate from Douglis and Nirenberg, [63]. Denote by $\mathcal{D}_1$ the subspace of $\mathcal{D}$ consisting of all functions in $\mathcal{D}$ with support in the unit ball. We write $|\cdot|_{k,p}$ for the norm defined by

$$|u|_{k,p}^p = \sum_{|\alpha| \leq k} |D^\alpha u|_{L^p(B_1(0))}^p.$$

Here $k \in \mathbb{N}$ and $1 < p < \infty$. The fundamental estimate given in [63] is the following

Theorem 6. For given $k \in \mathbb{N}$ and $1 < p < \infty$ there exists a positive constant $c = c(k, p)$ satisfying

$$|\bar{\partial} u|_{k,p} \geq c |u|_{k+1,p}$$

for all $u \in \mathcal{D}_1$.

In order to obtain some Sobolev type estimates for the section map $\bar{\partial} : C^\infty(S^2 \times \mathbb{C}^n) \to C^\infty(X)$ we take a finite covering $U_1, \ldots, U_m$ of S^2 so that U_j is biholomorphic to the open unit disc B around 0. Let $\beta_1, \ldots, \beta_n$ be a subordinate partition of unity. Denoting by $\varphi_j : U_j \to B$ suitable biholomorphic diffeomorphisms we define a norm $\| \|_{k,p}$ for $k \in \mathbb{N}, 1 < p < \infty$, by

$$\|u\|_{k,p}^p = \sum_{j=1}^m \left|(\beta_j \circ \varphi_j^{-1})(u \circ \varphi_j^{-1})\right|_{k,p}^p.$$

It is easy to show that all the norms on $C^\infty(S^2 \times \mathbb{C}^n)$ obtained by this procedure are equivalent. For the bundle $X \to S^2$ we take smooth complex linear trivialisations

A.4 THE CAUCHY-RIEMANN OPERATOR ON THE SPHERE

$\psi_j : X|U_j \to B \times \mathbb{C}^n$ covering φ_j. This means that the following diagramm is commutative

$$\begin{array}{ccc} X|U_j & \xrightarrow{\psi_j} & B \times \mathbb{C}^n \\ \downarrow & & \downarrow {pr_1} \\ U_j & \xrightarrow{\varphi_j} & B \end{array}$$

If f is a section of X we define the norm $||f||_{k,p}$ by

$$||f||_{k,p}^p = \sum_{j=1}^m \left| \psi_j(\varphi_j^{-1}) \left((\beta_j \circ \varphi_j^{-1})(f \circ \varphi_j^{-1}) \right) \right|_{k,p}^p.$$

Again all norms obtained this way are equivalent. Let us abbreviate $\hat{\psi}_j = \psi_j(\varphi_j^{-1})$ and $\hat{\beta}_j = \beta_j \circ \varphi_j^{-1}$. Then

$$\hat{\psi}_j(z) \left(\hat{\beta}_j(z) (\bar{\partial} u)(\varphi_j^{-1}(z)) \right)$$
$$= \hat{\beta}_j(z) \hat{\psi}_j(z) \left((Tu)(\varphi_j^{-1}(z)) + i(Tu)(\varphi_j^{-1}(z))i \right)$$
$$= \hat{\beta}_j(z) \hat{\psi}_j(z) \left[\left(T(u \circ \varphi_j^{-1})(z) + i(T(u \circ \varphi_j^{-1})(z)i \right) T\varphi_j(z) \right]$$
$$= \hat{\beta}_j(z) \hat{\psi}_j(z) \left[\left(T\hat{u}_j(z) + iT\hat{u}_j(z))i \right) T\varphi_j(z) \right],$$

where $\hat{u}_j = u_j \circ \varphi_j^{-1}$. This gives immediately the estimate

$$\left| \hat{\psi}_j \left(\hat{\beta}_j(\bar{\partial} u)(\varphi_j^{-1}) \right) \right|_{0,p}^p \geq d_j^p \left| \hat{\beta}_j \left(\frac{\partial}{\partial s} + i\frac{\partial}{\partial t} \right) \hat{u}_j \right|_{0,p}^p,$$

for a suitable constant d_j independent of u. If $d = \inf\{d_1, \ldots, d_m\}$, then

$$||\bar{\partial} u||_{0,p} \geq d \left(\sum_{j=1}^m \left| \hat{\beta}_j \left(\frac{\partial}{\partial s} + i\frac{\partial}{\partial t} \right) \hat{u}_j \right|_{0,p}^p \right)^{1/p}$$
$$= d \left(\sum_{j=1}^m \left| \left(\frac{\partial}{\partial s} + i\frac{\partial}{\partial t} \right)(\hat{\beta}_j \hat{u}_j) - \left(\left(\frac{\partial}{\partial s} + i\frac{\partial}{\partial t} \right)\hat{\beta}_j \right) \hat{u}_j \right|_{0,p}^p \right)^{1/p}$$
$$\geq d \left(\sum_{j=1}^m \left| \left(\frac{\partial}{\partial s} + i\frac{\partial}{\partial t} \right)(\hat{\beta}_j \hat{u}_j) \right|_{0,p}^p \right)^{1/p} - a \left(\sum_{j=1}^m |\chi_j \hat{u}_j|_{0,p}^p \right)^{1/p}$$

for a suitable constant $a > 0$ independent on the choice of u. Here χ_j is the characteristic function of the support of $\hat{\beta}_j$. Hence employing Theorem 6

$$||\bar{\partial} u||_{0,p} \geq d \left(\sum_{j=1}^m c^p |\hat{\beta}_j \hat{u}_j|_{1,p}^p \right)^{1/p} - a \left(\sum_{j=1}^m |\chi_j \hat{u}_j|_{0,p}^p \right)^{1/p}$$
$$\geq c' ||u||_{1,p} - a' ||u||_{0,p}$$

for suitable constants $a', c' > 0$, independent of u. We have proved

Theorem 7. There exist constants $a, c > 0$ such that for all $u \in C^\infty(S^2 \times \mathbb{C}^n)$

$$||\bar{\partial} u||_{0,p} \geq c||u||_{1,p} - a||u||_{0,p} \, .$$

Define $W^{1,p}(S^2 \times \mathbb{C}^n)$ and $L^p(X)$ as the completions of the spaces $C^\infty(S^2 \times \mathbb{C}^n)$ and $C^\infty(X)$ with respect to the norms $||\,||_{1,p}$ and $||\,||_{0,p}$. By the compact embedding of $W^{1,p} \hookrightarrow L^p$ we find in view of Theorem 7 that $\bar{\partial} : W^{1,p}(S^2 \times \mathbb{C}^n) \to L^p(X)$ has closed range. Since in view of Theorem 5 the range contains the dense set $C^\infty(X)$ we deduce $\bar{\partial}(W^{1,p}(S^2 \times \mathbb{C}^n)) = L^p(X)$ and hence $\bar{\partial}$ is surjective.

On the other hand, by interior elliptic regularity theory or by the well-known fact that a holomorphic distribution is a holomorphic function, we conclude from $\bar{\partial} u = 0$ that u is holomorphic and bounded and consequently constant.

We have proved

Theorem 8. The operator $\bar{\partial} : W^{1,p}(S^2 \times \mathbb{C}^n) \to L^p(X)$ is a surjective Fredholm operator whose kernel consists of the constant functions. Hence the index of $\bar{\partial}$ as a real operator is $2n$.

With some more efforts and rudimentary functional analysis one can prove

Theorem 9. For every $k \in \mathbb{N}$ and $1 < p < \infty$ there are positive constants $c = c(k,p)$ and $a = a(k,p)$ such that

$$||\bar{\partial} u||_{k,p} \geq c||u||_{k+1,p} - a||u||_{0,p} \, ,$$

for all $u \in C^\infty(S^2 \times \mathbb{C}^n)$.

As an operator from $W^{k+1,p}(S^2 \times \mathbb{C}^n)$ to $W^{k,p}(X)$ the linear operator $\bar{\partial}$ is again a surjective Fredholm operator whose kernel consists of the constant functions. For a general treatment of first order elliptic systems, we refer the reader to Vekua's book, [215], or its English translation.

A.5 Elliptic estimates near the boundary and an application

Denote by B_ε^+ the half ball given by

$$B_\varepsilon^+ = \{z \in \mathbb{C} \,|\, |z| < \varepsilon, Re(z) \geq 0\}$$

with boundary $\partial B_\varepsilon^+ = (-\varepsilon, \varepsilon)i$.

Let $||\cdot||_{k,p,\Omega}$ for $k \in \mathbb{N}$, $1 < p < \infty$ denote the usual Sobolev norms, i.e.,

$$||u||_{k,p,\Omega}^p = \sum_{|\alpha| \leq k} ||D^\alpha u||_{L^p(\Omega)}^p \, .$$

The classical estimate, which for example can be found in Vekua [215], for $\bar{\partial} = \frac{\partial}{\partial s} + i\frac{\partial}{\partial t}$ is

A.5 Elliptic estimates near the boundary and an application

Theorem 10. Given $\varepsilon > 0, k \in \mathbb{N}, 1 < p < \infty$, there exists a constant $c = c(k, p, \varepsilon) > 0$ such

$$\|\bar{\partial}u\|_{k,p,B_\varepsilon^+} \geq c\|u\|_{k+1,p,B_\varepsilon^+}$$

for all smooth $u : B_\varepsilon^+ \to \mathbb{C}^n$ with compact support in B_ε^+ satisfying the boundary condition

$$u(\partial B_\varepsilon^+) \subset \mathbb{R}^n .$$

Using this estimate and the usual differential quotient technique from standard elliptic regularity theory we obtain the following regularity result.

Let J be a smooth almost complex structure defined on $\mathbb{C}^n$ (or at least near 0) such that $J(0) = i$. Assume

$$u : B_\varepsilon^+ \to \mathbb{C}^n$$

is of class $W^{1,p}$ with $p > 2$ and satisfies

$$u(0) = 0$$

$$u\Big((-\varepsilon, \varepsilon)i\Big) \subset \mathbb{R}^n$$

$$\frac{\partial u}{\partial s} + J(u)\frac{\partial u}{\partial t} = 0 \text{ on } \mathring{B}_\varepsilon^+$$

where $\mathring{B}_\varepsilon^+ = B_\varepsilon^+ \setminus (-\varepsilon, \varepsilon)i$. We have the following

Theorem 11. Under the above assumptions there exists $\delta \in (0, \varepsilon)$ such that $u|B_\delta^+$ is smooth.

Our next aim is to give an application to a regularity question raised in Chapter 6. Let

$$w : [-\varepsilon, \varepsilon] \times [-\varepsilon, \varepsilon] \longrightarrow \mathbb{C}^n$$

be of class $W^{1,p}$, $2 < p < \infty$. Assume $w(0) = 0$, and J is an almost complex structure defined on a sufficiently large neighborhood of zero such that $J(0) = i$. Assume $H : [-\varepsilon, \varepsilon] \times U \to \mathbb{C}^n$ is a smooth map, where U is a sufficiently large neighborhood of 0 in $\mathbb{C}^n$. Moreover, let $H'(0,0) = 0$, where H' is the gradient with respect to the second variable for some Riemannian metric on U.

We consider the partial differential equation

$$(A.42) \qquad \frac{\partial w}{\partial s} + J(w)\frac{\partial w}{\partial t} + \chi_{[0,\infty)}(s)H'(t,w) = 0$$

on $(-\varepsilon, \varepsilon) \times (-\varepsilon, \varepsilon)$. Here $\chi_{[0,\infty)}$ is the characteristic function of $[0, \infty)$. The amazing fact is that

$$w\big|\Big([-\delta, 0] \times [-\delta, \delta]\Big)$$

and

$$w\big|\Big([0, \delta] \times [-\delta, \delta]\Big)$$

are smooth for some sufficiently small $\delta \in (0, \varepsilon)$. In particular,
$$w \mid \Big(\{0\} \times [-\delta, \delta] \Big) \longrightarrow \mathbb{C}^n$$
is smooth. Note that $w|([-\delta, \delta] \times [-\delta, \delta])$ is not smooth since the derivatives in the s-direction do not match. The above proof of the smoothness assertion will be reduced by a series of transformations to Theorem 11.

For simplicity of the argument we assume that all data H and J are defined on the whole of $\mathbb{C}^n$. Define an almost complex structure $\hat{J}$ on $[0, \varepsilon] \times [-\varepsilon, \varepsilon] \times \mathbb{C}^n \times \mathbb{C}^n$ by
$$\hat{J}(s, t, u, v)(a, b, h, k) = \Big(-b, a, J(u)h + bH'(t, u) + aJ(u)H'(t, u), -J(v)k \Big),$$

The definition of $\hat{J}$ is motivated by Gromov's trick and something that one could call the "inverse reflection principle". It will turn out that a solution w of (A.42) gives a holomorphic map α on $[0, \varepsilon) \times [-\varepsilon, \varepsilon)$ for the structure $\hat{J}$.

We leave it to the reader to verify that $\hat{J}^2 = -Id$. Let
$$j = \hat{J}(0, 0, 0, 0) .$$

We observe that
$$j(a, b, h, k) = \Big(-b, a, ih + bH'(0, 0) + aiH'(0, 0), -ik \Big) .$$

Define a linear real subspace of $\mathbb{R} \times \mathbb{R} \times \mathbb{C}^n \times \mathbb{C}^n$ by
$$T = \{(0, b, h, h,) \mid b \in \mathbb{R}, \ h \in \mathbb{C}^n\} .$$

Clearly $\dim_{\mathbb{R}}(T) = 2n + 1 =: N$. We note that
(A.43)
$$T \cap jT = \{0\} .$$

Indeed, if
$$(0, b_1, h_1, h_1) = j(0, b_2, h_2, h_2)$$
$$= \Big(-b_2, 0, ih_2 + b_2 H'(0, 0), -ih_2 \Big),$$
we find $b_1 = b_2 = 0$ implying
$$ih_2 = h_1$$
$$-ih_2 = h_1,$$
and, therefore, $h_1 = h_2 = 0$. Define a map $\alpha : [0, \varepsilon] \times [-\varepsilon, \varepsilon] \to \mathbb{R} \times \mathbb{R} \times \mathbb{C}^n \times \mathbb{C}^n$ by
$$\alpha(s, t) = \Big(s, t, u(s, t), u(-s, t) \Big) .$$

A.5 Elliptic estimates near the boundary and an application

Then
$$\alpha(0,t) = \Big(0, t, u(0,t), u(0,t)\Big) \in T$$
for $t \in [-\varepsilon, \varepsilon]$. Moreover α is in $W^{1,p}([0,\varepsilon] \times [-\varepsilon, \varepsilon], \mathbb{R} \times \mathbb{R} \times \mathbb{C}^n \times \mathbb{C}^n)$, since u, by assumption, belongs to this space. On $(0, \varepsilon) \times (-\varepsilon, \varepsilon)$ we compute

$$\alpha_s(s,t) + \hat{J}\Big(\alpha(s,t)\Big)\Big(\alpha_t(s,t)\Big)$$
$$= \Big(1, 0, u_s(s,t), -u_s(-s,t)\Big)$$
$$+ \Big(-1, 0, J(u(s,t))u_t(s,t) + H'(t, u(s,t)), -J(u(-s,t))u_t(-s,t)\Big)$$
$$= (0,0,0,0).$$

In view of (A.43) the set T is a totally real subspace of $(\mathbb{R} \times \mathbb{R} \times \mathbb{C}^n \times \mathbb{C}^n, j)$. It is a simple exercise to find a complex linear map

$$\Phi : (\mathbb{R} \times \mathbb{R} \times \mathbb{C}^n \times \mathbb{C}^n, j) \longrightarrow (\mathbb{C}^N, i)$$

mapping T onto $\mathbb{R}^N$. Recall that $N = 2n + 1$. Define $\beta = \Phi \circ \alpha$ and an almost complex structure $\tilde{J}$ on $\mathbb{C}^N$ by

$$\tilde{J}(A) = \Phi \hat{J}\Big(\Phi^{-1}(A)\Big)\Phi^{-1}.$$

We find that
$$\tilde{J}(0) = \Phi j \Phi^{-1}$$
$$= i\Phi\Phi^{-1}$$
$$= i.$$

Moreover, on $(0, \varepsilon) \times (-\varepsilon, \varepsilon)$,

$$\beta_s + \tilde{J}(\beta)\beta_t$$
$$= \Phi\alpha_s + \Phi\hat{J}(\Phi^{-1}\Phi\alpha)\Phi^{-1}\Phi\alpha_t$$
$$= \Phi[\alpha_s + \hat{J}(\alpha)\alpha_t]$$
$$= 0.$$

Finally, we observe that
$$\beta(0,t) \in \mathbb{R}^N$$
for $t \in (-\varepsilon, \varepsilon)$. Applying Theorem 11 we deduce, since $\beta \in W^{1,p}([0,\varepsilon] \times [-\varepsilon, \varepsilon], \mathbb{C}^N)$ and $\beta(i(-\varepsilon, \varepsilon)) \subset \mathbb{R}^N$, that β is smooth on $[0, \delta] \times [-\delta, \delta]$ for some $\delta \in (0, \varepsilon)$. Hence

$$[0, \delta] \times [-\delta, \delta] \to \mathbb{C}^n : (s, t) \to u(s, t)$$

and
$$[0,\delta] \times [-\delta,\delta] \to \mathbb{C}^n : (s,t) \to u(-s,t)$$
are smooth, as we wanted to prove.

We conclude with an observation used in Chapter 6. Recall the definition of S_T in Chapter 6 and assume $w : S_T \to M$ solves

(A.44) $\qquad 0 = \bar{\partial}w + \chi_{[-T,T] \times S^1}\left(H'(t,w)dt - J(u)H'(t,w)ds\right).$

Taking suitable complex coordinates near a point (T_0, t_0), respectively, $(-T_0, t_0)$ in $Z_T \subset S_T$ and suitable symplectic coordinates near $w(T_0, t_0)$, respectively, $w(-T_0, t_0)$ in M, we see that (A.44) looks, in these coordinates, like (A.42). We deduce, in particular, since T_0 was arbitrary, that the loops $t \mapsto w(-T, t)$ and $t \mapsto w(T, t)$ are smooth.

A.6 The generalized similarity principle

Let D be the closed unit disk $D = \{z \in \mathbb{C} | \, |z| \leq 1\}$. We denote by $A \in L^\infty(D, \mathcal{L}_\mathbb{R}(\mathbb{C}^n))$ a mapping which associates with $z \in D$ a real linear map $A(z) : \mathbb{C}^n \to \mathbb{C}^n$, so that the assignment $z \mapsto A(z)$ is essentially bounded. Let $z \mapsto J(z)$ be a smooth map associating with $z \in D$ a complex structure on $\mathbb{C}^n$, i.e $J(z)^2 = -1$, and $J(z) : \mathbb{C}^n \to \mathbb{C}^n$ is a real linear map. We consider a solution $w : D \to \mathbb{C}^n$ of the first order elliptic system

(A.45) $\qquad \dfrac{\partial w}{\partial s} + J(z)\dfrac{\partial w}{\partial t} + A(z)w = 0$ and $w(0) = 0$,

where $z = s + it$. We shall show that w is represented locally near 0 by a holomorphic map. Define $D_\delta = \{z \in D | \, |z| \leq \delta\}$. Generalizing a result of T. Carleman [37] to systems we shall prove

Theorem 12. There exist $0 < \delta < 1$, a holomorphic map $\sigma : D_\delta \to \mathbb{C}^n$ and a continuous map $\Phi : D_\delta \to \mathcal{Gl}_\mathbb{R}(\mathbb{C}^n)$,

$$\Phi \in \bigcap_{2<p<\infty} W^{1,p}\left(D_\delta, \mathcal{Gl}_\mathbb{R}(\mathbb{C}^n)\right),$$

satisfying on D_δ,
$$J(z)\Phi(z) = \Phi(z)i$$
and
$$w(z) = \Phi(z)\sigma(z).$$

It follows, in particular, that if 0 is a cluster point of zeros of the solution w, then it is also a cluster point of zeros of the holomorphic map σ, implying $\sigma \equiv 0$ and therefore $w \equiv 0$ on D_δ. This was used in the proof of the Arnold conjecture (Lemma 2 in 6.4).

A.6 THE GENERALIZED SIMILARITY PRINCIPLE

Proof. Reducing the local statement to a global one, the theorem will be obtained as a corollary of Theorem 8 in A.4. We shall view the disk D as a subset of $\mathbb{C} \subset \mathbb{C} \cup \{\infty\} = S^2$ and begin by rewriting equation (A.45). We choose a smooth map $z \mapsto \Psi(z)$ from D into $\mathcal{G}l_{\mathbb{R}}(\mathbb{C}^n)$ satisfying $J(z)\Psi(z) = \Psi(z)i$. Then we define $v : D \to \mathbb{C}^n$ by
$$w(z) = \Psi(z)v(z),$$
and compute:
$$\begin{aligned}
0 &= \tfrac{\partial w}{\partial s} + J(z)\tfrac{\partial w}{\partial t} + A(z)w \\
&= \Psi\left(\tfrac{\partial v}{\partial s} + i\tfrac{\partial v}{\partial t}\right) + A\Psi v + \left(\tfrac{\partial \Psi}{\partial s} + J\tfrac{\partial \Psi}{\partial t}\right)v \\
&= \Psi\left[\tfrac{\partial v}{\partial s} + i\tfrac{\partial v}{\partial t} + \Psi^{-1}A\Psi v + \Psi^{-1}\left(\tfrac{\partial \Psi}{\partial s} + J\tfrac{\partial \Psi}{\partial t}\right)v\right] \\
&= \Psi\left[\tfrac{\partial v}{\partial s} + i\tfrac{\partial v}{\partial t} + Bv\right].
\end{aligned}$$

We see that v solves on $\mathring{D}$ the equation

(A.46) $$\frac{\partial v}{\partial s} + i\frac{\partial v}{\partial t} + B(z)v = 0 \text{ and } v(0) = 0,$$

where $B \in L^\infty(D, \mathcal{L}_{\mathbb{R}}(\mathbb{C}^n))$. Now we represent $B(z)$ as follows:
$$B(z) = B_l(z) + B_a(z),$$
where $B_l(z)$ is complex linear and $B_a(z)$ complex anti-linear. We denote by $\Gamma : \mathbb{C}^n \to \mathbb{C}^n$ the anti-linear complex conjugation map $h \mapsto \bar{h}$, and choose $C \in L^\infty(D, \mathcal{L}_{\mathbb{C}}(\mathbb{C}^n))$ in such a way that
$$C(z)v(z) = \overline{v(z)}, \quad z \in D$$
where v is the given solutions of (A.46). Define $E \in L^\infty(D, \mathcal{L}_{\mathbb{C}}(\mathbb{C}^n))$ by
$$E(z) = B_l(z) + B_a(z) \circ \Gamma \circ C(z).$$
Then $E(z)$ is complex linear and
$$E(z)v(z) = B(z)v(z), \quad z \in D.$$
Therefore v satisfies on $\mathring{D}$,

(A.47) $$\frac{\partial v}{\partial s} + i\frac{\partial v}{\partial t} + E(z)v = 0 \text{ and } v(0) = 0,$$

where now $E \in L^\infty(D, \mathcal{L}_{\mathbb{C}}(\mathbb{C}^n))$. Observe that $\mathcal{L}_{\mathbb{C}}(\mathbb{C}^n) = \mathcal{L}(\mathbb{C}^n) \cong \mathbb{C}^{n^2}$. With the notations introduced in Appendix 4, we consider, for $1 < p < \infty$, the Cauchy-Riemann operator between sections,
$$\bar{\partial} : W^{1,p}\left(S^2 \times \mathcal{L}(\mathbb{C}^n)\right) \to L^p(X),$$

where the fibre $X_z, z \in S^2$ of the vector bundle $X \to S^2$ is the vector space consisting of all complex anti-linear maps $T_z S^2 \to \mathcal{L}(\mathbb{C}^n)$. By Theorem 8 in Appendix 4, the mapping

$$G : W^{1,p}\left(S^2 \times \mathcal{L}(\mathbb{C}^n)\right) \to L^p(X) \times \mathcal{L}(\mathbb{C}^n)$$

$$G(Z) := \left(\bar{\partial} Z, Z(0)\right),$$

is a linear isomorphism if $2 < p < \infty$. Note that in the case $2 < p < \infty$ we can estimate the C^0-norm in terms of the $W^{1,p}$-norms on 2-dimensional domains (Sobolev embedding theorem). If $G(Z) = (0, Id)$ we infer that $Z(z) = Id$ for $z \in S^2$. Hence if a linear map $\triangle : W^{1,p}(S^2 \times \mathcal{L}(\mathbb{C}^n)) \to L^p(X)$ is small (in the operator norm) then the perturbed map $G_\triangle$, defined by

$$G_\triangle(Z) = \left(\bar{\partial} Z + \triangle(Z), Z(0)\right),$$

is still an isomorphism, and $Z = G^{-1}(0, Id)$ is a map $S^2 \to \mathcal{G}l(\mathbb{C}^n)$ satisfying $Z(0) = Id$. If $z \in S^2$ we define $\hat{E}(z) : \mathcal{L}(\mathbb{C}^n) \to X_z$ as follows. If $z \in D$ we set

$$\hat{E}(z)(Z) = E(z) \circ Z \, ds - i \, E(z) \circ Z \, dt,$$

and if $z \in S^2 \backslash D$ we set $\hat{E}(z) = 0$. Then

$$\hat{E}(z) \in \mathcal{L}\left(\mathcal{L}(\mathbb{C}^n), X_z\right);$$

moreover, $z \mapsto \hat{E}(z)$ belongs to L^∞. For $0 < \varepsilon < 1$ we denote by χ_ε the characteristic function of the closed disk D_ε, and define $\triangle_\varepsilon \in \mathcal{L}(W^{1,p}(S^2 \times \mathcal{L}(\mathbb{C}^n)), L^p(X))$ by

$$(\triangle_\varepsilon Z)(z) := \chi_\varepsilon(z) \hat{E}(z) Z(z).$$

By the Sobolev embedding theorem we can estimate, if $2 < p < \infty$,

$$\|\triangle_\varepsilon Z\|_{0,p} \leq (\pi \varepsilon^2)^{\frac{1}{p}} \|\hat{E}\|_{L^\infty} \|Z\|_{L^\infty}$$

$$\leq c(\varepsilon) \|Z\|_{1,p},$$

where $c(\varepsilon) \to 0$ if $\varepsilon \to 0$. Consider now the special perturbation $G_\varepsilon := G_{\triangle_\varepsilon}$ and define $Z^* : S^2 \to \mathcal{L}(\mathbb{C}^n)$ by

(A.48) $$Z^* = G_\varepsilon^{-1}(0, Id).$$

If $\varepsilon > 0$ is sufficiently small, then $Z^*(z) \in \mathcal{G}l(\mathbb{C}^n)$ for all $z \in S^2$, hence is invertible; moreover, $Z^*(0) = Id$. Finally we define $\sigma : D_\varepsilon \to \mathbb{C}^n$ by

$$v(z) = Z^*(z) \sigma(z),$$

A.7 The Brouwer degree

and compute, for $z \in D_\varepsilon$,

$$\begin{aligned}
0 &= \tfrac{\partial v}{\partial s} + i\tfrac{\partial v}{\partial t} + E(z)v \\
&= \left(\tfrac{\partial Z^*}{\partial s} + i\tfrac{\partial Z^*}{\partial t} + EZ^*\right)\sigma + Z^*\left(\tfrac{\partial \sigma}{\partial s} + i\tfrac{\partial \sigma}{\partial t}\right) \\
&= \left[\left((\bar{\partial} + \hat{E})Z^*\right)(z)(1)\right]\sigma(z) + Z^*\left(\tfrac{\partial \sigma}{\partial s} + i\tfrac{\partial \sigma}{\partial t}\right) \\
&= Z^*\left(\tfrac{\partial \sigma}{\partial s} + i\tfrac{\partial \sigma}{\partial t}\right).
\end{aligned}$$

Here 1 is the obvious tangent vector in $T_z S^2 = T_z \mathbb{C} \cong \mathbb{C}$ at $z \in D_\varepsilon \subset S^2$. Since $Z^*(z)$ is invertible if $z \in D_\varepsilon$, we conclude that the map $\sigma : D_\varepsilon \to \mathbb{C}^n$ is holomorphic. Defining $\Phi(z) := \Psi(z)Z^*(z)$, for $z \in D_\varepsilon$, we obtain, using the definitions, the required representation:

$$\begin{aligned}
w(z) &= \Psi(z)v(z) \\
&= \Psi(z)Z^*(z)\sigma(z) \\
&= \Phi(z)\sigma(z).
\end{aligned}$$

From $w(0) = 0$ we deduce $\sigma(0) = 0$. In addition,

$$\begin{aligned}
\Phi(z)i &= \Psi(z)Z^*(z)i \\
&= \Psi(z)iZ^*(z) \\
&= J(z)\Psi(z)Z^*(z) \\
&= J(z)\Phi(z),
\end{aligned}$$

as desired.

So far, however, the δ constructed does depend on the p chosen. In order to find a δ independent of p we carry out the above procedure for a fixed p_0 to find a corresponding $\delta_0 > 0$. Then we employ inner elliptic regularity theory in order to conclude that Z^* belongs to $W^{1,p}_{\text{loc}}(B_{\delta_0}, \mathcal{L}_{\mathbb{R}}(\mathbb{C}))$ for every p. Now we can take any $\delta < \delta_0$ and the proof of Theorem 12 is complete. ∎

Related results and applications to the transversality questions involved in the construction of Floer homology are contained in A. Floer, H. Hofer and D. Salamon [87].

A.7 The Brouwer degree

In the following we shall describe the so called Brouwer degree or mapping degree used in Chapter 2. It serves as a qualitative criterion for the solvability of a nonlinear equation

$$f(x) = y$$

for x. This degree is an integer which "algebraically counts" the number of solutions. To be more precise we consider open bounded sets $U \subset \mathbb{R}^n$ and continuous maps $f : \bar{U} \to \mathbb{R}^n$. Given $y \in \mathbb{R}^n$, the triplet (f, U, y) is called admissible, if $y \notin f(\partial U)$. The aim is to associate with every admissible triplet an integer, denoted by $d(f, U, y)$ such that the assignment

$$(f, U, y) \mapsto d(f, U, y)$$

satisfies some natural properties formulated in the following axioms.

Axiom 1. (Solution property)
Assume (f, U, y) is admissible. If $d(f, U, y) \neq 0$ then there exists a solution $x \in U$ of $f(x) = y$.

The equations $f(x) = y$ and $f(x) - y = 0$ have the same solutions x, and we therefore require

Axiom 2. (Naturality)
If (f, U, y) is admissible, then

$$d(f, U, y) = d(f - y, U, 0).$$

Assume (f, U, y) is admissible and $V \subset U$ is an open set containing all the solutions, i.e, $(f|U)^{-1}(y) \subset V$. Then (f, V, y) is also admissible and has the same solutions, hence we postulate:

Axiom 3. (Excision)
If (f, U, y) is admissible and if the open set $V \subset U$ satisfies $f^{-1}(y) \subset V$, then

$$d(f, U, y) = d(f, V, y).$$

We wish the degree to "count" the solutions and hence require an additivity property:

Axiom 4. (Additivity)
Assume (f, U, y) and (f, V, y) are admissible and $\bar{U} \cap \bar{V} = \emptyset$. Then (clearly $(f, U \cup V, y)$ is admissible)

$$d(f, U \cup V, y) = d(f, U, y) + d(f, V, y).$$

The next axiom is the most important one and crucial for the applications. We want the solution criterion to be stable under perturbation and require that d is invariant under continuous deformations:

Axiom 5. (Homotopy invariance)
Let $t \mapsto (f_t, U, y), 0 \leq t \leq 1$ be an admissible deformation, i.e, $f : [0,1] \times \bar{U} \to \mathbb{R}^n$ is continuous and (f_t, U, y) is admissible for every $0 \leq t \leq 1$. Then

$$d(f_0, U, y) = d(f_1, U, y).$$

Finally, we normalize using a simple model equation.

A.7 THE BROUWER DEGREE

Axiom 6. (Normalization)
$$d\Big(Id, B_1(0), 0\Big) = 1,$$

where $B_1(0)$ is the open unit ball centered at the origin 0 in $\mathbb{R}^n$.

For the moment, let us assume that such a d, satisfying these axioms exists. We prove that the axioms uniquely determine d. It is called the Brouwer degree or mapping degree.

Assume first, that U is open and $0 \notin \bar{U}$. Then $(Id, U, 0)$ is admissible. Since there is no solution of $Id(x) = 0$ in U we conclude that $d(Id, U, 0) = 0$ in view of Axiom 1. Next we consider the linear map $f = T \in \mathcal{Gl}(\mathbb{R}^n)$.

Lemma 6. Assume U is a bounded open neighborhood of 0 and $T \in \mathcal{Gl}(\mathbb{R}^n)$. Then
$$d(T, U, 0) = \text{sign det}(T).$$

Proof. Recall that $\mathcal{Gl}(\mathbb{R}^n) = G^+ \cup G^-$ where $G^{\pm} = \{T \in \mathcal{Gl}(\mathbb{R}^n) | \pm \det(T) > 0\}$; G^+ and G^- are pathwise connected. Define $\vartheta_{\pm 1} \in \mathcal{Gl}(\mathbb{R}^n)$ by
$$\vartheta_{\pm 1} = \text{diag}(\pm 1, 1, \ldots, 1).$$

Then $\vartheta_+ \in G^+$ and $\vartheta_- \in G^-$. Put $\varepsilon = \text{sig det}(T)$ and choose a continuous path $t \to T_t$ in G^ε connecting $T_0 = \vartheta_\varepsilon$ with $T_1 = T$. Then $t \mapsto (T_t, U, 0)$ is an admissible deformation and we obtain by Axiom 5
$$d(\vartheta_\varepsilon, U, 0) = d(T, U, 0).$$

Since 0 is the unique solution we find by applying the excision Axiom twice that $d(T, U, 0) = d(\vartheta_\varepsilon, U, 0) = d(\vartheta_\varepsilon, B_1(0), 0)$. If $\varepsilon = 1$ we employ Axiom 6 to find
$$d(T, U, 0) = d\Big(Id, B_1(0), 0\Big) = 1$$
$$= \text{sign det}(T).$$

In order to deal with the case $\varepsilon = -1$, we define $\varphi_\tau : [-1, 1] \to \mathbb{R}$ by $\varphi_\tau(t) = |t| - \frac{1}{2} + \tau$, for all $0 \leq \tau \leq 1$ and introduce $f_\tau : [-1, 1] \times B_1^{n-1}(0) \to \mathbb{R}^n$ by
$$f_\tau(x_1, \ldots, x_n) = \Big(\varphi_\tau(x_1), x_2, \ldots, x_n\Big).$$

Then $(f_\tau, (-1, 1) \times B_1^{n-1}(0), 0)$ is an admissible homotopy. If $\tau = 1$ the equation
$$f_1(x) = 0$$
has no solutions in the associated open set. Hence by Axiom 5
$$d\Big(f_\tau, (-1, 1) \times B^{n-1}, 0\Big) = 0,$$

for every $\tau \in [0,1]$. If $\tau = 0$, then the equations $|x_1| - \frac{1}{2} = 0$, $x_j = 0$ for $2 \leq j \leq n$, have precisely the two solutions

$$x_- = (-\frac{1}{2}, 0, 0, \cdots, 0), \quad x_+ = (\frac{1}{2}, 0, \cdots, 0)$$

where $x_- \in V_- := (-\frac{3}{4}, -\frac{1}{4}) \times B^{n-1}$ and $x_+ \in V_+ := (\frac{1}{4}, \frac{3}{4}) \times B^{n-1}$. Consequently, by excision and additivity we obtain

$$0 = d(f_0, V_-, 0) + d(f_0, V_+, 0).$$

For $x \in \bar{V}_-$ we have $f_0(x) = (-x_1 - \frac{1}{2}, x_2, \ldots, x_n)$ and for $x \in \bar{V}_+$ we have $f_0(x) = (x_1 - \frac{1}{2}, x_2, \ldots, x_n)$, hence the previous equation is written

$$0 = d(\vartheta_- - x_+, V_-, 0) + d(\vartheta_+ - x_+, V_+, 0),$$

and by excision,

$$0 = d(\vartheta_- - x_+, B^n, 0) + d(\vartheta_+ - x_+, B^n, 0).$$

Using the homotopy invariance we can get rid of the term x_- to obtain

$$0 = d(\vartheta_-, B^n, 0) + d(\vartheta_+, B^n, 0)$$
$$= d(\vartheta_-, B^n, 0) + 1.$$

We deduce, that also for $\varepsilon = -1$,

$$d(T, U, 0) = d(\vartheta_-, B^n, 0) = -1$$
$$= \text{sign } \det(T),$$

and the proof of Lemma 6 is complete. ∎

Assume now that $(f, B_\varepsilon(x_0), 0)$ is admissible for some $\varepsilon > 0$. Assume, in addition that f is differentiable and satisfies

$$f^{-1}(0) = \{x_0\} \text{ and } f'(x_0) \in \mathcal{Gl}(\mathbb{R}^n).$$

Lemma 7.
$$d\Big(f, B_\varepsilon(x_0), 0\Big) = \text{sign } \det\Big(f'(x_0)\Big).$$

Proof. Define $f_t(x) := tf(x) + (1-t)f'(x_0)(x - x_0)$. Since $f'(x_0) \in \mathcal{Gl}(\mathbb{R}^n)$ we find a small $0 < \delta \leq \varepsilon$ such that $f_t(x) \neq 0$ if $|x| = \delta$, for all $0 \leq t \leq 1$. Using excision and homotopy invariance we conclude that

$$d\Big(f, B_\varepsilon(x_0), 0\Big) = d\Big(f, B_\delta(x_0), 0\Big)$$
$$= d\Big(f'(x_0)(x - x_0), B_\delta(x_0), 0\Big)$$
$$= d\Big(f'(x_0)(x - x_0), B_R(0), 0\Big),$$

A.7 THE BROUWER DEGREE

where $R > 0$ is so large, that $B_\delta(x_0) \subset B_R(0)$. Choosing R large enough we now get rid of the term $-f'(x_0)x_0$ through an admissible homotopy and obtain

$$d\Big(f, B_\varepsilon(x_0), 0\Big) = d\Big(f'(x_0), B_R(0), 0\Big).$$

But by excision the right hand side does not depend on the size of R; in view of Lemma 6, the proof of Lemma 7 is therefore complete. ∎

Now we appeal to two classical results in differential topology.

Lemma 8. *Let $\Omega \subset \mathbb{R}^n$ be a bounded open domain and $f : \bar{\Omega} \to \mathbb{R}^n$ a continuous map. Given $\varepsilon > 0$ there exists a smooth map $g : \bar{\Omega} \to \mathbb{R}^n$ satisfying*

$$|f(x) - g(x)| < \varepsilon \text{ for } x \in \bar{\Omega}.$$

Proof. In view of the classical Stone-Weierstrass theorem we can take for g a polynomial approximation of f. ∎

The second result we need is a version of the Morse-Sard theorem. Recall that $y \in \mathbb{R}^n$ is a regular value for f if $f'(x) \in Gl(\mathbb{R}^n)$ for all $x \in f^{-1}(y)$.

Lemma 9. *Let $U \subset \mathbb{R}^n$ be open and $f : U \to \mathbb{R}^n$ a smooth map. Then the set of regular values for f is residual in $\mathbb{R}^n$ and therefore dense.*

A proof can be found in the book on differential topology by M. Hirsch [113].

If (f, U, y) is an admissible pair we define

$$\sigma := \inf_{x \in \partial U} |f(x) - y|,$$

then $\sigma > 0$, and by Lemma 8 we can pick g smooth with $|f(x) - g(x)| < \varepsilon$ on U, where $\varepsilon < \sigma/2$. Then $f_t = (1-t)f + tg$ defines an admissible homotopy $t \mapsto (f_t, U, y)$. Hence by Axiom 5 and Axiom 2

$$d(f, U, y) = d(g, U, y) = d(g - y, U, 0).$$

For every $z \in B_\varepsilon(y)$ the homotopy $t \mapsto g_t := g - y + tz$ is admissible for $(U, 0)$ and hence

$$d(f, U, y) = d(g - z, U, 0),$$

for every $z \in B_\varepsilon(y)$. In view of Lemma 4 we can pick $z_0 \in B_\varepsilon(y)$ which is a regular value of g. Hence

$$d(f, U, y) = d(g, U, z_0).$$

Since $\bar{U}$ is compact and z_0 regular, the set $g^{-1} = \{x_1, x_2, \ldots, x_k\}$ is finite. By excision and additivity we, therefore, find for $\tau > 0$ sufficiently small, that

$$\begin{aligned} d(f, U, y) &= \sum_{j=1}^{k} d\Big(g, B_\tau(x_j), z_0\Big) \\ &= \sum_{j=1}^{k} d\Big(g - z_0, B_\tau(x_j), 0\Big) \\ &= \sum_{j=1}^{k} \text{sign det}\Big(g'(x_j)\Big), \end{aligned}$$

where we have used Lemma 7. Summarizing, under the assumption that a degree d satisfying the Axioms 1-6 exists we conclude

Proposition 7.

(i) Assume (f, U, y) is admissible, f is smooth and y a regular value. Then

$$d(f, U, y) = \sum_{x \in f^{-1}(y)} \text{sign det}\left(f'(x)\right).$$

(ii) If (f, U, y) is admissible, then there exists an admissible (g, U, y) homotopic to (f, U, y), where g is smooth and y is a regular value for g. In particular,

$$d(f, U, y) = \sum_{x \in g^{-1}(y)} \text{sign det}\left(g'(x)\right).$$

Proposition 7 shows that the degree d is uniquely determined by the Axioms 1-6, and we shall establish its existence. The usual approach is to start with the formula (i) as the definition of $d(f, U, y)$ if f is smooth with regular value y. The degree $d(f, U, y)$ of a general admissible triplet is then defined as $d(g, U, y)$ for a smooth g close to f having the regular value y. Then one verifies the axioms. The crucial step is to verify at first that this definition is independent of the choice of the approximation g. The argument is as follows. If g_0 and g_1 are two smooth approximations of f one defines the admissible deformation $G : [0, 1] \times U \to \mathbb{R}^n$ by $G(t, x) = (1 - t)g_0(x) + tg_1(x)$. One can assume y to be a regular value for all three maps g_1, g_2 and G. Since G is admissible, $G^{-1}(y) \subset [0, 1] \times U$, and since y is a regular value of G, the solution set $G(t, x) = y$ is either empty or consists of finitely many disjoint, compact, connected 1-dimensional manifolds. According to the classification of 1-dimensional manifolds, see J. Milnor [157], each component is either a circle (if it has no boundary) or a closed interval in which case its two boundaries belong to the faces $\{0\} \times U, \{1\} \times U$ as illustrated in the following figure:

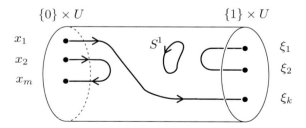

Fig. A7.1

A.7 The Brouwer degree

If the boundaries of a component belong to the same face, the corresponding solutions do not contribute to the degree in view of the orientation; also, a circle does not meet the boundary and hence does not contribute. Closed intervals connecting one face with the other one do contribute and, indeed, by the same amount again in view of the orientation, hence

$$\sum_{x \in g_0^{-1}(y)} \text{sign det } g_0'(x) = \sum_{\xi \in g_1^{-1}(y)} \text{sign det } g_1'(\xi) .$$

We do not carry out the details of this argument, and refer to H. Amann [5] or J. Milnor [157]. We shall establish the existence of the degree by an algebraic approach using homology theory.

In order to define the degree of an admissible triplet (f, U, y) we recall that $U \subset \mathbb{R}^n$ is open and bounded, $f : \bar{U} \to \mathbb{R}^n$ continuous and $y \notin f(\partial U)$. Viewing the sphere S^n as the 1-point compactification of $\mathbb{R}^n$, i.e., $S^n = \mathbb{R}^n \cup \{\infty\}$, we now consider f as a continuous map of the sphere, defined on $\bar{U} \subset S^n$. Denote by $U^c = S^n \setminus U$ the complement of U in S^n. Observing that $S^n \setminus \{y\}$ is homeomorphic to $\mathbb{R}^n$ (e.g., by means of the stereographic projection) we consider the diagram

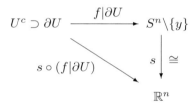

By the Tietze extension theorem we can extend the map $s \circ (f|\partial U)$ to a continuous map $\tilde{f} : U^c \to \mathbb{R}^n$. Since $\mathbb{R}^n$ is convex, all such extensions are homotopic. Define now the continuous extension $\tilde{F}$ of $f|\partial U$ by $\tilde{F} := s^{-1} \circ \tilde{f} : U^c \to S^n \setminus \{y\}$. Then any two such extensions of $f|\partial U$ are homotopic with boundary values fixed. Finally, we define the continuous map $F : S^n \to S^n$ extending $f : U \subset S^n \to S^n$ by

$$F(x) := \begin{cases} f(x) & \text{if } x \in U \\ \tilde{F}(x) & \text{if } x \in U^c . \end{cases}$$

The homotopy class of F is independent of the choices involved in the construction.

Considering the homology of S^n with $\mathbb{Z}$ coefficients we know that $H_n(S^n) = \mathbb{Z}$. Pick any generator of $H_n(S^n)$, say o_{S^n}. Then there exists an integer $d(F) \in \mathbb{Z}$ defined by

$$F_* o_{S^n} = d(F) o_{S^n} .$$

This integer is the homological degree of $F : S^n \to S^n$, and we now define

$$d(f, U, y) := d(F) \in \mathbb{Z} .$$

Since the homotopy class of F only depends on (f, U, y) and not on the choices involved, this integer is well-defined and we shall verify that $d(f, U, y)$ meets the Axioms 1-6, by using homology theory.

Axiom 1: Assume $d(f, U, y) \neq 0$. Then F is surjective, because otherwise its image would be in a subset homeomorphic to $\mathbb{R}^n$; hence F would be homotopic to a constant map so that $d(F) = 0$. By construction $y \notin F(U^c)$ and therefore $y \in F(U) = f(U)$ implying Axiom 1.

Axiom 2: The maps F associated with (f, U, y) and G associated with $(f - y, U, 0)$ are homotopic implying Axiom 2 by the homotopy axiom of homology.

Axiom 3: If F belongs to (f, U, y) then $F|V = f|V$ and $y \in F(V^c)$ implying Axiom 3.

Axiom 6: For the triplet $(Id, B^{2n}(1), 0)$ we can take $F = Id : S^n \to S^n$ and hence conclude $d(Id_{S^n}) = 1$ as desired.

Axiom 5: Let $(f_t, U, y), 0 \leq t \leq 1$ be an admissible homotopy. Then we can construct an associated family F_t depending continuously on t. Hence $t \mapsto d(F_t)$ is constant by the homotopy invariance of the homology implying the assertion.

We conclude, in particular, that $d(f, U, y)$ only depends on $f|\partial U$. Indeed if $f : \bar{U} \to \mathbb{R}^n$ and $g : \bar{U} \to \mathbb{R}^n$ agree on the boundary: $f|\partial U = g|\partial U$ then $tf(x)+(1-t)g(x)$, $0 \leq t \leq 1$ defines an admissible homotopy implying $d(f, U, y) = d(g, U, y)$.

Axiom 4: Consider first any admissible triplet (f, U, y) with associated map F on S^n. Then we have the following commutative diagram (assuming $n > 0$), with $S \equiv S^n$

$$H_n(S) \xrightarrow{i_*} H_n\left(S, S\setminus f^{-1}(y)\right) \xleftarrow{j_*} H_n\left(U, U\setminus f^{-1}(y)\right)$$

$$F_* \downarrow \qquad F_* \downarrow \qquad \swarrow f_*$$

$$H_n(S) \xrightarrow{k_*} H_n(S, S\setminus\{y\})$$

The maps j_* and k_* are isomorphisms: the first one by the excision property of homology and the second one as a consequence of the long exact sequence for the pair $(S, S\setminus\{y\})$. Hence we have

(A.49) $$d(F)o_S = k_*^{-1} f_* j_*^{-1} i_*(o_S) .$$

In order now to prove Axiom 4 we assume that $U = U_1 \cup U_2$ with $\bar{U}_1 \cap \bar{U}_2 = \emptyset$ so that $f_j = f|\bar{U}_j$, $j = 1, 2$ give admissible triplets (f_j, U_j, y). We have to prove that $d(f, U, y) = d(f_1, U_1, y) + d(f_2, U_2, y)$. We shall show that

(A.50) $$f_* j_*^{-1} i_*(o_S) = f_{1*} j_{1*}^{-1} i_{1*}(o_S) + f_{2*} j_{2*}^{-1} i_{2*}(o_S) .$$

A.7 THE BROUWER DEGREE

This implies in view of (A.49) that $d(F)o_S = d(F_1)o_S + d(F_2)o_S$ as desired. In order to prove (A.50) we consider the commutative diagram

$$\begin{array}{ccccc}
H_n(S) & \xrightarrow{i_*} & H_n\left(S, S\backslash f^{-1}(y)\right) & \xleftarrow{j_*} & H_n\left(U, U\backslash f^{-1}(y)\right) \\
\Delta \downarrow & & a \downarrow & & b \uparrow \\
H_n(S) \oplus H_n(S) & \longrightarrow & \bigoplus_{i=1}^{2} H_n\left(S, S\backslash f_i^{-1}(y)\right) & \longleftarrow & \bigoplus_{i=1}^{2} H_n\left(U_i, U_i\backslash f^{-1}(y)\right)
\end{array}$$

and

$$\begin{array}{ccc}
H_n\left(U, U\backslash f^{-1}(y)\right) & \xrightarrow{f_*} & H_n(S, S\backslash\{y\}) \\
b \uparrow & \nearrow (f_1)_* + (f_2)_* & \\
\bigoplus_{i=1}^{2} H_n\left(U_i, U_i\backslash f^{-1}(y)\right) & &
\end{array}$$

Here the maps j_* and $j_{1*} \oplus j_{2*}$ are isomorphisms by the excision in homology. The map $\Delta : \beta \mapsto (\beta, \beta)$ is the diagonal map. Moreover, $a(\sigma) = (a_{1*}(\sigma), a_{2*}(\sigma))$ and $b(\sigma_1, \sigma_2) = b_{1*}(\sigma_1) + b_{2*}(\sigma_2)$ where $a_j : (S, S\backslash F^{-1}(y)) \to (S, S\backslash f_j^{-1}(y))$ and $b_j : (U_j, U_j\backslash f_j^{-1}(y)) \to (U, U\backslash f^{-1}(y))$. Note that $f \circ b_j = f_j$ for $j = 1, 2$. If $\beta \in H_n(S)$ we can therefore compute

$$\begin{aligned}
f_* j_*^{-1} i_*(\beta) &= f_* b(j_{1*}^{-1} \oplus j_{2*}^{-1})(i_{1*} \oplus i_{2*})\Delta(\beta) \\
&= f_*\left(b_{1*} j_{1*}^{-1} i_{1*}(\beta) + b_{2*} j_{2*}^{-1} i_{2*}(\beta)\right) \\
&= f_{1*} j_{1*}^{-1} i_{1*}(\beta) + f_{2*} j_{2*}^{-1} i_{2*}(\beta)
\end{aligned}$$

as claimed in (A.50). We have verified that Axiom 4 holds. To sum up: we have demonstrated that d meets all the axioms, hence the existence proof of the mapping degree is complete.

Remarks: The Brouwer degree can be extended to arbitrary normed vector spaces E however, for a restricted class of continuous maps only. We call a triplet (f, Ω, y) admissible, if $\Omega \subset E$ is a bounded open subset, $f : \bar{\Omega} \to E$ is a continuous map of the form $f = id - k$ with a compact map k (i.e., the closure of $k(\Omega)$ is compact in E), and $y \in E$ is a point satisfying $y \notin f(\partial\Omega)$. There is a unique map associating to every admissible triplet an integer $d(f, \Omega, y) \in \mathbb{Z}$ satisfying the Axioms 1-6. It is called the Leray-Schauder degree; we made use of this degree in Chapter 3 in order to establish the existence of the symplectic capacity c_0.

The crucial observation in constructing this extension is the fact, that a compact map can be approximated (in the uniform topology) by compact mappings having their image in finite dimensional subspaces of E. Take such a map k satisfying $k(\bar{\Omega}) \subset E_n \subset E$, where dim $E_n = n$. If $x \in E$ is a solution of

$$(id - k)(x) = 0,$$

then $x \in E_n$ and hence one defines

$$d(id - k, \Omega, 0) = d(id - k, \Omega \cap E_n, 0),$$

where, on the right hand side, we have the Brouwer degree. This is well-defined. Indeed choosing a larger finite dimensional space $E_m \supset E_n$ then $d(id - k, \Omega \cap E_n, 0) = d(id - k, \Omega \cap E_m, 0)$. This follows by approximating $(id - k)|\Omega \cap E_m$ by a smooth map having 0 as a regular value and simply computing the index using the representation (:) of Proposition 1. Since the finite dimensional maps are dense in the set of compact maps (in the uniform topology) the continuous map $(id - k, \Omega, 0) \to d(id - k, \Omega, 0) \in \mathbb{Z}$ defined above has a unique extension, called the Leray-Schauder degree. For the details we refer the reader to H. Amann [5], K. Deimling [60] and E. Zeidler [232], Chapter 12.

A.8 Continuity property of the Alexander-Spanier cohomology

The mathematical tools needed for Chapter 6 are becoming more sophisticated. In particular, homology and cohomology theory is being used. Besides the basic properties of a (co-) homology theory the reader needs to know cup products and Poincaré duality. All this can be found in standard texts on algebraic topology. We suggest, for example, the book by J. Vick [216], which describes the standard material with great care and also offers an introduction to fixed point theory introducing the fixed point index closely related to degree theory. More comprehensive treatments on algebraic topology are the books by A. Dold [62], W.S. Massey [148, 149, 150], E. Spanier [203] and R. Switzer [209]. In particular the appendix in [62] concerned with the Kan- and Čech-extension of homotopy invariant functors is interesting for the following discussion of continuity properties of homotopy invariant functors.

Nonlinear analysis is — somewhat exaggerated — the study of nonlinear equations of the form

$$F(x) = y.$$

If the map $F : X \to Y$ and the topological spaces X and Y have some nice properties, for example, there is some compactness and the solution set is generically discrete, degree theory is a very powerful tool. Sometimes, however, the solution sets are generically expected to be manifolds and one would like, of course, to know what they are. This is in general a difficult task. But, one can try to compute some of their invariants, using, for example, homology theories. A situation which then occurs quite frequently is the following. If we consider the equation $F(x) = y$ we

A.8 Continuity property of the Alexander-Spanier cohomology

do not know whether F is generic so that the solution set might be quite badly behaved, for example, it might not be a manifold. On the other hand, it might be possible to approximate the solution set $F^{-1}(y)$ by solution sets $F^{-1}(y_k)$ which are generic, for example, manifolds. In this situation it is sometimes only possible to study $F^{-1}(y_k)$ by algebraic means (using homology theory) and the question then is: under which circumstances are we able to conclude something about $F^{-1}(y)$?

The important tool will be a continuity property for the homology theory, which we discuss in the following. We begin with an abstract discussion which is applicable in many circumstances and focus on the particular theory we need in Chapter 6, namely the Alexander-Spanier cohomology theory, see E. Spanier [203].

Let X be a topological space and $\mathcal{C}$ the category of all subspaces, the morphisms $\varphi : A \to B$ being the continuous maps. Let $\mathcal{A}$ be an algebraic category. We consider one of the following cases: $\mathcal{A}$ is the category of Abelian groups, or the category of commutative rings, or the category of modules over a fixed ring. Assume a contravariant functor $h : \mathcal{C} \to \mathcal{A}$ is given, i.e., to every $A \in \mathcal{C}$ we associate an algebraic object $h(A) \in \mathcal{A}$ and to a continuous map $\varphi : A \to B$ a homomorphism $\varphi^* = h(\varphi) : h(B) \to h(A)$. If $A \subset B$ we have a natural inclusion map and we denote by $i_{AB} : h(B) \to h(A)$ the induced map. Since h is a contravariant functor we have

$$i_{AB} i_{BC} = i_{AC} \quad \text{for } A \subset B \subset C$$

$$i_{AA} = \text{Id}_{h(A)}.$$

Given $A \subset X$ we denote by $\mathcal{O}_A$ the collection of all open neighborhoods of A. We denote for $U \in \mathcal{O}_A$ the homomorphism induced by the inclusion $A \subset U$ by $i_U : h(U) \to h(A)$.

Definition 1. Let h be as described above and $A \subset X$ fixed. We call h continuous at A if the following conditions hold

(i) Given $\alpha \in h(A)$ there exist $U \in \mathcal{O}_A$ and $\alpha_U \in h(U)$ such that $i_U(\alpha_U) = \alpha$

(ii) Given $\alpha_U \in h(U)$, with $U \in \mathcal{O}_A$ satisfying $i_U(\alpha_U) = 0$, there exists $V \in \mathcal{O}_A$ satisfying $A \subset V \subset U$ and $i_{VU}(\alpha_U) = 0$.

In other words any element $\alpha \in h(A)$ lives already on a neighborhood of A and any class $\alpha \in h(U), U \in \mathcal{O}_A$ which "dies" on A, "dies" already on some neighborhood. As we shall see in the following we can reconstruct $h(A)$ from the $h(U), U \in \mathcal{O}_A$ provided h is continuous at A.

Observe that $\mathcal{O}_A$ is a directed set in the following sense. We can define an order structure $\leq$ by

$$U \leq V :\Longleftrightarrow U \supset V$$

("smaller sets are larger"). Moreover given $U, V \in \mathcal{O}_A$ there exists $W \in \mathcal{O}_A$ satisfying

$$W \geq U, W \geq V.$$

Indeed, just take $W = U \cap V$. Hence we have the following structure. For $U \leq V$ we have a homomorphism
$$i_{VU} : h(U) \longrightarrow h(V)$$
and for every $U \in \mathcal{O}_A$ a homomorphism
$$i_U : h(U) \longrightarrow h(A) \,.$$
Moreover, the following holds: If $U \leq V \leq W$ then
$$i_{WV} i_{VU} = i_{WU} \text{ and } i_{UU} = Id \,.$$
Moreover, if $U \leq V$, then
$$i_V i_{VU} = i_U \,.$$
We first carry out a construction called the direct limit and then show that it is isomorphic to $h(A)$ provided h is continuous at A.

Define $M_A \in \mathcal{A}$ as the direct sum
$$M_A = \bigoplus_{U \in \mathcal{O}(A)} h(U) \,.$$
If A consists of abelian groups, M_A will be an abelian group, etc. Denote by φ_U the homomorphism
$$\varphi_U : h(U) \longrightarrow M_A \,,$$
which is the inclusion into the U-th factor.

Consider the subgroup, subring or submodule K_A generated by elements of the form
$$\varphi_U(\alpha_U) - \varphi_V i_{VU}(\alpha_U)$$
where $U, V \in \mathcal{O}_A$ satisfy $U \leq V$ and where $\alpha_U \in h(U)$. Define L_A by
$$L_A = M_A / K_A \,.$$
Depending on $\mathcal{A}$ this quotient is either an Abelian group, or a commutative ring or a module.

Elements of L_A are equivalence classes $[(\alpha_U)]$ where $(\alpha_U) \in M_A$ with $\alpha_U = 0$ up to finitely many. If $\alpha_{U_1}, \ldots, \alpha_{U_k}$ are the non zero elements, then
$$(\alpha_U) = \sum_{i=1}^{k} \varphi_{U_i}(\alpha_{U_i}) \,.$$
Let $U_\infty = \bigcap_{i=1}^{k} U_i$. Pick any $V \in \mathcal{O}_A, V \geq U_\infty$. Then for $i = 1, \ldots, k$
$$\beta_i := \varphi_{U_i}(\alpha_{U_i}) - \varphi_V i_{VU_i}(\alpha_{U_i})$$

A.8 Continuity property of the Alexander-Spanier cohomology

belongs to K_A. Consequently,

$$(\alpha_U) = \sum_{i=1}^{k} \beta_i + \varphi_V\left(\sum_{i=1}^{k} i_{VU_i}(\alpha_{U_i})\right)$$
$$=: \beta + \varphi_V(\gamma)$$

where $\beta \in K_A$ and $\gamma \in h(V)$. Hence for every class $a \in L_A$ there exists $U_a \in \mathcal{O}_A$ such that for every $V \geq U_a$, a is represented by $\varphi_V(\alpha_V)$ for a suitable element $\alpha_V \in h(V)$. For an element $(\alpha_U) \in M_A$ we consider the element $\hat{\Phi}((\alpha_U)) := \sum_{U \in \mathcal{O}_A} i_U(\alpha_U) \in h(A)$. We note that the sum is finite. For the element

$$\beta = \varphi_U(\alpha_U) - \varphi_V i_{VU}(\alpha_U),$$

with $V \geq U$ we compute

$$\hat{\Phi}(\beta) = i_U(\alpha_U) - i_V i_{VU}(\alpha_U)$$
$$= i_U(\alpha_U) - i_U(\alpha_U)$$
$$= 0.$$

Clearly $\hat{\Phi} : M_A \to h(A)$ defines a homomorphism. By the preceding discussion its kernel contains K_A. Hence we obtain a homomorphism

$$\Phi : L_A \longrightarrow h(A).$$

In a precise sense Φ gives us what the $h(U), U \in \mathcal{O}_A$, know about the $h(A)$. The above construction is the well-known direct limit denoted by

$$\check{h}(A) := L_A = \mathrm{dir}\lim_{U \in \mathcal{O}_A} h(U).$$

It is an algebraic measure of what the neighborhoods of A know about A with respect to the contravariant functor h.

Theorem 13. Assume h is continuous at A. Then

$$\Phi : \check{h}(A) \longrightarrow h(A)$$

is an isomorphism.

Proof. Assume $\Phi(a) = 0$. We can represent $a \in \check{h}(A)$ by an element $\varphi_V(\alpha_V)$ for $V \geq U_a$ where $\alpha_V \in h(V)$ by the preceding discussion. Hence

$$0 = \Phi(a) = i_V(\alpha_V).$$

By property (ii) of the continuity definition we find $W \geq V$ such that

$$\alpha_W := i_{WV}(\alpha_V) = 0.$$

Note that
$$\varphi_V(\alpha_V)$$
$$= \varphi_V(\alpha_V) - \varphi_W i_{WV}(\alpha_V)$$
$$= \varphi_V(\alpha_V) - \varphi_W(\alpha_W)$$
belongs to K_A. Hence $a = 0$. Next let $\beta \in h(A)$. By property (i) there exists $U \in \mathcal{O}_A$ and $\alpha_U \in h(U)$ with $i_U(\alpha_U) = \beta$. Let $a = [\varphi_U(\alpha_U)]$. Then $\Phi(a) = \beta$. ∎

Assume next that h is a contravariant functor on the category of topological spaces. Let X and A be as before and assume Y is another topological space. If $f : X \to Y$ is a continuous map we would like to study $f|A : A \to Y$ and say something about its properties once we are able to study them for $f|U : U \to Y$, $U \in \mathcal{O}_A$. We have the commutative diagram

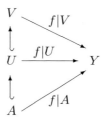

giving
$$i_U(f|U)^* = (f|A)^*$$
$$i_{UV}(f|V)^* = (f|U)^* .$$

We define a homomorphism
$$\check{f} : h(Y) \longrightarrow \check{h}(A)$$
by
$$\check{f}(a) = [\varphi_X f^*(a)] .$$
We observe that
$$[\varphi_X f^*(a)] = [\varphi_U (f|U)^*(a)]$$
$$= [\varphi_U i_{UX} f^*(a)] ,$$
so that for all $U \in \mathcal{O}_A$
$$\check{f}(a) = [\varphi_U (f|U)^*(a)] .$$
We compute for $a \in h(U)$
$$\Phi \circ \check{f}(a) = i_U(f|U)^*(a)$$
$$= (f|A)^*(a) ,$$

so that we obtain the commutative diagram

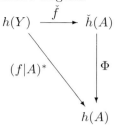

Note also that we have the commutative diagram

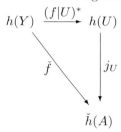

where $j_U(\alpha_U) = [\varphi_U(\alpha_U)]$.

Proposition 8. For every $a \in \check{h}(A)$ there exists $U \in \mathcal{O}_A$ and $\alpha_U \in h(U)$ with $j_U(\alpha_U) = a$. Given $\alpha_U \in h(U)$ with $j_U(\alpha_U) = 0$ there exists $V \geq U$ with $i_{VU}(\alpha_U) = 0$.

Proof. By the previous discussion an element $a \in \check{h}(A)$ can be written in the form $[\varphi_V(\gamma)]$ for some $\gamma \in h(V)$ and some suitable $V \in \mathcal{O}_A$. Hence

$$j_V(\gamma) = a \, .$$

Assume $j_U(\alpha_U) = 0$. Then we find $U_1, \ldots, U_k, V_1, \ldots, V_k$, with $V_i \geq U_i$ and $\alpha_{U_i} \in h(U_i)$ with

(A.51) $$\varphi_U(\alpha_U) = \sum_{l=1}^{k} \varphi_{U_i}(\alpha_{U_i}) - \varphi_V\left(i_{VU_i}(\alpha_{U_i})\right) .$$

Pick any $W \in \mathcal{O}_A$ with

(A.52) $$W \geq U, U_1, \ldots, U_k, V_1, \ldots, V_k \, .$$

Then applying $\sum_{C \leq W} i_{WC}$ to (A.51) we obtain

$$\begin{aligned} i_{WU}(\alpha_U) &= \sum_{l=1}^{k} i_{WU_i}(\alpha_{U_i}) - i_{WV_i} i_{V_i U_i}(\alpha_{U_i}) \\ &= \sum_{l=1}^{k} i_{WU_i}(\alpha_{U_i}) - i_{WU_i}(\alpha_{U_i}) \quad = 0 \, . \end{aligned}$$

This is true for every W satisfying (A.52) proving our result. ∎

Theorem 14. Let $f : X \to Y$ be continuous and $A \subset X$. Assume for every $U \in \mathcal{O}_A$ that $(f|U)^*$ is injective. Then $\check{f}$ is injective.

Proof. Assume $\check{f}(a) = 0$. This implies for every $U \in \mathcal{O}_A$ by the definition of $\check{f}$,
$$0 = [\varphi_U(f|U)^*(a)] = j_U\left((f|U)^*(a)\right).$$
By Proposition 8 there exists $V \geq U$ such that
$$0 = i_{VU}\left((f|U)^*(a)\right).$$
But $i_{VU}(f|U)^*(a) = (f|V)^*(a)$ implying that $a = 0$ since $(f|V)^*$ is injective. ∎

Corollary 1. Assume $f : X \to Y$ is continuous and $A \subset X$. If h is continuous at A and if for every $U \in \mathcal{O}_A$ the map $(f|U)^*$ is injective, then $(f|A)^*$ is injective.

Proof. We have the commutative diagram

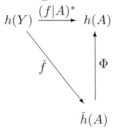

By Theorem 14, the map $\check{f}$ is injective and, by Theorem 13, the map Φ is an isomorphism. Hence $(f|A)^*$ is injective. ∎

Next we apply the previous discussion to the Alexander-Spanier cohomology theory. We refer the reader to the original articles by E. Spanier [202], the book by E. Spanier [203] and the excellent book by W.S. Massey [149].

Definition 2. (see also [202]) Let X be a topological space and A a subspace. We say A is taut in X if for every coefficient ring R the Alexander-Spanier cohomology functors $\bar{H}^i(\cdot, R)$, for every i, are continuous at A.

The following very useful theorem is due to E. Spanier and we refer the reader to [204]

Theorem 15. In each of the following cases A is taut in X

(i) A is compact and X is Hausdorff.
(ii) A is closed and X is paracompact and Hausdorff.
(iii) A is arbitrary and every open set in X is paracompact and Hausdorff.
(iv) A is a neighborhood retract.

So we note that in a metrizable space every subset A is taut, and hence, by Theorem 13,
$$\bar{H}^*(A) \cong \mathrm{dir}\lim_{U \in \mathcal{O}_A} \bar{H}^*(U).$$

Index

Abraham, R. and Robbin, J., 291
action, 7, 203, 228
action of a fixed point, 151
action of a periodic solution, 99
action spectrum, 151, 152, 186, 269
action spectrum, compactness, 152
action, bifurcation diagram, 187
action-angle coordinates, 273, 278
admissible Hamiltonians, 70
Alexander-Spanier cohomology, 208, 212, 213, 219
almost complex structure, 13–15, 200, 242
Ambrosetti, A., 80
anti-holomorphic disc, 230
Arnold conjecture, 193, 198, 199
Arnold conjecture, torus, 202
Arnold, V.I., 48, 196, 198, 199, 216, 250, 279
Arnold-Jost theorem, 135, 273, 279
Arzelà-Ascoli theorem, 235, 241
asymptotic boundary conditions, 258
auxiliary structures, 13
Avez, A., 48

Banach manifold $W^{1,p}$, 232
Bangert, V., 132
Banyaga, A., 198
Beckner, W., 290
Benci, V., 131, 133
Bennequin, D., 126
Betti numbers, 199, 222
bi-invariant metric, 149
bi-invariant pseudo-metrics, 150
Bialy, M., 186, 187, 190
Bialy-Polterovich theorem, 186
bifurcation diagram, 187
bifurcation diagram, simple, 187, 188
Birkhoff invariants, 43, 263
Birkhoff normal form, 42, 43, 46
Birkhoff's theorem, 43, 47
Birkhoff, G., 42, 47, 131, 194
Birkhoff-Lewis theorem, 263

Blanchard, P., 81
Bolotin, S.V., 131
Bordeaux bottle, 99
boundary operator, 256
bounded orbits, 205, 234, 236, 253
bounded solution, unregularized flow, 227
Brüning, E., 81
Brezis, H., 211
Brouwer, L., 195, 211
Brunella, M., 216
bubbling off analysis, 240
bundle of anti-linear maps, 232

Calabi invariant, 150
canonical basis, 3
canonical bi-invariant metric, 149
canonical one-form, 128
Cantor set, 48
cap product, 213
capacity preserving homeomorphisms, 64
Cartan's identity, 11, 13
Cauchy-Riemann operator, 224, 242, 253, 291
Cauchy-Riemann operator on sections, 295
celestial mechanics, 18
chain complex, 256, 261
chain homomorphism, 259, 267
chain homotopy, 259
Chang, K.C., 81
Chang, S.Y., 290
Chaperon, M., 215, 250
characteristic line bundle, 22, 113
Chekanov, Yu. V., 216
Chern-Simons functional, 264
Clarke, F., 24
classical Hamiltonian systems, 127
classical systems, 136
classification of symplectic forms, 12
closed characteristic, 22, 23
closedness of $\mathcal{D}$ in C^0, 59

Closing lemma, 21, 105
Closing problem, 21
cohomology ring, 199
commutator, 148
compactness of the action spectrum, 152
compatible almost complex
    structure, 227
compatible complex structure, 2
compatible inner product, 2
complementary symplectic subspaces, 2
complex structure, 2, 231
computation of Floer homology, 260
configuration space, 127
conformality axiom, 51
conformality of c_0, 70
Conley's index theory, 257
Conley, C., 198, 200, 203, 215, 221, 258
Conley-index, 221
Conley-Morse equation, 221
Conley-Zehnder theorem, 203
connected simple system of C. Conley, 258
connecting orbit, 205, 206, 253, 254, 261
contact form, 125
contact structure, 119, 125
contact type, 269
contact type hypersurface, 119
continuity property of
    A.-S. cohomology, 212, 220
convex energy surface, 23
Courant-Hilbert minimax principle, 39
critical points, 205
critical value, 79
Croke, C., 99
cup-length, 199, 216
cylinder, 231

Dacorogna-Moser theorem, 31
Darboux's normal form, 12
Darboux's theorem, 10
de Rham cohomology group, 12
de Rham, G., 12
definition of c_0, 70
definition of c_0-Lipschitz type, 116
definition of the universal minimax level
    γ, 156
degree, homological definition, 213
determinant, 5

differential forms, 4
direct method in the calculus of variations, 23
directed set of admissible pairs, 267
displacement energy, 150
displacement energy as a special capacity, 172
distinguished line bundle, 22
Dold, A., 213
Douglis, A., 296
dual action principle, 31
Duistermaat, J.J., 286

Ekeland, I., 31, 33, 51, 59, 61, 66, 99
elementary symplectic
    transformation, 275
Eliashberg conjecture, 169
Eliashberg, Y., 59, 123, 126, 150, 201, 216, 269
Eliashberg-Gromov theorem, 59
Eliasson, H., 232
elliptic methods, 222
elliptic operator, first order, 225
elliptic regularity theory, 227
energy, 253
energy estimate in C^0, 173
energy of a symplectic map, 146, 165
energy surface, 19
energy-capacity inequality, 147, 165
Euler characteristic, 198
Euler equation, 25
Euler-Maupertuis-Jacobi principle, 130
existence problem for periodic solutions, 18, 21, 31, 41, 76, 98, 183
existence results on prescribed energy surfaces, 105

Fet, A.I., 131
first Chern class, 228, 255
flat space, locally, 190
Floer complex, 77
Floer groups, 262
Floer homology, 193, 256, 257
Floer, A., 56, 77, 103, 200, 216, 250, 253, 257, 259, 265, 269
Floquet multipliers, 110, 111, 135
Fomenko, A.T., 286
Fortune, B., 201

Fourier integral operators, 216
Franks, J., 132
Fredholm index, 209, 210
Fredholm map, 209
Fredholm map, proper, 210
Fredholm operator, 298
Fredholm theory, 254
Fukaya, K., 265
fundamental class, 213

Genecand, C., 49
generating function, 13, 189, 273
geodesic flow, 130
geodesic path in $\mathcal{D}$, 184
geodesic path, regular, 185
geodesic, length, 186
geodesic, minimal, 186, 190
Givental, A.B., 216
Gluck, H., 131
gluing construction, 256
gradient flow, 204, 206
gradient-like flow, 206, 217
Green's function, 291
Gromov's squeezing theorem, 33, 55
Gromov, M., 7, 17, 33, 59, 126, 144, 240
Gromov-width, 17, 56
group of compactly supported symplectic diffeomorphism, 143

Hörmander, L., 216
Hadamard, J., 131
Hamilton-Jacobi equation, 189, 190
Hamiltonian diffeomorphism, 198
Hamiltonian principle, 23
Hamiltonian vector field, 1, 6, 8, 9, 13, 15
Hamiltonian, quasi autonomous, 186
Herman's non-closing lemma, 137, 139
Herman, M., 105
higher order capacities, 66
Hilbert, D., 131
Hofer, H., 33, 51, 56, 59, 61, 66, 99, 102, 103, 106, 115, 119, 126, 132, 133, 150, 169, 174, 216, 228, 250, 253, 262, 265, 269
holomorphic disc, 229
holomorphic disc, area, 229

homotopy invariance of Floer groups, 257
Hurewicz homomorphism, 228
hypersurface, 19
hypersurface of c_0-Lipschitz type, 115

inner capacity, 57
inner regularity of c_0, 70
inner regularity of a capacity, 57
integrable Hamiltonian system, 46, 134, 278
integrals in involution, 46
invariance of Hamiltonian vector fields, 13
invariance of the symplectic form, 13
invariant Lagrangian tori, 48
isolated invariant set, 222
isotropic subspace, 54

Jacobi metric, 130, 131
Jiang, M.Y., 136
Joker, 240
Jost, R., 279

K.A.M.-theory, 48, 135
Künzle, A., 99
kinetic energy, 129
Klingenberg, W., 131
Kolmogorov, A.N., 48
Koslov, V.V., 136
Kronecker flow, 47, 48

Lagrangian mechanics, 127
Lagrangian submanifold, 216
Le Calvez, P., 263
Lefschetz fixed point theory, 195
Lefschetz, S., 195
Legendre transformation, 25
length of a symplectic arc, 184
Leray-Schauder degree, 96
Lewis, D.C., 47
Lie derivative, 11
limit sets, 217, 218
linear canonical map, 4
linear symplectic group $Sp(n)$, 5
linear symplectic map, 4, 6
Liouville form, 128
Liouville theorem, 277

Liouville, J., 135
little trick, 236
Ljapunov function, 218, 219
Ljapunov-Schmidt reduction, 215
Ljusternik Schnirelman theory, 131
Ljusternik-Schnirelman, 197, 199, 206, 208, 218, 243
Ljusternik-Schnirelman theory, 77
localization of Sikorav's estimate, 178
Loewner, C., 196
loop space, 203
Lutz, R., 126

Ma, R., 103
Martinet, J., 126
Mather set, 49
Mather, J., 49
Maupertuis-Jacobi principle, 136
Mawhin, J., 81
McDuff, D., 267
Milnor, J., 211
minimax characterisation, 77
Minimax lemma, 79
Mishchenko, A.S., 286
mod 2-degree, 210
monotonicity axiom, 51
monotonicity of c_0, 70
monotonicity of the Gromov width, 18
Morse complex, 261
Morse decomposition, 219, 221
Morse equation, 222, 261
Morse homology, 261
Morse index, 221
Morse theory, 31, 77, 131
Morse-Smale flow, 257, 261
Moser's deformation method, 12
Moser's theorem, 16
Moser, J., 12, 42, 47–49, 135, 198, 263, 286
Mountain Pass Lemma, 80

Nekhoroshev, N., 286
Nikishin, N., 195
Nirenberg, L., 211, 296
non-flatness of $\mathcal{D}$, 185
nonresonant of order s, 43
nontriviality axiom, 51, 57
nontriviality of c_0, 70

Novikov, S., 262

Offin, D., 134
Ono, K., 228, 262
orbit of a point, 182
order structure on D, 149
orthogonal complement, 2
orthogonal sum, 3
orthogonality, 2
$osc(H)$, 146
oscillation, 146

P.S.-condition, 78
Pöschel, J., 48
Palais, R., 80
Palais-Smale condition, 78
parametrized family of energy surfaces, 106
parametrized family of hypersurfaces, 114
partial differential equation, 252
partial ordering of admissible pairs, 267
period group, 228
periodic orbit, 18, 23
periodic solution, 18
perturbation methods, 47
perturbations of integrable systems, 47
phase function, 216
Poincaré continuation method, 110
Poincaré duality, 213
Poincaré inequality, 27
Poincaré lemma, 7, 11, 239
Poincaré polynomial, 199, 221, 261
Poincaré recurrence theorem, 1, 139
Poincaré section map, 111
Poincaré's recurrence, 263
Poincaré, H., 19, 21, 110, 194, 197
Poincaré-Birkhoff fixed point theorem, 194
Poincaré's recurrence theorem, 20
Poincaré, H., 194
Poincaré-Birkhoff twist theorem, 194
Poisson bracket, 43
Poisson bracket of two functions, 277
Polterovich, L., 150, 186, 187, 190
potential energy, 129
proper homotopy, 211
Pugh, C.C., 21, 139

pullback of a form, 4, 6, 12
pullback of a vector field, 8

quadratic form, 35
quasi autonomous, 186, 187, 190
quasi-autonomous, 188, 191
quasi-periodic solutions, 47, 48, 50, 135

Rabinowitz, P., 23, 24, 77, 80, 81, 123, 133
real filtration of complex, 266
recurrence, 19
Reeb vector field, 125, 127
regular pair (H, J), 255
regular value, 211, 212
regularity of critical points, 88
representation of the gradient flow, 90
rescaling, 237
rest points, 208, 218
Riemannian metric, compatible, 14
Riemannian sphere, 231, 232, 294
Riesz' theorem, 77
Riesz, F., 77
rigidity of symplectic diffeomorphisms, 58
rigidity of symplectic embeddings, 51
rigidity phenomena, 34
Robinson, C., 21, 112, 139
Rybakowski, K., 222

Salamon, D., 222, 228, 253, 255, 258, 261–263
Sard's theorem, 119
Sard-Smale theorem, 211, 212, 255
Schwarz, M., 254, 257, 261
Seifert, H., 21, 131
set of compatible almost complex structures, 15
set of distinguished Hamiltonians, 70
Siburg, K.F., 52, 100
Siegel, C.L., 47
Sikorav's estimate, 177
Sikorav's estimate, localized form, 178
Sikorav's theorem, 177, 178
Sikorav, J.C., 177, 200, 216, 265
Simon, C., 195
Smale degree, 210
Sobolev class, 232

Sobolev embedding theorem, 241
Sobolev space, 82, 203
space of compatible almost complex structures, 15
special symplectic capacity, 172
standard symplectic space, 4
Stein, E., 242
Stokes' theorem, 12, 240
Struwe, M., 81, 119
symplectic 2-form, 10
symplectic action, 7
symplectic atlas, 10
symplectic basis, 3
symplectic capacities of higher order, 66
symplectic capacity, 18, 51, 52, 56, 58, 66, 70
symplectic capacity $c_0(M, \omega)$, 265
symplectic capacity, definition, 51
symplectic capacity, special, 172
symplectic classification of open polydiscs, 269
symplectic classification of quadratic forms, 35
symplectic coordinates, 4, 10
symplectic diffeomorphism, 6
symplectic embeddings, 31
symplectic energy, properties, 147
symplectic fixed points, 222
symplectic group, 5
symplectic homeomorphisms, 51
symplectic homology, 193, 265
symplectic homology groups, 267
symplectic homology groups, exact triangle, 268
symplectic homology theory, 56, 216
symplectic invariants, 18
symplectic manifolds, 1, 9
symplectic map, linear, 4
symplectic maps between manifolds, 12
symplectic polar coordinates, 46
symplectic structure, 9
symplectic structure on a cotangent bundle, 128
symplectic subspace, 3
symplectic vector space, 1

Taubes, C., 257
Terng, C., 80

three-body problem, 21
trace-theorem, 287
transformation formula for vector fields, 8
transformation of a Hamiltonian vector field, 9
transversally hyperbolic, 205
Traynor, L., 216

uniformization theorem, 231
universal minimax family $\mathcal{F}$, 156
universal minimax level γ, 155
universal minimax set, 155
universal variational principle, 154
unparameterized periodic solution, 113
unregularized gradient flow equation, 224, 227

variation of constant formula, 214
Vekua, I.N., 298
Viterbo's homological special symplectic capacity, 183
Viterbo's theorem, 123
Viterbo, C., 23, 102, 103, 123, 132, 169, 182, 216
volume form, 5

weak nontriviality axiom, 51, 66
Weierstrass, K., 35
Weinstein conjecture, 23, 119, 120, 123, 126, 132
Weinstein, A., 23, 24, 42, 99, 120, 123, 198, 201
Willem, M., 81
Wysocki, K., 269

Yang-Mills action, 264

Zehnder, E., 99, 106, 115, 119, 137, 198, 200, 203, 215, 255, 258, 263
Ziller, W., 131

Bibliography

[1] C. Abbas and H. Hofer. Holomorphic curves and global questions in contact geometry. To appear in Birkhäuser.

[2] R. Abraham and J. Robbin. *Transversal mappings and flows*. W.A. Benjamin, Inc., 1967.

[3] B. Aebischer, M. Borer, Ch. Leuenberger, M. Kälin, and H.M. Reimann. Symplectic geometry, an introduction based on the Seminar in Berne 1992. Birkhäuser, 1994.

[4] S. Alpern and V.S. Prasad. Combinatorial proofs of the Conley-Zehnder-Franks theorem on a fixed point for torus homeomorphisms. *Advances in Mathematics*, 99:238–247, 1993.

[5] H. Amann. Lectures on some fixed point theorems. Monografias de matemática IMPA, Rio de Janeiro.

[6] A. Ambrosetti and P. Rabinowitz. Dual variational methods in critical point theory and applications. *J. Funct. Anal.*, 14:349–381, 1973.

[7] V.I. Arnold. Proof of a theorem of A.N. Kolmogorov on the invariance of quasiperiodic motions under small perturbations of the Hamiltonian. *Uspeki Mat. Nauk*, 18:13–40, 1963. Russian Math. Surveys 18 (1963), 9–36.

[8] V.I. Arnold. Small denominators and problems of stability of motions in classical and celestial mechanics. *Uspeki Mat. Nauk*, 18:91–196, 1963. Russian Math. Surveys 18 (1963), 85–193.

[9] V.I. Arnold. Sur une propriété topologique des applications globalement canoniques de la méchanique classique. *C. R. Acad. Paris*, 261:3719–3722, 1965.

[10] V.I. Arnold. Proceedings of Symposia in Pure Mathematics. Mathematical developments arising from Hilbert problems. *AMS, Providence*, 28:66, 1976.

[11] V.I. Arnold. *Mathematical methods of classical mechanics*. Springer-Verlag, Berlin und New York, 1978. In particular: Appendix 9.

[12] V.I. Arnold and A. Avez. *Ergodic problems of classical mechanics*. W.A. Benjamin, Inc., 1968. In particular Appendix 26.

[13] M. Audin and J. Lafontaine. Holomorphic curves in symplectic geometry. Birkhäuser, 1994

[14] V. Bangert. Geodätische Linien auf Riemannschen Mannigfaltigkeiten. *Jber. d. Dt. Math.-Verein.*, 87:39–66, 1985.

[15] V. Bangert. On the existence of closed geodesics on two spheres. *Intern. Journal of Mathematics*, 4:1:1–10, 1993.

[16] A. Banyaga. Sur la structure du groupe des difféomorphismes qui presérvent une forme symplectique. *Comment. Math. Helvetici*, 53:174–227, 1978.

[17] W. Beckner. Sharp Sobolev inequalities on the sphere and the Moser-Trudinger inequality. *Annals of Mathematics*, 138:213–242, 1993.

[18] V. Benci. Closed geodesic for the Jacobi metric and periodic solutions of prescribed energy of natural Hamiltonian systems. *Ann. Inst. H. Poincaré, Analyse non linéaire*, 1:401–412, 1984.

[19] V. Benci, H. Hofer, and P. Rabinowitz. A remark on a-priori bounds and existence for periodic solutions of Hamiltonian systems. *NATO ASI Series, Series C*, In: Periodic Solutions of Hamiltonian Systems and Related Topics,(209):85–88, 1987.

[20] D. Bennequin. Entrelacements et équations de Pfaff. *Astérisque*, 107–108:83–161, 1983.

[21] H. Berestycki, J.M. Lasry, G. Mancini, and B. Ruf. Existence of multiple periodic orbits on star-shaped Hamiltonian surfaces. *Comm. Pure Appl. Math.*, 38:253–289, 1985.

[22] M. Bialy and L. Polterovich. Geodesics of Hofer's metric on the group of Hamiltonian diffeomorphisms. To appear: Duke Math. Journal, 1993.

[23] G.D. Birkhoff. Proof of Poincaré's geometric theorem. *Trans. Amer. Math. Soc.*, 14:14–22, 1913.

[24] G.D. Birkhoff. The restricted problem of three bodies. *Rend. Circ. Mat. Palermo*, 39:265–334, 1915.

[25] G.D. Birkhoff. *Dynamical Systems*. Colloquium Publ. A.M.S., Vol. 9, 1927.

[26] G.D. Birkhoff. Stability and the equations of dynamics. *Am. J. Math.*, 49:1–38, 1927.

[27] G.D. Birkhoff and D.C. Lewis. On the periodic motions near a given periodic motion of a dynamical system. *Annali di Math.*, 12:117–133, 1933.

[28] Ph. Blanchard and E. Brüning. *Variational methods in mathematical physics*. Springer, 1992.

[29] S.V. Bolotin. Libration motions of natural dynamical systems. *Moscow University Bulletin*, 3, 1978.

[30] Yu.G. Borisovich, V.G. Zvyagin, and Yu.I. Sapronov. Non-linear Fredholm maps and the Leray-Schauder theory. *Russian Math. Surveys*, 32(4):1–54, 1977.

[31] J. Bourgain. Approximation of solutions of the cubic nonlinear Schrödinger equations by finite-dimensional equations and nonsqueezing properties. *IMRN*, 2:79–90, 1994.

[32] H. Brezis and L. Nirenberg. Nonlinear functional analysis. In preparation.

[33] L.E.J. Brouwer. Beweis des ebenen Translationssatzes. *Math. Ann.*, 72:37–54, 1912.

[34] M. Brown. A new proof of Brouwer's lemma on translation arcs. *Houston J. of Math.*, 10:35–41, 1984.

[35] M. Brown and W.D. Neumann. Proof of the Poincaré-Birkhoff fixed point theorem. *Michigan Math. J.*, 24:21–31, 1977.

[36] M. Brunella. On a theorem of Sikorav. *L'Enseignement Mathématique*, 37:83–87, 1991.

[37] T. Carleman. Sur un problème d'unicité pour les systèmes d'équations aux dérivées partielles à deux variables indépendantes. *Ark. Mat. Asts. Fys.*, 26B, No. 17:1–9, 1939.

[38] P.H. Carter. An improvement of the Poincaré-Birkhoff theorem. *Trans. Amer. Math. Soc.*, 269:285–299, 1982.

[39] K.-C. Chang. *Infinite dimensional Morse theory and multiple solution problems*. Birkhäuser, 1993.

[40] S-Y. A. Chang. Extremal functions in a sharp form of Sobolev inequality. *ICM Berkeley*, 1:715–723, 1986.

[41] M. Chaperon. Phases génératrices en géométrie symplectique. In: D. Bennequin, Rencontre entre physiciens théoriciens et mathématiciens en l'honneur de René Thom. Publication of the Université de Strasbourg.

[42] M. Chaperon. Quelques questions de géométrie symplectique. *Séminaire Bourbaki 1982–83, Astérisque*, 105–106:231–249, 1983.

[43] M. Chaperon. Une idée du type "géodésique brisée" pour les systémes Hamiltoniens. *C.R. Acad. Sci. Paris Sér. I Math.*, 298:293–296, 1984. Série I, no. 13.

[44] M. Chaperon. Lois de conservation et géométrie symplectique. *C.R. Acad. Sci. Paris*, 312:345–348, 1991. Série I.

[45] M. Chaperon and B. D'Onofrio. Sur un théorème de Claude Viterbo. *C.R. Acad. Sci. Paris*, 306:597–600, 1988. Série I.

[46] M. Chaperon and E. Zehnder. Quelques résultat en géométrie symplectique. In *Séminaire Sud-Rhodanien de Géométrie III: Autour du Théorème de Poincaré-Birkhoff*, pages 51–121, 1984. Travaux en cours. Hermann Paris.

[47] Yu.V. Chekanov. Legandrova teorija morsa. *Uspekhi Mat. Naul.*, 42:4:129–141, 1987.

[48] A. Chenciner. Sur un énoucé dissipatif du théorème géométrique de Poincaré-Birkhoff. *C.R. Acad. Sci. Paris*, 294, I:243–245, 1982.

[49] A. Chenciner. La dynamique on voisinage d'un point fixe elliptique conservatif: de Poincaré et Birkhoff à Aubry et Mather. *Séminaire Bourbaki Astérisque*, 121/122:147–170, 1984.

[50] A. Floer, H. Hofer and K. Wysocki. Applications of symplectic homology I, to appear in *Math. Zeit.*

[51] K. Cieliebak, A. Floer, and H. Hofer. Symplectic homology II: A general construction. To appear in Math. Zeit.

[52] K. Cieliebak, A. Floer, H. Hofer, and K. Wysocki. Applications of symplectic homology II. Preprint E.T.H. Zürich (1994).

[53] F. Clarke. A classical variational principle for periodic Hamiltonian trajectories. *Proc. Amer. Math. Soc.*, 76:186–188, 1979.

[54] C. Conley. *Isolated invariant sets and Morse index*. CBMS Reg. Conf. Series in Math. 38. AMS, 1978.

[55] C. Conley and E. Zehnder. The Birkhoff-Lewis fixed point theorem and a conjecture of V. I.Arnold. *Invent. Math.*, 73:33–49, 1983.

[56] C. Conley and E. Zehnder. *An index theory for periodic solutions of a Hamiltonian system*, volume 1007 of *Geometric Dynamics, Lecture Notes in Mathematics*. Springer, 1983. 132–145.

[57] C. Conley and E. Zehnder. Morse type index theory for flows and periodic solutions of Hamiltonian equations. *Comm. Pure Appl. Math.*, 37:207–253, 1984.

[58] C.B. Croke and A. Weinstein. Closed curves on convex hypersurfaces and periods of nonlinear oscillations. *Invent. Math.*, 64:199–202, 1981.

[59] B. Dacorogna and J. Moser. On a partial differential equation involving the Jacobian determinant. *Ann. Inst. Henri Poincaré, Analyse non linéaire*, 7:1–26, 1990.

[60] K. Deimling. *Nonlinear functional analysis*. Springer-Verlag, Berlin Heidelberg, 1985.

[61] W.-Y. Ding. A generalization of the Poincaré-Birkhoff theorem. *Proc. AMS*, 88:341–346, 1983.

[62] A. Dold. *Lectures on Algebraic Topology*. Grundlagen der mathematischen Wissenschaften 200, Springer 1980. second edition.

[63] A. Douglis and L. Nirenberg. Interior estimates for elliptic systems of partial differential equations. *Comm. Pure Appl. Math.*, 8:503–538, 1955.

[64] J.J. Duistermaat. On global action-angle coordinates. *Commun. on Pure and Appl. Math.*, 33:687–706, 1980.

[65] I. Ekeland. Une théorie de Morse pour les systèmes hamiltoniens convexes. *Ann. Inst. Henri Poincaré, Analyse non linéaire*, 1:19–78, 1984.

[66] I. Ekeland. *Convexity methods in Hamiltonian mechanics*. Springer, 1990. Ergebnisse der Mathematik und ihrer Grenzgebiete, 3. Folge.

[67] I. Ekeland and H. Hofer. Convex Hamiltonian energy surfaces and their closed trajectories. *Comm. Math. Physics*, 113:419–467, 1987.

[68] I. Ekeland and H. Hofer. Symplectic topology and Hamiltonian dynamics. *Math. Zeit.*, 200:355–378, 1990.

[69] I. Ekeland and H. Hofer. Symplectic topology and Hamiltonian dynamics II. *Math. Zeit.*, 203:553–567, 1990.

[70] I. Ekeland and J.M. Lasry. On the number of periodic trajectories for a Hamiltonian flow on a convex energy surface. *Ann. Math*, 112:283–319, 1980.

[71] Y. Eliashberg. A theorem on the structure of wave fronts and its application in symplectic topology. *Funct. Anal. and its Appl.*, 21:227–232, 1987.

[72] Y. Eliashberg. Topological characterization of Stein manifolds of dimension > 2. *Intern. J. Math.*, 1:29–46, 1990.

[73] Y. Eliashberg. Filling by holomorphic discs and its applications. *London Math. Society Lecture Notes*, pages 45–67, 1991. Series 151.

[74] Y. Eliashberg. Contact 3-manifolds, twenty year since J. Martinet's work. *Ann. Inst. Fourier*, 42:165–192, 1992.

[75] Y. Eliashberg and H. Hofer. Unseen symplectic boundaries. In *Proceedings of a Conference in honour of E. Calabi*, 1991.

[76] Y. Eliashberg and L. Polterovich. Bi-invariant metrics on the group of Hamiltonian diffeomorphisms. International Journal of Mathematics Vol. 4, No. 5 (1993) 727–738.

[77] Y.M. Eliashberg. Estimates on the number of fixed points of area preserving transformations. Preprint Syktyvkar 1978 (Russian).

[78] Y.M. Eliashberg. Rigidity of symplectic and contact structures. Preprint, 1981.

[79] H. Eliasson. Geometry of manifolds of maps. *J. Diff. Geom.*, 1:169–194, 1967.

[80] A. Fahti. An orbit closing proof of Brouwer's lemma on translation arcs. *Enseignement Math.*, 33:315–322, 1987.

[81] A.I. Fet and L.A. Lusternik. Variational problems on closed manifolds. *Dokl. Akad. Nauk. S.S.S.R.*, 81:17–18, 1951.

[82] A. Floer. Proof of the Arnold conjecture for surfaces and generalizations to certain Kähler manifolds. *Duke Math. J.*, 53:1–32, 1986.

[83] A. Floer. An instanton invariant for 3-manifolds. *Comm. Math. Phys.*, 118:215–240, 1988.

[84] A. Floer. Morse theory for Lagrangian intersection theory. *J. Diff. Geom.*, 18:513–517, 1988.

[85] A. Floer. The unregularised gradient flow of the symplectic action. *Comm. Pure Appl. Math.*, 41:775–813, 1988.

[86] A. Floer. Cuplength estimates on Lagrangian intersections. *Comm. Pure Appl. Math.*, 42:335–356, 1989.

[87] A. Floer, Hofer H., and D. Salamon. Transversality in the elliptic Morse theory for the symplectic action. Preprint E.T.H. Zürich 1994.

[88] A. Floer, Hofer H., and C. Viterbo. The Weinstein conjecture in $P \times \mathbb{C}^l$. *Math. Zeit.*, 203:355–378, 1989.

[89] A. Floer and H. Hofer. Coherent orientations for periodic orbit problems in symplectic geometry. *Math. Zeit.*, 212:13–38, 1993.

[90] A. Floer and H. Hofer. Symplectic homology I: Open sets in $\mathbb{C}^n$. *Math. Zeit.*, 215:37–88, 1994.

[91] M. Flucher. Fixed points of measure preserving torus homeomorphisms. *manuscripta math.*, 68:271–293, 1990.

[92] O. Forster. *Riemannsche Flächen*. Heidelberger Taschenbücher. Springer, 1977.

[93] B. Fortune. A symplectic fixed point theorem for CP^n. *Invent. Math.*, 81:22–46, 1985.

[94] B. Fortune and A. Weinstein. A symplectic fixed point theorem for complex projective spaces. *Bulletin AMS*, 12:128–130, 1985.

[95] J. Franks. Periodic points and rotation numbers for area preserving diffeomorphisms of the plane. Preprint, Northwestern University.

[96] J. Franks. Generalizations of the Poincaré-Birkhoff theorem. *Ann. of Math.*, 128:139–151, 1988.

[97] J. Franks. Recurrence and fixed points of surface homeomorphisms. *Ergod. Th. and Dynam. Sys.*, 8:99–107, 1988. Charles Conley Memorial Issue.

[98] J. Franks. *A variation of the Poincaré-Birkhoff theorem*. Amer. Math. Soc., Providence, 1988. Contemp. Math., 81.

[99] J. Franks. Periodic points and rotation numbers for area preserving diffeomorphisms of the plane. *IHES-Publ.Math.*, 71:105–120, 1990.

[100] J. Franks. Geodesics on S^2 and periodic points of annulus homeomorphisms. *Invent. Math.*, 108:403–418, 1992.

[101] J. Franks. A new proof of the Brouwer plane translation theorem. *Ergod. Th. and Dynam. Sys.*, 12:217–226, 1992.

[102] K. Fukaya. Floer homology for oriented 3-manifolds. *Advanced Studies in Pure Mathematics*, 20:1–92, 1992.

[103] J.M. Gambaudo and P. Le Calvez. Infinité d'orbites périodiques pour les difféomorphismes conservatifs de l'anneau. Preprint, 1991.

[104] C. Genecand. Transversal homoclinic orbits near elliptic fixed points of area preserving diffeomorphisms of the plane. *Dynamics Reported, New Series*, 2:1–30, 1993.

[105] A.B. Givental. Global properties of Maslov index and Morse theory. *Funct. Anal. and its Appl.*, 22:141–142, 1988.

[106] H. Gluck and W. Ziller. *Existence of periodic motions of conservative systems*. Princeton University Press, 1983. In: Seminar on minimal submanifolds, E. Bombieri, ed.

[107] M. Gromov. Pseudoholomorphic curves in symplectic manifolds. *Invent. Math.*, 82:307–347, 1985.

[108] M. Gromov. *Partial differential relations*. Ergebnisse der Mathematik. Springer, 1986.

[109] L. Guillon. Le théorème de translation de Brouwer: une démonstration simplifiée menant à une nouvelle preuve du théorème de Poincaré-Birkhoff. 1991.

[110] M.-R. Herman. Differentiabilite optimale et contre-exemples à la fermeture en topologie C^∞ des orbites recurrentes de flots Hamiltoniens. *Comptes Rendus de l'Académie des Sciences. Serie-I.-Mathématique*, 313:49–51, 1991. No. 1.

[111] M.-R. Herman. Exemples de flots Hamiltoniens dont aucune perturbation en topologie C^∞ n'a d'orbites periodiques sur un ouvert de surfaces d'energies. *Comptes Rendus de l'Académie des Sciences. Serie-I.-Mathématique*, 312:989–994, 1991. No. 13.

[112] D. Hilbert. Über das Dirichlet'sche Prinzip. *Jhrb. der Deutschen Math. Vereinigung*, VIII, Erstes Heft:184–188, 1900.

[113] M.W. Hirsch. *Differential topology*. Graduate Texts in Math. 33. Springer-Verlag, 1976.

[114] H. Hofer. Lagrangian embeddings and critical point theory. *Ann. Inst. H. Poincaré, Anal. Non Linéaire*, 2:407–462, 1985.

[115] H. Hofer. Ljusternik-Schnirelman theory for Lagrangian intersections. *Ann. Inst. H. Poincaré*, 5(5):465–499, 1988.

[116] H. Hofer. On the topological properties of symplectic maps. *Proceedings of the Royal Society of Edinburgh*, 115 A:25–38, 1990.

[117] H. Hofer. Estimates for the energy of a symplectic map. *Comm. Math. Helv.*, 68:48–72, 1993.

[118] H. Hofer. Pseudoholomorphic curves in symplectisations with application to the Weinstein conjecture in dimension three. *Invent. Math.*, 114:515–563, 1993.

[119] H. Hofer and D. Salamon. Floer homology and Novikov rings. To appear A. Floer Memorial Volume.

[120] H. Hofer, C. Taubes, A. Weinstein, and E. Zehnder, editors. *Floer Memorial Volume*. To appear, Birkhäuser, 1994.

[121] H. Hofer and C. Viterbo. The Weinstein conjecture in cotangent bundles and related results. *Annali Sc. Norm. Sup. Pisa*, 15:411–445, 1988. Serie IV, Fasc III.

[122] H. Hofer and C. Viterbo. The Weinstein conjecture in the presence of holomorphic spheres. *Comm. Pure Appl. Math.*, 45(5):583–622, 1992.

[123] H. Hofer and E. Zehnder. *A new capacity for symplectic manifolds*, pages 405–428. Analysis et cetera. Academic press, 1990. Edit: P. Rabinowitz and E. Zehnder.

[124] L. Hörmander. *The analysis of linear partial differential operators*, volume I–IV. Springer, Grundlehren Edition, 1983–1985.

[125] M.-Y. Jiang. Hofer-Zehnder symplectic capacity for two dimensional manifolds. *Proc. Royal Soc. Edinb.*, 123A:945–950, 1993.

[126] R. Jost. Winkel- und Wirkungsvariable für allgemeine mechanische Systeme. *Helvetica Physica Acta*, 41:965–968, 1968.

[127] W. Klingenberg. *Lectures on closed geodesics*. Springer, Grundlehren Math. Wiss. 230, 1978.

[128] A.N. Kolmogorov. *The general theory of dynamical systems and classical mechanics*. Proc. Int. Congress of Math. Amsterdam 1954. W.A. Benjamin, Inc., 1967. see Appendix D in R. Abraham: Foundations of mechanics.

[129] V.V. Koslov. Calculus of variations in the large and classical mechanics. *Russian Math. Survey*, 40:2:37–71, 1985.

[130] S. Kuksin. Infinite-dimensional capacities and a squeezing theorem for Hamiltonian PDE's. Preprint F.I.M., ETH Zürich (1993).

[131] A. Künzle. The least characteristic action as symplectic capacity. F.I.M. report, ETH Zürich (1991).

[132] K. Kuperberg. A smooth counter example to the Seifert conjecture. To appear in Annals of Mathematics.

[133] F. Lalonde. Energy and capacities in symplectic topology. Preprint, 1994.

[134] F. Lalonde and D. McDuff. The geometry of symplectic energy. Preprint, 1993.

[135] F. Lalonde and D. McDuff. Hofer's L^∞-geometry: energy and stability of Hamiltonian flows I. Preprint, 1994.

[136] F. Laudenbach and J.C. Sikorav. Persistence d'intersections avec la section nulle au cours d'une isotopie Hamiltonienne dans un fibre cotangent. *Invent. Math.*, 82:349–357, 1985.

[137] H.V. Lê and Kaoru Ono. Symplectic fixed points, Calabi invariant and Novikov homology. Max-Planck-Institut Bonn MPI/93–27.

[138] P. Le Calvez. Sur les diffeomorphismes de l'anneau et du tore. Preprint, Université Paris-Sud.

[139] P. Le Calvez. Une généralisation du théorème de Conley-Zehnder aux homéomorphismes du tore de dimension deux. Prépublications Mathématiques de l'Université Paris-Nord, 93–02, 1993.

[140] P. Le Calvez. Phases génératrices discrètes de difféomorphismes du tore ou de l'anneau. Preprint, Unversité Paris-Sud, 1991.

[141] P. Le Calvez. Un théorème de translations pour les difféomorphismes du tore ayant des points fixes. Prépublications Université de Paris-Sud, mathématiques, 92–85 (1992), 1991.

[142] E. Lima. Orientability of smooth hypersurfaces and the Jordan-Brouwer separation theorem. *Expo. Math.*, 5:283–286, 1987.

[143] L. Lusternik and L. Schnirelman. *Méthodes topologiques dans les problèmes variationnels.* Herman, Paris, 1934.

[144] R. Lutz. Structures de contact sur les fibrés principaux en circles de dimension 3. *Ann. Inst. Fourier*, 3:1–15, 1977.

[145] A.M. Lyapunov. Problème général de la stabilité du movement. *Ann. Fac. Sci. Toulouse*, 2:203–474, 1907.

[146] R. Ma. A remark on Hofer-Zehnder symplectic capacity in $M \times \mathbb{R}^{2n}$, 1992. Preprint, Tsinghua University.

[147] J. Martinet. Formes de contact sur les variétés de dimension 3. *Lecture Notes in Math.*, 209:142–163, 1971.

[148] W.S. Massey. *Algebraic topology: an introduction.* Graduate Texts in Mathematics 56. Springer Verlag, 1967.

[149] W.S. Massey. *Homology and cohomology theory.* Marcel Dekker, Incl. New York and Basel, 1978.

[150] W.S. Massey. *Singular homology theory.* Graduate Texts in Mathematics 70. Springer Verlag, 1980.

[151] J. Mather. Existence of quasiperiodic orbits for twist homeomorphisms of the annulus. *Topology*, 21:457–467, 1982.

[152] J. Mawhin and M. Willem. *Critical point theory and Hamiltonian systems.* Springer, 1989.

[153] D. McDuff. Elliptic methods in symplectic geometry. *Bulletin AMS*, 23(2):311–358, 1990.

[154] D. McDuff. Symplectic manifolds with contact type boundaries. *Invent. Math.*, 103:651–671, 1991.

[155] D. McDuff and L. Polterovich. Symplectic packings and algebraic geometry. Preprint, 1993.

[156] D. McDuff and D. Salamon. Introduction to symplectic topology. To appear.

[157] J. Milnor. *Topology from the differentiable viewpoint*. The University Press of Virginia, Charlottesville, 1972.

[158] A.S. Mishchenko and A.T. Fomenko. Generalized Liouville method of integration of Hamiltonian systems. *Funct. Anal. and its Appl.*, 12:113–121, 1978.

[159] J. Moser. On invariant curves of area preserving mappings of the annulus. *Nachr. Akad. Wiss. Göttingen, Math. Phys. Kl.*, pages 1–20, 1962.

[160] J. Moser. On the volume elements on a manifold. *Trans. Amer. Math. Soc.*, 120:286–294, 1965.

[161] J. Moser. A rapidly convergent iteration method and non-linear partial differential equations I and II. *Ann. Scuola Norm. Sup. Pisa*, 20:265–315, 499–535, 1966.

[162] J. Moser. *Stable and random motions in dynamical systems*. Princeton University Press, 1973. In: Annals of Math. Studies, Vol. 77.

[163] J. Moser. Periodic orbits near an equilibrium and a theorem of A. Weinstein. *Comm. Pure Appl. Math.*, 29:727–747, 1976.

[164] J. Moser. Proof of a generalized form of a fixed point theorem due to G.D. Birkhoff. *Geometry and Topology, Lecture Notes in Mathematics*, 597:464–494, 1977.

[165] J. Moser. A fixed point theorem in symplectic geometry. *Acta Mathematica*, 141:17–34, 1978.

[166] J. Moser. Integrable Hamiltonian systems and spectral theory. In *Lezioni Fermiane, Accademia nazionale dei Lincai Scuola Normale Superiore, Pisa*, 1981.

[167] J. Moser and E. Zehnder. *Notes on Dynamical systems*. Courant Institute, 1980.

[168] N. Nekhoroshev. Action-angle variables and their generalization. *Trans. Mosc. Math. Soc.*, 26 181–198, 1972.

[169] N. Nikishin. Fixed points of diffeomorphisms on the two sphere that preserve area. *Funkts. Anal. i Prilozhen*, 8:84–85, 1974.

[170] S.P. Novikov. Multivalued functions and functionals — an analogue of the Morse theory. *Soviet Math. Dokl.*, 24:222–225, 1981.

[171] D. Offin. A class of periodic orbits in classical mechanics. *J. Diff. Eq.*, 66:90–116, 1987.

[172] K. Ono. On the Arnold conjecture for weakly monotone symplectic manifolds. Preprint Max-Planck-Institut für Mathematik MPI/93-4.

[173] R.S. Palais. Critical point theory and the minimax principle. In: Global Analysis. *Proc. AMS*, XV:185–212, 1970.

[174] R.S. Palais and Terng C.-L. *Critical point theory and submanifold geometry*, volume 1353. Springer Lecture Notes in Mathematics, 1988.

[175] R.S. Palais and S. Smale. A generalized Morse theory. *Bull. AMS*, 70:165–171, 1964.

[176] H. Poincaré. *Methodes nouvelle de la méchanique céleste*, volume 3. Gauthier-Villars, Paris, 1899.

[177] H. Poincaré. Sur un théorème de géométrie. *Rend. Circolo Math. Palermo*, 33:375–407, 1912.

[178] J. Pöschel. Integrability of Hamiltonian systems on Cantor sets. *Comm. Pure Appl. Math.*, 35:653–696, 1982.

[179] C.C. Pugh and R.C. Robinson. The C^1-closing lemma, including Hamiltonians. *Erg. Th. and Dynam. Sys.*, 3:261–313, 1983.

[180] P. Rabinowitz. Periodic solutions of Hamiltonian systems. *Comm. Pure Appl. Math.*, 31:157–184, 1978.

[181] P. Rabinowitz. Periodic solutions of Hamiltonian systems on a prescribed energy surface. *J. Diff. Equ.*, 33:336–352, 1979.

[182] P. Rabinowitz. *Minimax methods in critical point theory with applications to differential equations*. CBMS, Regional Conf. Ser. in Math. 65. AMS, 1986.

[183] C. Robinson. A global approximation theorem for Hamiltonian systems. *Proc. Symposium in Pure Math.*, XIV:233–244, 1970. Global Analysis, AMS.

[184] K.P. Rybakowski. On the homotopy index for infinite-dimensional semiflows. *Trans. Amer. Math. Soc.*, 269:351–382, 1982.

[185] K.P. Rybakowski. *The homotopy index and partial differential equations.* Springer, 1987.

[186] K.P. Rybakowski and E. Zehnder. A Morse equation in Conley's index theory for semiflows on metric spaces. *Ergod. Th. and Dynam. Sys.*, 5:123–143, 1985.

[187] D. Salamon. Connected simple system and the Conley index of isolated invariant sets. *Trans. Amer. Math. Soc.*, 291:1–41, 1985.

[188] D. Salamon. Morse theory, the Conley index and Floer homology. *Bull. London Math. Soc.*, 22:113–140, 1990.

[189] D. Salamon and E. Zehnder. KAM theory in configuration space. *Comm. Math. Helvetici*, 64:84–132, 1989.

[190] D. Salamon and E. Zehnder. Morse theory for periodic solutions of Hamiltonian systems and the Maslov index. *Comm. Pure Appl. Math.*, 45:1303–1360, 1992.

[191] M. Schwarz. *Morse homology.* Progress in Mathematics, 111. Birkhäuser, 1993.

[192] P.A. Schweitzer. Counter examples to the Seifert conjecture and opening closed leaves of foliations. *Ann. Math.*, 100:386–400, 1974.

[193] H. Seifert. Periodische Bewegungen mechanischer Systeme. *Math. Zeit.*, 51:197–216, 1948.

[194] K.F. Siburg. Symplectic capacities in two dimensions. *manuscripta math.*, 78:149–163, 1993.

[195] C.L. Siegel. über die Existenz einer Normalform analytischer Hamiltonischer Differentialgleichungen in der Nähe einer Gleichgewichtslösung. *Math. Ann.*, 128:144–170, 1954.

[196] C.L. Siegel and J. Moser. *Lectures on Celestial Mechanics.* Springer Grundlehren, Band 187, 1971.

[197] J.-C. Sikorav. Problèmes d'intersections et de points fixes en géométrie Hamiltonienne. *Comm. Math. Helv.*, 62:62–73, 1987.

[198] J.-C. Sikorav. Systèmes Hamiltoniens et topologie symplectique. Dipartimento di Matematica dell' Università di Pisa, 1990. ETS, EDITRICE PISA.

[199] J.-C. Sikorav. Homologie associée à une fonctionelle (d'après A. Floer). *Séminaire Bourbaki Exp. 730–744 Astérisque*, 201–202–203:115–143, 1991.

[200] C.P. Simon. A bound for the fixed point index of an area preserving map with applications in mechanics. *Invent. Math.*, 26:187–200, 1974.

[201] S. Smale. An infinite dimensional version of Sard's theorem. *Amer. J. Math.*, 87:861–866, 1965.

[202] E. Spanier. Cohomology theory for general spaces. *Ann. of Math. (2)*, 49:407–427, 1948.

[203] E. Spanier. *Algebraic topology*. Springer Verlag, New York, 1966.

[204] E. Spanier. Tautness for Alexander-Spanier cohomology. *Pacific Journal of Mathematics*, (75) 2:561–563, 1978.

[205] E.M. Stein. *Singular integrals and differentiability properties of functions*. Princeton University Press, 1970.

[206] A. Stephen. On homeomorphisms of the plane which have no fixed points. *Abh. Math. Seminar Univ. Hamburg*, 30:61–74, 1967.

[207] M. Struwe. Existence of periodic solutions of Hamiltonian systems on almost every energy surface. *Bol. Soc. Bras. Mat.*, 20:49–58, 1990.

[208] M. Struwe. *Variational methods*. Springer, 1990.

[209] R. Switzer. *Algebraic topology — homotopy and homology*. Grundlehren, 212. Springer Verlag, Berlin, 1975.

[210] C.H. Taubes. Self-dual Yang-Mills connections on non-self-dual 4-manifolds. *J. Diff. Geom.*, 17:139–170, 1982.

[211] C.H. Taubes. A framework for Morse theory for the Yang-Mills functionals. Preprint Harvard University, 1986.

[212] L. Traynor. Symplectic homology via generating functions. Preprint Stanford University, 1993.

[213] L. Traynor. Symplectic packing constructions. Preprint, 1993.

[214] I. Ustilovsky. Conjugate points on geodesics of Hofer's metric. Preprint, 1993.

[215] I.N. Vekua. *Verallgemeinerte analytische Funktionen*. Akademie-Verlag, Berlin, 1963.

[216] J. Vick. *Homology theory, an introduction to algebraic topology*. Academic Press, 1973.

[217] C. Viterbo. A proof of the Weinstein conjecture in $\mathbb{R}^{2n}$. *Ann. Inst. H. Poincaré, Anal. non linéaire*, 4:337–357, 1987.

[218] C. Viterbo. Symplectic topology as the geometry of generating functions. *Math. Annalen*, 292:685–710, 1992.

[219] F. Warner. *Foundations of differentiable manifolds and Lie groups*. Scott, Foresman and Company, 1971.

[220] K. Weierstrass. *Mathematische Werke*. 1858. Berlin, Band I: 233–246, Band II: 19–44, Nachtrag: 139–148.

[221] A. Weinstein. Contact surgery and symplectic handlebodies. *Hokkaido Math. Journal* 20 (1990) 241–251.

[222] A. Weinstein. Lagrangian submanifolds and Hamiltonian systems. *Ann. Math.*, 98:377–410, 1973.

[223] A. Weinstein. Normal modes for nonlinear Hamiltonian systems. *Invent. Math.*, 20:47–57, 1973.

[224] A. Weinstein. *Lectures on symplectic manifolds*, volume 29 of *CBMS, Reg. Conf. Series in Math*. AMS, 1977.

[225] A. Weinstein. Periodic orbits for convex Hamiltonian systems. *Ann. Math.*, 108:507–518, 1978.

[226] A. Weinstein. On the hypothesis of Rabinowitz's periodic orbit theorems. *J. Diff. Equ.*, 33:353–358, 1979.

[227] A. Weinstein. On extending the Conley-Zehnder fixed point theorem to other manifolds. *Proc. Sympos. Pure Math.*, 45, part 2:541–544, 1986. AMS.

[228] F.W. Wilson. On the minimal sets of nonsingular vector fields. *Ann. Math.*, 84:529–536, 1966.

[229] J.C. Yoccoz. Travaux de Herman sur les tores invariants. *Bourbaki Exp. 745-579, Astérisque*, 206:311–345, 1992.

[230] E. Zehnder. Homoclinic points near elliptic fixed points. *Comm. Pure Appl. Math.*, 26:131–182, 1973.

[231] E. Zehnder. Remarks on periodic solutions on hypersurfaces. *NATO ASI Series, Series C*, In: Periodic Solutions of Hamiltonian Systems and Related Topics,(209):267–279, 1987.

[232] E. Zeidler. *Nonlinear functional analysis and its applications I*. Springer-Verlag, New York Inc., 1986.

Birkhäuser Advanced Texts /
Basler Lehrbücher

Managing Editors:

H. Amann (Universität Zürich)
H. Kraft (Universität Basel)

This series presents, at an advanced level, introductions to some of the fields of current interest in mathematics. Starting with basic concepts, fundamental results and techniques are covered, and important applications and new developments discussed. The textbooks are suitable as an introduction for students and non-specialists, and they can also be used as background material for advanced courses and seminars.

We encourage preparation of manuscripts in TeX for delivery in camera-ready copy which leads to rapid publication, or in electronic form for interfacing with laser printers or typesetters. Proposals should be sent directly to the editors or to: Birkhäuser Verlag, P. O. Box 133, CH–4010 Basel, Switzerland

M. Brodmann, **Algebraische Geometrie**
1989. 296 Seiten. Gebunden. ISBN 3-7643-1779-5

E.B. Vinberg, **Linear Representations of Groups**
1989. 152 pages. Hardcover. ISBN 3-7643-2288-8

K. Jacobs, **Discrete Stochastics**
1991. 296 pages. Hardcover. ISBN 3-7643-2591-7

S.G. Krantz, **H.R. Parks**, **A Primer of Real Analytic Functions**
1992. 194 pages. Hardcover. ISBN 3-7643-2768-5

L. Conlon, **Differentiable Manifolds: A First Course**
1992. 369 pages. Hardcover. *First edition, 2nd revised printing.*
ISBN 3-7643-3626-9

M. Artin, **Algebra**
1993. 723 Seiten. Gebunden. ISBN 3-7643-2927-0

H. Hofer / E. Zehnder, **Symplectic Invariants and Hamiltonian Dynamics**
1994. Approx. 348 pages. Hardcover. ISBN 3-7643-5066-0

Progress in Mathematics

Edited by:

J. Oesterlé
Départment de Mathématiques
Université de Paris VI
4, Place Jussieu
75230 Paris Cedex 05, France

A. Weinstein
Department of Mathematics
University of California
Berkeley, CA 94720
U.S.A.

Progress in Mathematics is a series of books intended for professional mathematicians and scientists, encompassing all areas of pure mathematics. This distinguished series, which began in 1979, includes authored monographs, and edited collections of papers on important research developments as well as expositions of particular subject areas.

We encourage preparation of manuscripts in such form of TeX for delivery in camera-ready copy which leads to rapid publication, or in electronic form for interfacing with laser printers or typesetters.
Proposals should be sent directly to the editors or to: Birkhäuser Boston, 675 Massachusetts Avenue, Cambridge, MA 02139, U.S.A.

1 GROSS. Quadratic Forms in Infinite-Dimensional Vector Spaces
2 PHAM. Singularités des Systèmes Differentiels de Gauss-Manin
4 AUPETIT. Complex Approximation
5 HELGASON. The Radon Transform
6 LION/VERGNE. The Weil representation Maslov index and Theta series
7 HIRSCHOWITZ. Vector bundles and differential equations
10 KATOK. Ergodic Theory and Dynamical Systems I
11 BALSLEY. 18th Scandinavian Congress of Mathematicians
12 BERTIN. Séminaire de Théorie de Nombres, Paris 79-80
13 HELGASON. Topics in Harmonic Analysis on Homogeneous Spaces
14 HANO. Manifolds and Lie Groups
15 VOGAN JR. Representations of Real Reductive Lie Groups
16 GRIFFITHS/MORGAN. Rational Homotopy Theory and Differential Forms
17 VOVSI. Triangular Products of Group Representations and Their Applications
18 FRESNEL/VAN DER PUT. Géometrie Analytique Rigide et Applications
19 ODA. Periods of Hilbert Modular Surfaces
20 STEVENS. Arithmetic on Modular Curves
21 KATOK. Ergodic Theory and Dynamical Systems II
22 BERTIN. Séminaire de Théorie de Nombres, Paris 80-81
23 WEIL. Adeles and Algebraic Groups
24 LE BARZ. Enumerative Geometry and Classical Algebraic Geometry
25 GRIFFITHS. Exterior Differential Systems and the Calculus of Variations
27 BROCKETT. Differential Geometric Control Theory
28 MUMFORD. Tata Lectures on Theta I
29 FRIEDMANN. The Birational Geometry of Degenerations
30 YANO/KON. Submanifolds of Kaehlerian and Sasakian Manifolds
31 BERTRAND. Approximations Diophantiennes et Nombres Transcendant
32 BROOKS. Differential Geometry
33 ZUILY. Uniqueness and Non-Uniqueness in the Cauchy Problem
34 KASHIWARA. Systems of Microdifferential Equations
35/36 ARTIN/TATE. Vol. 1 Arithmetic. Vol. 2 Geometry
37 BOUTET. Mathématique et Physique

38 BERTIN. Séminaire de Théorie de Nombres, Paris 81-82
39 UENO. Classification of Algebraic and Analytic Manifolds
40 TROMBI. Representation Theory of Reductive Groups
41 STANLEY. Combinatorics and Commutative Algebra
42 JOUANOLOU. Théorèmes de Bertini et Applications
43 MUMFORD. Tata Lectures on Theta II
45 BISMUT. Large Deviations and the Malliavin Calculus
47 TATE. Les Conjectures de Stark sur les Fonctions L d'Artin en $s=0$
48 FRÖHLICH. Classgroups and Hermitian Modules
49 SCHLICHTKRULL. Hyperfunctions and Harmonic Analysis on Symetric Spaces
50 BOREL ET AL. Intersection Cohomology
51 Séminaire de Théorie de Nombres, Paris 82-83
52 GASQUI/GOLDSCHMIDT. Déformations Infinitésimales desStructures Conformes Plates
53 LAURENT. Théorie de la 2ième Microlocalisation dans le Domaine Complexe
54 VERDIER. Module des Fibres Stables sur les Courbes Algébriques
55 EICHLER/ZAGIER. The Theory of Jacobi Forms
56 SHIFFMAN/SOMMESE. Vanishing Theorems on Complex Manifolds
57 RIESEL. Prime Numbers and Computer Methods for Factorization
58 HELFFER/NOURRIGAT. Hypoellipticité Maximale pour des Operateurs Polynomes de Champs de Vecteurs
59 GOLDSTEIN. Séminaire de Théorie de Nombres, Paris 83-84
60 ARBARELLO. Geometry Today
62 GUILLOU. A la Recherche de la Topologie Perdue
63 GOLDSTEIN. Séminaire de Théorie des Nombres, Paris 84-85
64 MYUNG. Malcev-Admissible Algebras
65 GRUBB. Functional Calculus of Pseudo-Differential Boundary Problems
66 CASSOU-NOGUES/TAYLOR. Elliptic Functions and Rings of Integers
67 HOWE. Discrete Groups in Geometry and Analysis
68 ROBERT. Autour de l'Approximation Semi-Classique
69 FARAUT/HARZALLAH. Analyse Harmonique: Fonctions Speciales et Distributions Invariantes
70 YAGER. Analytic Number Theory and Diophantine Problems
71 GOLDSTEIN. Séminaire de Théorie de Nombres, Paris 85-86
72 VAISMAN. Symplectic Geometry and Secondary Characteristic Classes
73 MOLINO. Riemannian Foliations
74 HENKIN/LEITERER. Andreotti-Grauert Theory by Integral Formulas
75 GOLDSTEIN. Séminaire de Théorie de Nombres, Paris 86-87
76 COSSEC/DOLGACHEV. Enriques Surfaces I
77 REYSSAT. Quelques Aspects des Surfaces de Riemann
78 BORHO/BRYLINSKI/MCPHERSON. Nilpotent Orbits, Primitive Ideals, and Characteristic Classes
79 MCKENZIE/VALERIOTE. The Structure of Decidable Locally Finite Varieties
80 KRAFT/ SCHWARZ/PETRIE (eds.) Topological Methods in Algebraic Transformation Groups
81 GOLDSTEIN. Séminaire de Théorie des Nombres, Paris 87–88
82 DUFLO/PEDERSEN/VERGNE (eds.) The Orbit Method in Representation Theory
83 GHYS/DE LA HARPE (eds.) Sur les Groupes Hyperboliques d'après M. Gromov
84 ARAKI/KADISON (eds.) Mappings of Operator Algebras
85 BERNDT/DIAMOND/HALBERSTAM/ HILDEBRAND (eds.) Analytic Number Theory
89 VAN DER GEER/OORT/STEENBRINK (eds.) Arithmetic Algebraic Geometry
90 SRINIVAS. Algebraic K-Theory
91 GOLDSTEIN. Séminaire de Théorie des Nombres, Paris 1988-89

- 92 CONNES/DUFLO/JOSEPH/RENTSCHLER. Operator Algebras, Unitary Representations, Enveloping Algebras, and Invariant Theory. A Collection of Articles in Honor of the 65th Birthday of Jacques Dixmier
- 93 AUDIN. The Topology of Torus Actions on Symplectic Manifolds
- 94 MORA/TRAVERSO (eds.) Effective Methods in Algebraic Geometry
- 95 MICHLER/RINGEL (eds.) Representation Theory of Finite Groups and Finite–Dimensional Algebras
- 96 MALGRANGE. Equations Différentielles à Coefficients Polynomiaux
- 97 MUMFORD/NORMAN/NORI. Tata Lectures on Theta III
- 98 GODBILLON. Feuilletages, Etudes géométriques
- 99 DONATO/DUVAL/ELHADAD/ TUYNMAN. Symplectic Geometry and Mathematical Physics. A Collection of Articles in Honor of J.-M. Souriau
- 100 TAYLOR. Pseudodifferential Operators and Nonlinear PDE
- 101 BARKER/SALLY. Harmonic Analysis on Reductive Groups
- 102 DAVID. Séminaire de Théorie des Nombres, Paris 1989-90
- 103 ANGER/PORTENIER. Radon Integrals
- 104 ADAMS/BARBASCH/VOGAN. The Langlands Classification and Irreducible Characters for Real Reductive Groups
- 105 TIRAO/WALLACH. New Developments in Lie Theory and Their Applications
- 106 BUSER. Geometry and Spectra of Compact Riemann Surfaces
- 107 BRYLINSKI. Loop Spaces, Characteristic Classes and Geometric Quantization
- 108 DAVID. Séminaire de Théorie des Nombres, Paris 1990-91
- 109 EYSSETTE/GALLIGO. Computational Algebraic Geometry
- 110 LUSZTIG. Introduction to Quantum Groups
- 111 SCHWARZ. Morse Homology
- 112 DONG/LEPOWSKY. Generalized Vertex-Algebras and Relative Vertex Operators
- 113 MOEGLIN/WALDSPURGER. Décomposition Spectrale et Series d'Eisenstein
- 114 BERENSTEIN/GAY/VIDRAS/YGER. Residue Currents and Bezout Identities
- 115 BABELON/CARTIER/KOSMANN-SCHWARZBACH (eds.) Integrable Systems. The Verdier Memorial Conference
- 116 DAVID (ed.) Séminaire de Théorie des Nombres, Paris, 1991-1992
- 117 AUDIN/LAFONTAINE (eds.) Holomorphic Curves in Symplectic Geometry
- 118 VAISMAN. Lectures on the Geometry of Poisson Manifolds
- 122 GUILLEMIN. Moment Maps and Combinatorial Invariants of Hamiltonian T^n-spaces
- 123 BRYLINSKI/BRYLINSKI/GUILLEMIN/KAC (eds.) Lie Theory and Geometry: In Honor of Bertram Kostant
- 124 AEBISCHER/BORER/KÄLIN/ LEUENBERGER/REIMANN. Symplectic Geometry. An Introduction based on the Seminar in Bern, 1992
- 125 LUBOTZKY. Discrete Groups, Expanding Graphs and Invariant Measures
- 126 RIESEL. Prime Numbers and Computer Methods for Factorization

Monographs in Mathematics

Managing Editors:
H. Amann / K. Grove / H. Kraft / P.-L. Lions

Editorial Board:
H. Araki / J. Ball / F. Brezzi / K.C. Chang / N. Hitchin / H. Hofer / H. Knörrer / K. Masuda / D. Zagier

The foundations of this outstanding book series were laid in 1944. Until the end of the 1970s, a total of 77 volumes appeared, including works of such distinguished mathematicians as Carathéodory, Nevanlinna and Shafarevich, to name a few. The series came to its name and present appearance in the 1980s. According to its well-established tradition, only monographs of excellent quality will be published in this collection. Comprehensive, in-depth treatments of areas of current interest are presented to a readership ranging from graduate students to professional mathematicians. Concrete examples and applications both within and beyond the immediate domain of mathematics illustrate the import and consequences of the theory under discussion.

Published in the series since 1983

Volume 78 **H. Triebel, Theory of Function Spaces I**
1983, 284 pages, hardcover, ISBN 3-7643-1381-1.

Volume 79 **G.M. Henkin/J. Leiterer, Theory of Functions on Complex Manifolds**
1984, 228 pages, hardcover, ISBN 3-7643-1477-X.

Volume 80 **E. Giusti, Minimal Surfaces and Functions of Bounded Variation**
1984, 240 pages, hardcover, ISBN 3-7643-3153-4.

Volume 81 **R.J. Zimmer, Ergodic Theory and Semisimple Groups**
1984, 210 pages, hardcover, ISBN 3-7643-3184-4.

Volume 82 **V.I. Arnold / S.M. Gusein-Zade / A.N. Varchenko, Singularities of Differentiable Maps – Vol. I**
1985, 392 pages, hardcover, ISBN 3-7643-3187-9.

Volume 83 **V.I. Arnold / S.M. Gusein-Zade / A.N. Varchenko, Singularities of Differentiable Maps – Vol. II**
1988, 500 pages, hardcover, ISBN 3-7643-3185-2.

Volume 84 **H. Triebel, Theory of Function Spaces II**
1992, 380 pages, hardcover, ISBN 3-7643-2639-5.

Volume 85 **K.R. Parthasarathy, An Introduction to Quantum Stochastic Calculus**
1992, 300 pages, hardcover, ISBN 3-7643-2697-2.

Volume 86 **M. Nagasawa, Schrödinger Equations and Diffusion Theory**
1993, 332 pages, hardcover, ISBN 3-7643-2875-4.

Volume 87 **J. Prüss, Evolutionary Integral Equations and Applications**
1993, 392 pages, hardcover, ISBN 3-7643-2876-2.

Volume 88 **R.W. Bruggeman, Families of Automorphic Forms**
1994, 328 pages, hardcover, ISBN 3-7643-5046-6.